Plant Image Analysis

Fundamentals and Applications

Plant Image Analysis

Fundamentals and Applications

Edited by
S. Dutta Gupta and Yasuomi Ibaraki

CRC Press
Taylor & Francis Group
Boca Raton London New York

CRC Press is an imprint of the
Taylor & Francis Group, an **informa** business

CRC Press
Taylor & Francis Group
6000 Broken Sound Parkway NW, Suite 300
Boca Raton, FL 33487-2742

Printed on acid-free paper
Version Date: 20140717

International Standard Book Number-13: 978-1-4665-8301-6 (Hardback)

Library of Congress Cataloging-in-Publication Data

Plant image analysis : fundamentals and applications / editors: S. Dutta Gupta, Yasuomi Ibaraki.
pages cm
Includes bibliographical references and index.
ISBN 978-1-4665-8301-6 (hardcover : alk. paper) 1. Plants--Analysis. 2. Image analysis. 3. Phenotype. I. Gupta, S. Dutta. II. Ibaraki, Yasuomi.

QK865.P59 2015
580--dc23 2014023648

Visit the Taylor & Francis Web site at
http://www.taylorandfrancis.com

and the CRC Press Web site at
http://www.crcpress.com

Contents

Preface

Image analysis is a useful tool for obtaining quantitative information for target objects. The application of imaging techniques to plant and agricultural sciences has previously been confined to images obtained through remote sensing techniques. Technological advancement in the development of powerful hardware, picture capturing tools, and robust algorithms in a cost-effective manner paves the path of image analysis toward nondestructive and objective evaluation of biological objects, and opens up a new window to look into the field of plant science. The complex, dynamic nature of plant responses to unexpected changes in the environment compelled scientists to contemplate the application of image analysis in high-throughput phenotyping for different purposes. Various types of imaging techniques, such as red, green, and blue (RGB) imaging, hyperspectral imaging, fluorescence imaging, and thermal imaging, have contributed significantly to different aspects of crop performance and improvement.

Predicting crop performance as a function of genome architecture is one of the major challenges for crop improvement in the twenty-first century to ensure agricultural production that will satisfy the needs of a human population likely to exceed 9 billion by 2050. Compared to the advancements made in the "next generation" genotyping tools, plant phenotyping technology has progressed slowly over the past 25 years. Constraints in plant phenotyping capability limit our approaches to dissect complex traits such as stress tolerance and yield potential. In recent years, phenomics facilities are popping up with the development of new methodological applications of nonconventional optical imaging coupled with computer vision algorithms and widening the set of tools available for automated plant phenotyping.

The present book provides a comprehensive treatise of recent developments in image analysis of higher plants. The book introduces readers to the fundamentals of images and image analysis and then features various types of image analysis techniques covering a diverse domain of plant sciences. It covers imaging techniques that include RGB imaging, hyperspectral imaging at the small canopy level, thermal imaging, photochemical reflectance index (PRI) imaging, chlorophyll fluorescence imaging, reactive oxygen species (ROS) imaging, and chromosome imaging. The book includes 16 chapters presenting a wide spectrum of applications of image analysis that are relevant to assessment of plant growth, nutrient status, and photosynthetic efficiency both in vivo and in vitro, early detection of diseases and stress, cellular detection of reactive oxygen species, plant chromosome analysis, fruit crop yield, and plant phenotyping. The chapters are written by international experts who are pioneers and have made significant contributions to this fascinating field.

The book is designed for graduate students, research workers, and teachers in the fields of cell and developmental biology, stress physiology, precision agriculture, and agricultural biotechnology, as well as professionals involved in areas that utilize machine vision in plant science.

We express our deep sense of gratitude to all the contributors for their kind support and cooperation in our humble approach to present the current status, state of the art, and future outlook of plant image analysis. Thanks are also due to Dr. Rina Dutta Gupta for her support and encouragement throughout the preparation of this volume. Finally, we thank CRC Press for giving us the opportunity to bring out this book.

S. Dutta Gupta
Kharagpur, India

Y. Ibaraki
Yamaguchi, Japan

Contributors

Ferhan Ayaydin
Cellular Imaging Laboratory
Biological Research Center
Szeged, Hungary

Étienne Belin
Laboratoire d'Ingénierie des
Systèmes Automatisés (LISA)
Université d'Angers
Angers, France

Landry Benoit
Laboratoire d'Ingénierie des
Systèmes Automatisés (LISA)
Université d'Angers
Angers, France

Tristan Boureau
Université d'Angers
Institut de Recherche en
Horticulture et Semences
INRA, Agrocampus-Ouest
Université d'Angers
Beaucouzé, France

Valérie Caffier
INRA
Institut de Recherche en
Horticulture et Semences
Beaucouzé, France

and

Agrocampus-Ouest
Université d'Angers
Angers, France

François Chapeau-Blondeau
Laboratoire d'Ingénierie des
Systèmes Automatisés (LISA)
Université d'Angers
Angers, France

Yann Chéné
Laboratoire d'Ingénierie des
Systèmes Automatisés (LISA)
Université d'Angers
Angers, France

Didier Demilly
GEVES
Station Nationale d'Essais de
Semences (SNES)
Beaucouzé, France

Sylvie Ducournau
GEVES
Station Nationale d'Essais de Semences (SNES)
Beaucouzé, France

Carolyne Dürr
INRA
Institut de Recherche en Horticulture et Semences
Beaucouzé, France

Nieves Fernandez-Garcia
Department of Abiotic Stress and Plant Pathology
Centro de Edafologia y Biologia Aplicada del Segura
Consejo Superior de Investigaciones Cientificas
Murcia, Spain

Andrew P. French
Centre for Plant Integrative Biology
University of Nottingham
Nottingham, UK

Kiichi Fukui
Department of Biotechnology
Graduate School of Engineering
Osaka University
Osaka, Japan

S. Dutta Gupta
Agricultural and Food Engineering Department
Indian Institute of Technology
Kharagpur, India

Éva Hideg
Institute of Biology
Faculty of Sciences
University of Pécs
Pécs, Hungary

Yasuomi Ibaraki
Faculty of Agriculture
Yamaguchi University
Yamaguchi, Japan

Seiji Kato
Yamanashi Prefectural Agritechnology Center
Yamanashi, Japan

Gloria Flor Mata-Donjuan
Department of Mechatronics
Polytechnic Queretaro University
Querétaro, México

Adán Mercado-Luna
Department of Biosystems
School of Engineering
Queretaro State University
Querétaro, México

Nobuko Ohmido
Graduate School of Human Development and Environment
Kobe University Nada Ku
Kobe, Japan

Enrique Olmos
Department of Abiotic Stress and Plant Pathology
Centro de Edafologia y Biologia Aplicada del Segura
Consejo Superior de Investigaciones Cientificas
Murcia, Spain

A. Payne
Central Queensland University
Queensland, Australia

Michael P. Pound
Centre for Plant Integrative Biology
University of Nottingham
Nottingham, UK

Enrique Rico-García
Department of Biosystems
School of Engineering
Queretaro State University
Querétaro, México

David Rousseau
Université de Lyon
Université Lyon 1
Villeurbanne, France

Kotaro Takayama
Ehime University
Matsuyama, Japan

P. Trivedi
Agricultural and Food Engineering Department
Indian Institute of Technology
Kharagpur, India

Marie-Hélène Wagner
GEVES—Station Nationale d'Essais de Semences
Beaucouzé, France

Toshiyuki Wako
Division of Plant Sciences
National Institute of Agrobiological Sciences
Tsukuba, Japan

K. Walsh
Central Queensland University
Centre for Plant and Water Science
Queensland, Australia

Yanyou Wu
Key Laboratory of Modern Agricultural Equipment and Technology
Chinese Ministry of Education
Jiangsu University
Zhenjiang, People's Republic of China

and

State Key Laboratory of Environmental Geochemistry
Institute of Geochemistry
Chinese Academy of Sciences
Guiyang, People's Republic of China

Kaiyan Zhang
State Key Laboratory of Environmental Geochemistry
Institute of Geochemistry
Chinese Academy of Sciences
Guiyang, People's Republic of China

chapter one

An introduction to images and image analysis

Michael P. Pound and Andrew P. French

Contents

1.1 Introduction

In this chapter, the reader will be presented with a basic introduction to images, image data, and some basic and widely used image processing techniques. When developing or understanding image analysis approaches in general, not just for the study of plant growth, it is necessary to have an understanding of the underlying data representations, within which is buried the information we wish to extract in the analysis stage. With a good understanding of the raw data, the reader will be well placed to comprehend the function and limitations of more specific image analysis methods. But understanding that an image is essentially a matrix of numbers that can represent different kinds of spatial information—dependent on the sensor type, image type, resolution, etc.—is the first step toward forming an image analysis solution.

Storing images for automated image analysis is a different technical problem than that of storing digital images for later manual analysis. Using high-quality raw data is crucial, as recapturing the data at a later stage is at best costly, at worst impossible, and in both cases clearly undesirable. Choice of image type, format, and compression is crucial here. Most people have heard of JPEG images, and some may know they compress the data, but do you know why you need to be careful of using them for data collected for scientific image analysis research? In this chapter, we hope to answer such fundamental questions.

Data storage these days is often thought of as prolific and cheap. Certainly, it is cheap to store terabytes of data. But over the course of an imaging-intensive research project, it may be that petabytes of data storage have to be allocated, and often in triplicate to allow for a backup strategy. An automated plant phenotyping setup using a variety of image sensors and capturing 3D data could easily end up accumulating this much data. Then, decisions relating to compression and bit depth of the images become serious considerations, so an understanding of these is essential.

A comparison of some commonly encountered color spaces is presented next. Some approaches in image analysis perform best on individual channels of an image (or components of a color space), and so an understanding of what is available and the differences between them will help the user to prepare data for a particular processing technique. For example, we may wish to segment an image into areas of the same color (hue), but we might not be interested in different brightness values (intensity) within the same color regions. Here, choosing a color space where we can separate hue from brightness would be a sensible choice.

Following a description of common data and file formats, and color spaces, some basic image analysis processing techniques are then presented. These techniques are often used as part of a preprocessing stage, prior to running more complex image analysis algorithms. Ways of removing different kinds of noise from an image are discussed, and useful terms are defined, including what *segmentation* means. Simple example segmentation approaches are described—in this case, related to binary thresholding. Morphological operations, which allow us to process geometric structures in the binary image, are then described in the context of cleaning up binary plant-related images. Finally, an introduction to some basic image features, such as edges and how to detect them, is presented.

If the reader becomes familiar with the concepts in this chapter, his or her understanding of the more involved image analysis techniques in the rest of the book should have a good foundation.

1.2 What is an image?

1.2.1 Image structure

An image is most often represented as a two-dimensional, rectangular grid of pixels. Images represented this way are called raster images. Each position in the image is located using positive integer values on a Cartesian coordinate system. The main distinguishing feature between images and regular Cartesian coordinates is that the origin of the image, pixel (0, 0), is found in the upper left corner of the image. At each coordinate, a pixel represents the color at that point. An example image can be seen in Figure 1.1.

This chapter will deal exclusively with raster images. Images captured by biologists using cameras, scanners, and microscopes will all use this representation, and image analysis algorithms assume that an image is in this form. However, it should be noted that there is a theoretical opposite of a raster image, often called a vector graphic. In these images, objects are represented as a series of points, lines, and more complex paths, generated using mathematical expressions. The benefits of vector graphics are scale and resolution, and device independence; if you scale a vector graphic, it

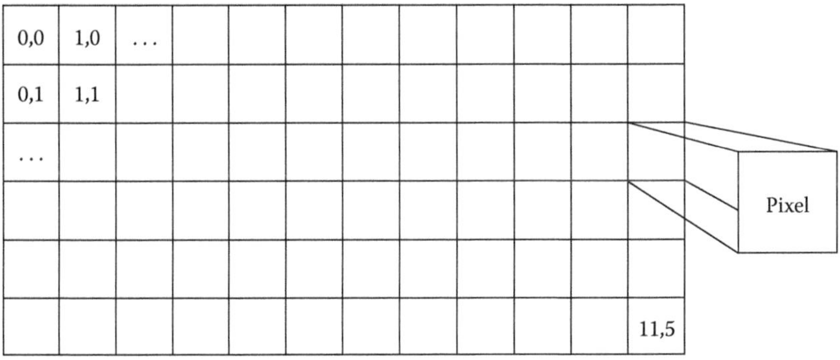

Figure 1.1 An example image of width 12 pixels and height 6 pixels.

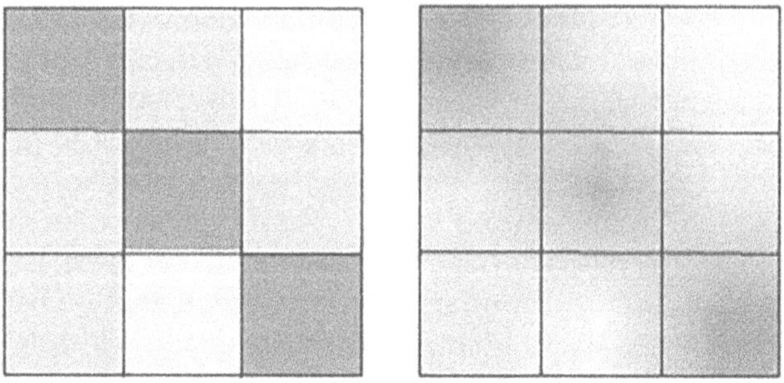

Figure 1.2 Two representations of the same image. Left: The common interpretation of pixels, as small squares. Computer monitors display pixels in this manner. Right: A smoothed image, treating each pixel as a point in the center of each square, and linearly interpolating between each value for positions between pixels.

simply becomes larger, rather than becoming pixelated due to inadequate resolution. Nevertheless, computer displays consist of a grid of pixels, like a raster image, so any vector graphics must first be converted into a raster image before they are shown. This process is called rasterization.

1.2.2 Pixels

As discussed above, at each position in the image a pixel represents the brightness and the color at that point. Although pixels are usually thought of as a small square section of the image, mainly because of the similarities between image pixels and display pixels in computer monitors, strictly speaking they represent some sampled point of the image. Thus, in reality, pixels represent a single point of color, or intensity level, assigned to a coordinate. Figure 1.2 shows this distinction.

In most cases the distinction between a pixel as a square and a pixel as a point is largely arbitrary. However, there are times in image analysis where it might be necessary to calculate the color or gray value *between* two pixels, in which case the rectangle representation would be inadequate, and interpolation should be used.

1.2.3 Bit depth and color channels

Along with width and height values in pixels, images are also described using their bit depth. This is the total number of bits, zeros, and ones that describe the color or intensity at each pixel location. As with all values stored on a computer, the number of bits used to store some information tells us how many different values can be stored. For example, 8 bits is a block of 8 zeros and ones, and can distinguish between $2^8 = 256$ different

Image data	00000000 01000000 00100110 11110000	10101100 01000110 11111111

Coordinate	(0,0)	(0,1)	(0,2)	(0,3)	(11,3)	(11,4)	(11,5)
Gray value	0	128	38	240	172	70	255
Intensity							

Figure 1.3 An example of grayscale image data, and the respective values that these data represent. Each row in the image is listed one after another in order from top to bottom. Pixel values range from 0 (black) to 255 (white).

values. Eight-bit images are usually, but not always, grayscale, and the pixels are usually stored consecutively in a list (see Figure 1.3).

For color images the theory is the same, except that additional channels are used to store separate values for each color component. For example, in a 24-bit color image, 8 bits are used for each value of red, green, and blue, and a tuple of RGB represents a single pixel. At each pixel, the combination of red, green, and blue produces the final color in the image. Table 1.1 provides details of some common bit and channel combinations for images used in image analysis.

1.2.4 *Image file formats*

While there are many file formats, most image analysis in the biological domain will encounter only a select few dedicated image formats. It is true that there are many proprietary microscope formats that contain image data, and include metadata such as microscope settings, etc., but for the purpose of image analysis, often we want to export from these files to image data in a more conventional format (that said, software such as Fiji (Schindelin et al., 2012) is capable of opening a wide variety of microscope formats and working with the data directly). The file format dictates not only how the image data are stored, but also what compression is used. Some file formats such as TIFF will also allow a user to tag an image with relevant information, such as date of capture, which can be particularly helpful when capturing images during an experiment.

Image compression is a technique whereby the raw image data are transformed in such a way as to make them more memory efficient. Compression can take one of two forms. Lossless compression aims to shrink the size of the image data, while preserving all of the information held within them. Most images will contain some degree of repetition, for example, a solid area in a single color. Where this occurs, these contiguous blocks can be compressed into a single instruction, allowing a decoding algorithm to reconstruct the entire image. For example, pixel data such as 000000111111112222223333 may be compressed using run length encoding in a form such as 6-0, 8-1, 6-2, 4-3, which requires less space.

Table 1.1 Descriptions of Notable Image Bit and Channel Combinations

Name	Number of channels	Bits per channel	Total possible color combinations	Description
1-bit binary	1	1	2	In a binary image, values can be either 0 (black) or 1 (white). There are no intermediate values. Cameras and other capture devices do not usually capture a binary image; rather, they are obtained by thresholding a higher-quality image into foreground and background pixels.
8-bit grayscale	1	8	256	As seen in Figure 1.3, an 8-bit grayscale image stores up to 256 levels of gray, referred to as intensity levels due to their correspondence with image brightness.
16-bit grayscale	1	16	65,536	Structurally identical to an 8-bit grayscale image, but with double the number of bits per pixel. This grayscale image can be used when 256 intensity levels is insufficient to capture the information required.
8-bit indexed	1	8	256 of 16 m	Contains 256 grayscale or color values, but these values are stored separately from the pixel data that index them. Can be used if some color is required, but only 8 bits per pixel is available.

24-bit RGB	3	8	16 m	3 bytes per pixel, allows for a total of 16 million color combinations. A very common image format. In practice, each pixel is usually stored in BGR order, but this makes little difference to image analysis algorithms.
32-bit ARGB	4	8	16 m + transparency	Much like 24 bpp RGB, this includes a separate alpha channel for transparency. This image format is very common in web applications. The similar 32 bpp RGB format simply ignores any transparency data in the last 8 bits.
RAW	—	—	—	This is a platform-dependent format that allows images to be captured without any alteration. This will usually require conversion into a more common format, using proprietary software included with the camera or capture device. This format is only beneficial if some aspect of the image would be lost if it was preconverted to another format.

Note: The only remaining consideration with regard to bits per pixel is the resulting image size. Assuming there is no compression in the image (a topic covered in the next section), high bit depths can cause file sizes to become increasingly large. In previous decades, where magnetic storage was smaller and more expensive, this was more of a concern than it is now. However, even now RAW and uncompressed image formats can be very large.

The obvious advantage of lossless compression is that images are unaltered, and all information is retained and can be used in any image analysis steps. The disadvantage is that this approach can still be memory inefficient where images are hard to compress. Images that are repetitive or have large blocks in the same color can be compressed easily, but photos of biological subjects may not have these properties. The common ZIP format is an example of lossless compression that can be used on any file type. TIFF file formats allow different types of compression behind the scenes, including ZIP-like lossless compression.

Lossy compression attempts to reduce the file size further than lossless approaches, but at the cost of some image information, which will be irrevocably lost. Traditionally, lossless compression is used in domains where memory efficiency is paramount, such as transmitting images across the Internet. Many compression algorithms also attempt to exploit properties of the human visual system such that the loss in quality is not noticeable to most observers. For example, the JPEG image format applies more lossless compression to the color information in an image than to the brightness information. This is because human eyes are more sensitive to contrast than they are to changes in color, and hence more color compression can be achieved before a human observer will notice.

In image analysis, the appearance of an image to a human is of little importance, and modern memory is very cheap. Users capturing images should think carefully before using lossy compression methods, and in most cases lossless compression will ensure that all information is retained for image analysis. Certainly if JPEG is used, the highest-quality setting should be used.

Table 1.2 shows a selection of common image file formats and their properties.

1.2.5 Color spaces

The color image data described above were concerned with 8-bit RGB color, that is, color separated into three separate red, green, and blue components. This is one of many ways to represent color, and not all image formats store RGB data. In addition, many image analysis routines can be used effectively in different color spaces, so knowledge of these spaces is helpful. Diagrammatic representations of common color spaces can be seen in Figure 1.4.

Biologists who are familiar with confocal microscopy should note that confocal images are often stored using arbitrary colors, sometimes correlating with the colors of the lasers used. In fact, the confocal microscope measures only intensity of the fluorescence at a point and at given wavelengths, so in reality confocal images are similar to a group of

Table 1.2 A Comparison of Common Image File Formats

Full name	Extension	Compression	Allows tagging	Description
Bitmap (BMP)	.bmp	Lossless	No	Can make use of limited compression, but is often uncompressed. This results in very large file sizes.
Portable Network Graphic (PNG)	.png	Lossless	No	Now very common, reasonable lossless encoding now makes PNG preferable to BMP in many situations.
Tagged Image File Format (TIFF)	.tiff	Lossless	Yes	Often used in scientific research. Uses lossless encoding but also allows a significant amount of extra information to be included in tags.
Joint Photographic Experts Group (JPEG)	.jpg, .jpeg	Lossy	Yes	Generally used to store photographs, the lossy compression in JPEG might be unsuitable for scientific use. However, the amount of compression can be altered, and at low levels a large decrease in file size can still be obtained, with minimal loss in quality.
Graphics Interchange Format (GIF)	.gif	Lossy	No	GIF compression uses a color palette of only 256 colors, and is unsuitable for scientific use in most cases.

grayscale intensity images, and do not adhere to any of the color spaces discussed below.

1.2.5.1 RGB

The RGB color space splits each pixel into three colors, representing the three primary color components, red, green, and blue. RGB can be visualized as a three-dimensional cube (Figure 1.4a), where each axis represents one of the color channels. The color black is found at the origin, where RGB values are (0, 0, 0). White is found at the opposite corner, with RGB values of (255, 255, 255) for an 8-bit image. Grayscale pixels are found along the line between the black and white corners, where R, G, and B have the same value. The RGB format is popular because it matches the structure of pixels in monitors and other displays. However, the main notable drawback of RGB is that it combines color and brightness into the same space. Conversions between RGB and all other color spaces exist; thus, the required color component can be separated from the brightness component by converting it into a color space that makes that distinction.

Although this representation is almost exclusively referred to as RGB, many bitmap files, in particular the Windows BMP format, actually store the data in BGR order. In most cases pixels are stored in 32-bit blocks; thus,

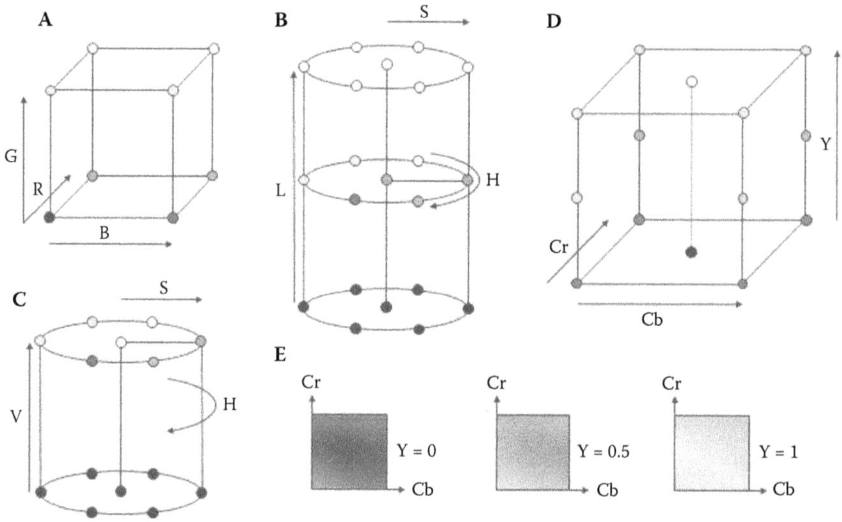

Figure 1.4 **(See color insert.)** Diagrammatic representations of common color spaces. (a) The RGB cube. (b) The HSL cylinder. (c) The HSV cylinder. (d) The YCbCr cube. (e) A selection of planes taken from the YCbCr cube, demonstrating how color changes with Y, Cb, and Cr.

a pixel will usually comprise 24 bits of BGR, followed by an additional unused 8 bits. In other formats such as PNG, these 8 bits are used as a transparency, or alpha, channel.

1.2.5.2 HSV

The HSV color space represents the separate hue, saturation, and value components of a pixel. HSV is most easily viewed as a cylinder (Figure 1.4c), with hue being the position around the edge, saturation the distance from the center to the edge, and value the position from the top to the bottom.

Hue represents the color of the pixel, and usually a value in degrees, from 0 to 360. At 0°, the hue color is red; by rotating about the hue wheel, the colors will pass through blue and then green, and then finally back to red. Saturation represents the intensity of the color, from strong through to grayscale. The closer to the center of the cylinder cross section the saturation value lies, the less the hue value will be expressed. Finally, the position up and down the cylinder represents value, or brightness. Toward the bottom there will be darker pixels, with lighter pixels above. Any position on the HSV cylinder can be matched by a pixel in RGB, and conversion between the two color spaces is simple. Because HSV separates color (H + S) from intensity (V), the HSV color space separates the color components of a pixel from the grayscale component in a way that RGB does not. HSV is therefore more stable during changing lighting conditions, which may occur when analyzing images over time.

1.2.5.3 HSL

Similar to HSV, the hue, saturation, and lightness (HSL) color space converts RGB into a distinct color and brightness components. However, there are slight differences, as can be seen in the color space diagram (Figure 1.4b). While the saturation value still influences whether a color is vibrant, or washed out, the lightness value now assigns the blackness and whiteness of a color. For image analysis purposes, HSV and HSL are similar. However, it is often thought that HSL is the more intuitive space, as a high value of lightness will produce a white pixel, rather than a pixel whose whiteness depends on the additional saturation variable.

1.2.5.4 YCbCr

YCbCr (sometimes referred to as YUV) exists to reduce the redundancy inherent in signals sent using RGB. The Y component, scaled between 0 and 1, represents the luminance of a pixel. The color components Cb and Cr represent the blue difference and red difference. Any RGB color can be found on the Cb and Cr axis, with the luminance Y specifying the shade of that color. A diagram of the color space is shown in Figure 1.4d;

however, it is often easier to visualize as only CbCr using constant Y values, as in Figure 1.4e. While the amount of information per pixel in YCbCr is not any greater than RGB, by separating the luminance and color information, much like with HSV, different compression algorithms can be applied to the color or to the intensity of an image. It is the YCbCr color space that the JPEG file format uses to highly compress the color information. Human observers cannot resolve color information as accurately as grayscale intensity, so video encoding for TV is also processed as YCbCr.

While compression of the color information is not ideal for images meant for use in image analysis, it should be noted that many algorithms, such as segmentation and stereo reconstruction, can work well on grayscale images. Other algorithms, like common edge filters, operate exclusively on grayscale images. It is therefore more important to preserve intensity information than color information when capturing images.

1.3 *Analyzing images*

Image analysis is the process of working from raw pixel data to obtaining some useful information from the image, typically a measurement of the objects within. While some image analysis algorithms can be quite complex (as will be seen in later chapters), there are a variety of simple techniques that see widespread use due to their varied applicability. For example, image filtering is often used to reduce or remove image noise prior to further processing. Image segmentation can be used to locate regions of interest, separating foreground from background, and ranges in complexity, from very simple techniques to very complex ones.

1.3.1 *Image filtering*

Image noise occurs in all captured images, regardless of the quality of the sensor. It is caused by a number of factors, but primarily by naturally occurring electronic noise. The different causes of image noise produce unwanted variations in pixel color or intensity, away from the true color of the object being viewed. These variations are drawn from different probability distributions, depending on the nature of the noise. For example, the majority of noise generated by a camera sensor will follow a Gaussian distribution.

Image filtering is an effective way of reducing noise in an image while preserving important aspects of the subject. The nature of the filter used should depend on the nature of the noise, and multiple filters can be used where there are multiple sources of noise. This section discusses a variety of noise types and suggests appropriate image filters to reduce them.

1.3.2 *Kernel convolution*

Many image filters use the mathematical convolution operation to convolve an image with a mask, often called a kernel. An example kernel may be structured like this:

1	1	1
2	3	2
1	1	1

where coordinates of values in the kernel are defined relative to the center, not the top corner like an image:

(−1, −1)	(0, −1)	(1, −1)
(−1, 0)	(0, 0)	(1, 0)
(−1, 1)	(0, 1)	(1, 1)

The kernel is altered depending on the effect that the filter requires. The discrete convolution operation at pixel coordinate is defined as

$$I'(u,v) = \frac{1}{\sum H(i,j)} \sum_{i=-\infty}^{\infty} \sum_{j=-\infty}^{\infty} I(u-i, v-j) H(i,j)$$

where I is the original image, I' is the filtered image, and H is the kernel to be applied to I. In other words, for each pixel in the source image, we apply the kernel at that point. We then multiply all neighboring pixels under the mask by the corresponding value in the kernel, and sum the result for all neighbors. How many pixels are considered at each location is dependent on the size of the kernel. The division by $\Sigma H(i, j)$, the sum of all elements in the kernel, ensures that the image intensity is not altered, should the sum of the values in the kernel not be 1. The following example aims to illustrate the process of convolution.

Given the following kernel:

1	1	1
2	3	2
1	1	1

and the following image:

1	3	5	2	8	8	5	1
6	4	6	9	3	1	9	9
2	7	1	5	3	6	8	2
1	3	8	2	4	9	7	3
7	3	5	6	4	1	5	4
3	1	4	7	9	2	9	9

The filtered value at coordinate (4, 2) is calculated as

$$\frac{1}{9(9*1+3*1+1*1+5*2+3*3+6*2+2*1+4*1+9*1)} = 6.55$$

This process is repeated for all values in the image, resulting in a new, filtered image as output.

1.3.2.1 Mean filter

The mean filter averages local pixel values in a region, reducing the magnitude of uniform noise. The filter also has the effect of blurring the image. The main form of noise in an image is quantization noise, that is, noise that forms where pixels have been quantized into a discrete range, for example, 0–255 in 8-bit images. If a sensor can measure intensity or color with a higher degree of accuracy than 8 bits, the encoding of the image will cause pixels to take the nearest appropriate value, and some additional information will be lost. This loss will cause a uniform error throughout the image.

The kernel for a mean filter is given as

1	1	1
1	1	1
1	1	1

or

1	1	1	1	1
1	1	1	1	1
1	1	1	1	1
1	1	1	1	1
1	1	1	1	1

Two examples are shown, a 3 × 3 kernel and a 5 × 5 kernel. A mean filter kernel can be any size, N × N. The larger the kernel, the more pixels are included in the mean calculation, and the larger the blur. Because a mean filter treats all pixels under the kernel as equal in weight, there is a disadvantage that pixels far from the center pixels can influence its color.

1.3.2.2 *Gaussian filter*

Unlike a mean filter, a Gaussian filter gives higher weight to pixels closer to the center of the kernel. These values are drawn from a two-dimensional normal distribution, where the mean of this distribution is the center of the kernel, and the standard deviation is varied depending on how strong the effect of the filter should be. The general formula for a 2D normal distribution is given as

$$f(x,y) = A\exp\left(-\left(\frac{(x-x_0)^2}{2\sigma^2}+\frac{(y-y_0)^2}{2\sigma^2}\right)\right)$$

where A is the amplitude, (x, y) is the center, and σ is the standard deviation (s.d.) of both directions of the Gaussian. The s.d. is kept identical for both directions, to ensure the same amount of blur is applied horizontally and vertically in the image.

The dimensions of the kernel necessary to represent a normal distribution are dependent on the standard deviation. While theoretically normal distributions are of infinite size, in practice the majority (98%) of the distribution can be represented as a kernel with a radius of σ * 2.5. However, for even relatively small σ, this produces kernels large enough to become computationally inefficient. The appropriate dimensions of a kernel based on a Gaussian with is 20 × 20. One benefit of using a normal distribution is that the x and y components can be separated into two separate passes over the image; an N × N kernel can be split into two kernels of size N × 1 and 1 × N. The result of convolution with the first kernel is convoluted with the second, to produce the same result that would be obtained if convoluting with the much less efficient N × N kernel. The x and y component kernels for a Gaussian filter of $\sigma = 3$ are

$f(x) =$

0.004	0.009	0.018	0.033	0.055	0.081	0.106	0.126	0.133	0.126	0.106	0.081	0.055	0.033	0.018	0.009	0.004

T

$f(y) =$

0.004	0.009	0.018	0.033	0.055	0.081	0.106	0.126	0.133	0.126	0.106	0.081	0.055	0.033	0.018	0.009	0.004

Compared to a mean filter, a Gaussian filter is much better at preserving edges, and distant pixels do not influence the target pixel value as much as closer pixels. An example of an image filtered using a Gaussian blur can be seen in Figure 1.5a and b.

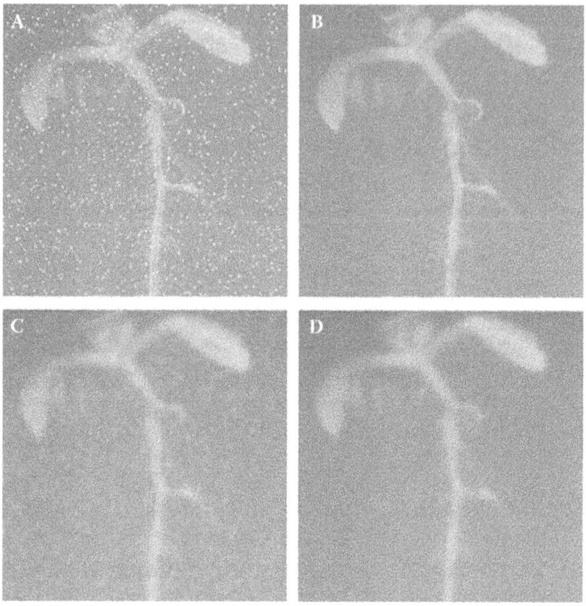

Figure 1.5 (a) Image of an *Arabidopsis* seedling; the image has had artificially added Gaussian noise, followed by salt-and-pepper noise. (b) The same image filtered with a median filter; the majority of the salt-and-pepper noise has been removed, but the Gaussian noise remains. (c) The same image filtered with a Gaussian blur. Much of the Gaussian noise has been removed, but the salt-and-pepper noise has been spread out to neighboring pixels. (d) A median filter followed by a Gaussian filter. Combining two image filters first removes the salt-and-pepper noise, and then reduces the Gaussian blur. (Courtesy of Ric Traini, Centre for Plant Integrative Biology, University of Nottingham.)

1.3.2.3 *Median filter*

A median filter works in a similar way to the operation of a mean filter, except that the median values of the pixels under the mask are calculated, rather than the mean value. A median filter does not use a specific kernel, simply a window size, below which the median is calculated. For example, given the following image, using a window size of 3 × 3:

1	3	5	2	8	8	5	1
6	4	6	9	3	1	9	9
2	7	1	5	3	6	8	2
1	3	8	2	4	9	7	3
7	3	5	6	4	1	5	4
3	1	4	7	9	2	9	9

the filtered value at coordinate (4, 2) is calculated as the median of {9, 3, 1, 5, 3, 6, 2, 4, 9}. The median of these values is 4.

A median filter is very effective at removing salt-and-pepper noise, where errors in the capture device lead to some pixels having extremely low or high intensities. High levels of Gaussian noise will also lead to some extremely high or low values. As the noise is normally distributed, these values will be uncommon, but a median filter will remove them where they do occur. Figure 1.5c shows a typical output of a median filter on an image containing both Gaussian and salt-and-pepper noise. Figure 1.5d shows a more effective approach of first using a median filter to remove salt-and-pepper noise, and following this with a Gaussian filter to smooth the remaining Gaussian noise.

1.3.3 Segmentation

Segmentation is the process of splitting the pixels of an image into groups, where each group has some meaningful distinction from the others. The simplest form of segmentation would split pixels into two classes, where one class represents the areas of interest in the image, and the other represents the background. Segmentation into two groups of pixels is often achieved using thresholding, the process of grouping pixels based on their intensity or color.

1.3.3.1 Binary thresholding

The simplest form of thresholding into two classes is called binary thresholding. In binary thresholding, some intensity or color value is chosen; all pixels below this level are classed as background, and all pixels above are classed as foreground. This approach essentially separates bright foreground from darker background. An example can be seen in Figure 1.6.

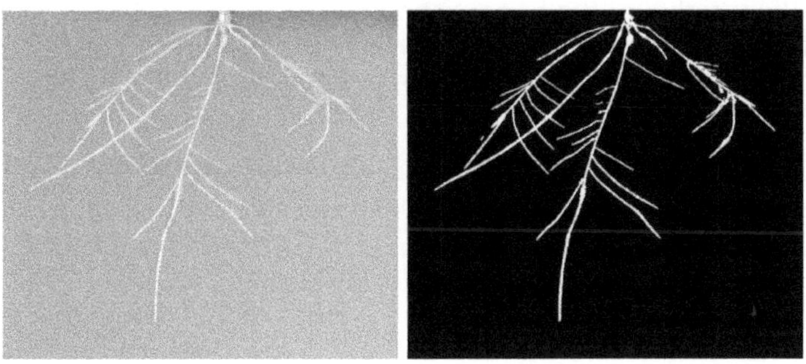

Figure 1.6 Grayscale image of a wheat root grown on germination paper. Right: A binary image resulting from thresholding the left image. A threshold value of 177 was manually chosen.

While binary thresholding can be appropriate for images where the foreground and background pixels are clearly defined, it can become problematic where intensity levels are not so clear. The user must also specify the level at which thresholding occurs. A level midway between 0 and 255 may work on many images, but not on images that are darker or brighter than usual. Binary thresholding over an entire image will also fail to adequately handle images that include gradual changes in intensity over the entire image. Where one side of the image is lighter or darker than another, the results of the segmentation will change in the different regions of the image.

1.3.3.2 *Adaptive thresholding*

One disadvantage inherent in binary thresholding is that the threshold itself must be set manually by the user. This process can be time-consuming and subjective; a better approach would be to automatically calculate the optimum threshold. A popular approach to so-called adaptive thresholding is the Otsu method (Otsu, 1979).

Otsu thresholding operates by minimizing the intraclass variance, calculated as a weighted sum of the variances of both the foreground and background classes. In other words, Otsu chooses a threshold level t that minimizes the variance of the pixels in both classes. This intraclass variance calculation is made easier by instead calculating the interclass variance, that is, the variance between the foreground and background pixels. Otsu showed that the threshold t that maximizes the interclass variance is the same value that minimizes the intraclass variance.

Interclass variance is given as

$$\sigma_B^2 = w_b * w_f * (\mu_b - \mu_f)^2$$

where the weights of the foreground and background classes, w_b and w_f, are calculated as the number of pixels in the respective classes, divided by the total number of pixels. μ_b and μ_f are the mean values of the background and foreground classes.

Otsu thresholding calculates σ_B^2 for each possible threshold value, and then selects the optimum threshold:

$$\text{Threshold} = \arg\max_{t\sigma_B^2(t)}$$

1.3.3.3 *Region-based segmentation*

While adaptive thresholding is preferable to manual thresholding in many cases, broad changes in image intensity over the entire image can still lead to poor results. To account for global changes in image intensity, it is often beneficial to split an image into smaller subregions, before separately applying an adaptive threshold to each region. If some regions appear brighter than others, local adaptive thresholding can treat these regions as distinct, and apply a different threshold value.

1.3.3.4 Advanced segmentation

The segmentation techniques described so far can be thought of as low-level pixel-based methods. While the average intensity of neighboring pixels is a factor in region-based segmentation, regions are generally large, and each pixel's influence is slight. More advanced segmentation algorithms will use the locations of pixels to better effect, by clustering groups of neighboring pixels into separate regions. This chapter will not cover such algorithms in detail; however, those interested are encouraged to read about watershed segmentation (Vincent and Soille, 1991).

1.3.4 Morphological operations

When an image is segmented into a binary result (such as foreground and background) without using any prior models of expected shape, it is likely that the initial results will need further processing to improve the segmentation. For example, due to variations in gray level, a thin foreground object may have missing sections (see Figure 1.7).

To remove such extraneous noise and holes from images, morphological processing can be used. Morphological operators are commonly used on binary images. Two of the most basic processes are performed with the erosion operator and the dilation operator, which will be explained here. Much of the other morphological operations can be thought of in terms of these two foundations operators.

The erosion operator has the effect of shrinking the foreground pixels representing an area. That is, the foreground pixel area is eroded away, leaving less foreground present. This is achieved using a structuring element, similar to the kernel used in filtering. A typical morphological kernel is a 3 × 3 square as follows:

1	1	1
1	1	1
1	1	1

This is placed over every pixel in the foreground, with the current pixel under test being placed in the (0, 0) central position. The neighboring pixels that fall under the kernel are considered, and if *all* of them are

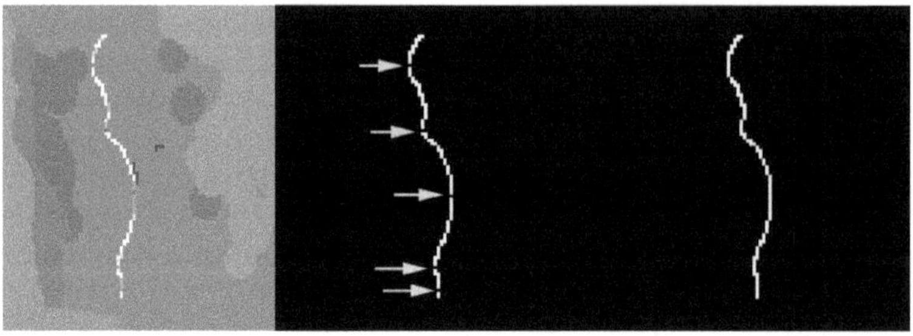

Figure 1.7 Example of morphological erosion and dilation. Original image of plant roots (top left), binary image produce by thresholding, inverted for clarity (top right), erosion applied to the binary image (bottom left), and dilation applied to the binary image (bottom right).

foreground pixels, then the pixel under consideration is set to the foreground in the output image; otherwise, it is set to the background.

The opposite of the erode operator, and the second of the two most basic operators, is the dilate operator. Its use proceeds with the same kernel and procedure as above, except that the output pixel is set to the foreground if *at least one* of the other pixels in the kernel is a foreground. As its name implies, this has the effect of dilating or growing the foreground segmentation.

By chaining a dilate operation and an erode operation together, we can fill small holes in the foreground. It is easy to imagine how a dilation operation can fill holes, by expanding the boundary of foreground shapes until a gap between them is filled. Clearly this affects the shape and size of the foreground object. Following the dilation with an erosion operation using the same structural element still allows small gaps to be bridged, but is less destructive to the shape of the original foreground element, as the boundary is shrunk again after the initial expansion (see Figure 1.8, right panel). The opening operator effectively produces the opposite effect to closing, and has the effect of removing small areas of foreground, such as speckle noise (rather than small holes).

1.3.5 Edge detection

Finding features in the image can be thought of as one conceptual level above finding groups of pixels, which the thresholding-based methods we have seen so far have produced. A common feature requirement is to segment the edges from the images. This allows us to find the boundaries of objects. A crude method is to use morphological operations on a binary image; dilation followed by subtracting away the original image will leave only the dilated pixels around the edges of

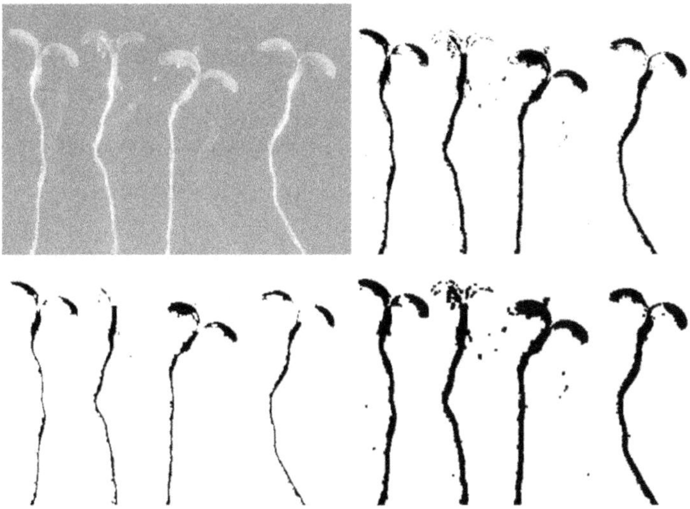

Figure 1.8 Example of a common problem with fixed-level global thresholding. Suppose we want to segment the central linear feature in the simulated image (left); using global thresholding at a cutoff value of 128 correctly categorizes all the background pixels as black, but misclassifies some of the darker foreground pixels into the background category (center). To fix this, a morphological dilation is performed, followed by an erosion. This is also referred to as morphological closing, and has the effect of filling small holes in the foreground (right).

the foreground areas. Often, though, the images in which we are trying to find an edge are not binary. An edge can represent a separation between regions of different colors, textures, 3D depths, etc., but is most simply introduced as a difference in intensity between neighboring regions in an image.

Figure 1.9 shows the first derivative of the intensity plot across the edge—note how it gives rise to a sharp peak on the strong edge, which we can clearly see in the figure. This suggests the first derivative will make a good basis for an edge detection operator. A kernel convolution operation can achieve the desired effect. A first derivative approximation in the x direction can be calculated by passing the following kernel across the image:

−1	0	+1

Hopefully it is clear to the reader why this will give a large response when sited over a light/dark vertical edge. One of the most widely known edge operators is the Sobel operator. It uses the same principles, but has two kernels each designed to search for edges in different directions:

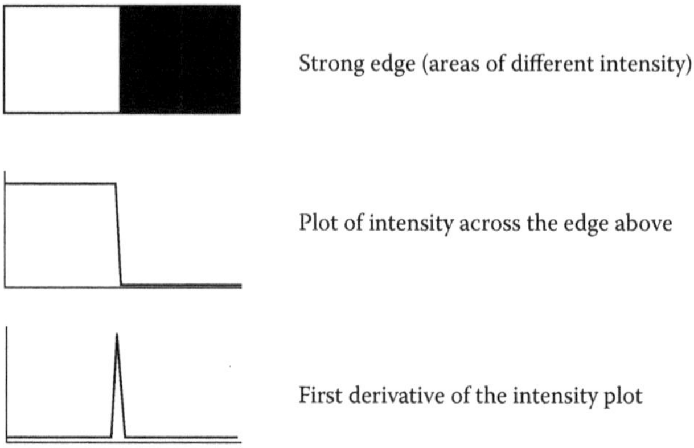

Figure 1.9 Top: A synthetic image illustrating two regions with a clear edge between them. Center: Intensity profile across the two regions. Bottom: First derivate (gradient) of the intensity profile line.

−1	0	+1	
−2	0	+2	
−1	0	+1	

Gx

+1	+2	+1	
0	0	0	
−1	−2	−1	

Gy

The results of the operators can be combined using Pythagoras' theorem to give a magnitude for the response to the edge detection:

$$|G| = \sqrt{Gx^2 + Gy^2}$$

It is also possible to combine the two responses to produce a dominant direction for the edge. For more details see Sonka et al. (1999).

1.4 Conclusion

In this chapter we have presented to the reader an introduction into the structure and design of images, and an overview of many common low-level image analysis operations. An understanding of the underlying data representations of image data is crucial if one hopes to design effective image analysis algorithms, or correctly make use of existing algorithms and tools.

An overview of the storage mechanisms behind pixel data was presented, followed by a comparison of some popular image formats used within plant science research and further afield. We presented a brief overview of the contrasts between lossless and lossy compression, in the hope that researchers will consider their image capture requirements before embarking on a new project. Finally, we covered many of the fundamental image analysis algorithms necessary for a researcher to begin analyzing his or her own images. While these techniques could be considered only an introduction to image analysis, they can be found throughout the literature, in many complex image analysis applications.

References

Otsu, N., A threshold selection method from gray-level histograms, *IEEE Trans. Syst. Man. Cybern.*, SMC-9, 62–66, 1979.

Schindelin, J., Arganda-Carreras, I., Frise, E., et al., Fiji: an open-source platform for biological-image analysis, *Nature Methods*, 9, 676–682, 2012.

Sonka, M., Hlavac, V., and Boyle, R., *Image processing, analysis, and machine vision*, 2nd ed., Brooks/Cole, California, 1999.

Vincent, L., and Soille, P., Watersheds in digital spaces: an efficient algorithm based on immersion simulations, *IEEE Trans. Pattern Anal. Mach. Intell.*, 13, 583–598, 1991.

chapter two

Image analysis for plants: Basic procedures and techniques

Yasuomi Ibaraki and S. Dutta Gupta

Contents

2.1 Introduction

Image analysis is a promising tool for nondestructive analysis of biological objects, and has been widely used in botanical research and practical agriculture. The technique is now readily available at low cost and is being widely applied to objects from the cell level to the plant and canopy levels. Advances in devices for digital image acquisition and personal computers have contributed to this progress. Software for image analysis is also now readily available. The main advantage of image analysis is its potential for nondestructive and objective analysis. The objectives of the analysis include measurement (of size, population, growth, etc.), quality evaluation, classification, and visualization. As a useful research tool, image analysis has been widely used in microscopy for improving the visual appearance of an image to a human viewer or for measurement of various features of organelles, cells, and organs from his or her images. In addition, it is possible to quantify elongation or expansion in roots or shoots using serially acquired images for plant growth analysis (Spalding and Miller, 2013). Image-based analysis of morphological features may also be an effective tool for phenomics in plant (Arvidsson et al., 2011; Iyer-Pascuzzi et al., 2010; Zhong et al., 2009; Keyser et al., 2013; Chapter 9 in this book). Furthermore, images are used for photosynthetic analysis, via chlorophyll fluorescence imaging or photochemical reflectance imaging, giving spatial information on photosynthetic properties within a leaf, plant, or canopy.

Image analysis is also promising for practical use in agriculture. Evaluation of plant status based on visual inspection is often performed for management of cultivation. Image analysis has the potential for objective evaluation of plant status and is expected to help in the management of cultivation. In particular, nondestructive evaluation of plant status using images permits monitoring a time course of plant status, yielding valuable information, including growth rates and developmental stages. In addition, image analysis is promising for acquiring information about the physiological state of plants, including leaf area index (Liu and Pattey, 2010), chlorophyll content (Yadav et al., 2010; Dutta Gupta et al., 2013), and disease severity (Corkidi et al., 2006; Wijekoon et al., 2008; Cui et al., 2010). Thermal imaging of leaves can also provide information about transpiration and can be used for stress detection, and numerous applications have been published to date (Jones et al., 2002). Moreover, it has been recently reported that light intensity distribution on a canopy surface can be estimated using images acquired through a specific optical filter (Ibaraki et al., 2012).

Given that readings of reflection and radiation of electromagnetic waves from a target object or area are acquired as images in remote sensing, image analysis is one of the basic components of remote sensing data analysis. Numerous applications for plants, mainly for large plant canopies, have been reported in this area of research. The potential uses of remote sensing for horticultural crops have been reviewed by Usha and Singh (2013).

In this chapter, procedures and techniques of image analysis along with its application in higher plants are discussed, mainly focusing on macroscopic imaging at the level of plant organs such as leaves and roots, whole plants, or small plant canopies.

2.2 *Procedures of image analysis for biological objects*

2.2.1 *Basic flow*

Image analysis of a biological target proceeds as follows: (1) acquisition of an image of the target object, (2) preprocessing of the image for facilitating further processing, (3) selection of pixels of interest, and (4) extraction of characteristic features.

After preprocessing of the acquired digital image, pixels corresponding to the target object in the image are selected. Characteristic features are then calculated for the selected area (the selected pixels) according to the purpose of analysis. For examples, the pixels are counted in case of measurement of size (length or area), and red, green, and blue (RGB) values may be extracted for color analysis.

For texture analysis, the flow is somewhat different and does not require selection of pixels corresponding to the target objects (Ibaraki, 2006). Details of texture analysis of plants are described with application examples in Section 2.7.

2.2.2 *Image acquisition*

Digital images of target objects can now be readily acquired with a digital camera, a digital video camera, or a camera attached to a personal computer or a cellular phone. Image data acquired with analog imaging devices, in the form of a printed picture or a video signal, may be converted into digital image data with special devices such as a scanner, a digitizer, or an analog–digital converter. An image is an expression of the spatial distribution of light intensity or color. A digital image consists of a number of small compartments referred to as picture elements or pixels, and a digital value expressing light intensity is assigned to each pixel. This value is often referred to as pixel value, pixel digital number, intensity, or gray level.

Properties of the acquired image, including spatial resolution, number of gray levels, color system, and the file format (compressed or uncompressed), should be given proper attention. A digital camera saves images in Joint Photographic Experts Group (JPEG) format, which is one kind of lossy compression image format. Part of the image information is lost in the compression process, particularly when a high compression rate is applied. Because images in uncompressed formats such as BMP, uncompressed TIFF, and raw data format are easy to analyze and introduce the highest optical performance in space and color resolution of the imaging device, they are desirable for precise image-based measurement. However, images of these types have greater file sizes and require more calculation time.

Most digital cameras are provided with automatic image control systems such as automatic gain control and automatic white balance control. Automatic gain control changes sensitivity of the camera to the input of light, i.e., the output (pixel value in an image), and automatic white balance results in color changes in the image according to lighting conditions. These automatic control systems change imaging conditions and, as a result, may make comparison among the images difficult. Moreover, when light intensity is to be measured from images, the linearity between input and output should be confirmed.

Another important aspect in imaging target objects for analysis is to acquire an image in which it is easy to select (extract) the target objects from the background. Therefore, attention should be paid to the background in imaging. It is effective to place behind the target object a board of a different color than the target. Materials used for supporting plants in cultivation, such as poles and nets, should have different colors than leaves, stems, or fruits in practical application of image analysis. Lighting conditions in imaging also influence the success of extraction of the target object. Special lighting devices are sometimes used for plant imaging to extract the objects of interest. For example, Keyser et al. (2013) used a black light to extract leaves.

In specialized imaging such as thermal, hyperspectral, and fluorescence imaging, suitable imaging devices are required for each specific purpose, although image analysis follows the same process as that of the normal image. In particular, thermal imaging cannot be performed with normal digital cameras, as the imaging device requires the detection of infrared rays. However, a general-purpose charge-coupled device (CCD) or digital cameras may be used for hyperspectral imaging and fluorescence imaging by combining optical filters including band-pass filters or long/short-pass filters, which transmit only specific wavelengths of light. Indices using spectral reflectance such as normalized difference vegetation index (NDVI) or photochemical reflectance index (PRI) may be estimated by changing the band-pass filters attached to a camera (Ibaraki et al., 2010). For chlorophyll fluorescence imaging, optical filters are used for both camera and light source for exciting fluorescence (Omasa and Takayama, 2003; Ibaraki and Matsumura, 2005). Modulation lighting systems enable us to image fluorescence emitted in response to constant excitation light intensity (fluorescent quantum yield). Commercially available chlorophyll fluorescence imaging systems (Fluorcam, PSI; Imaging PAM, Walz) are equipped with modulated lighting systems and permit readily to image PSII quantum yield (Figure 2.1), although they are still expensive. Lasers, which are narrow-band light sources, have also been used as lighting devices for imaging fluorescence (Novák, 2011). Recently, systems for imaging green fluorescent protein at the macroscopic level—in a whole leaf (Stephan et al., 2011) or root (Novák, 2011)—have been developed. In addition, multispectral fluorescence imaging can provide useful information about plants' state of health (Lenk et al., 2007).

2.2.3 *Preprocessing*

Preprocessing is a procedure for facilitating subsequent processing of the image. The purpose of preprocessing includes noise reduction, geometric

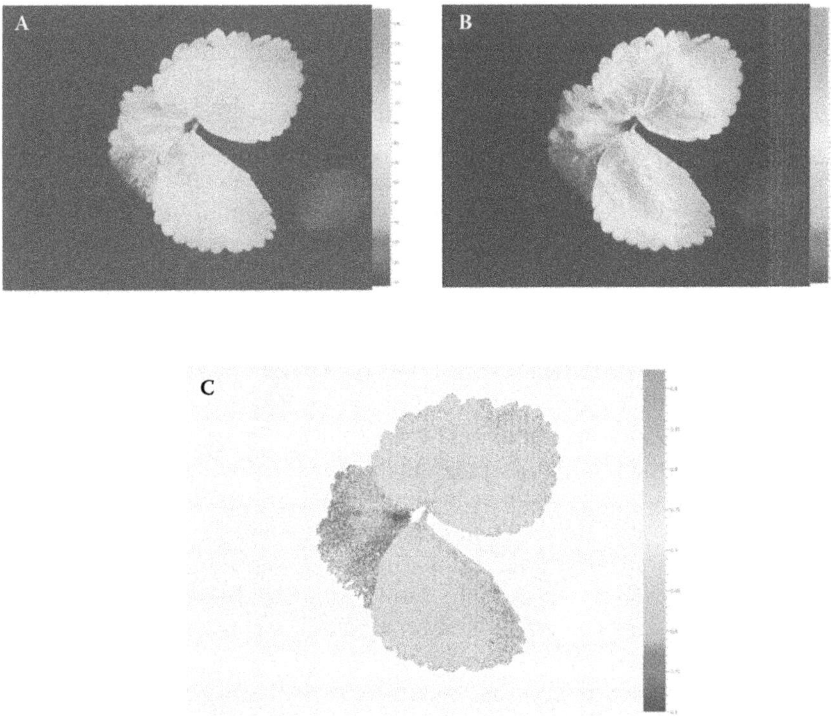

Figure 2.1 Chlorophyll fluorescence images of strawberry leaves acquired by a commercially available chlorophyll fluorescence imaging system. (a) F image, (b) Fm image, and (c) Φ_{PSII} image. Only the left leaf had been irradiated with high intensity of light, showing the reduced Φ_{PSII}.

correction, modification of spatial resolution and number of gray levels, and conversion of color mode. For noise reduction, a smoothing filter, which is a matrix used to calculate the average pixel value using several pixels around the target pixel, including a moving average filter, Gaussian filter, or other type, is often used. A median filter assigns to a target pixel the median value of several pixels around it and is also used for noise reduction.

Processes called opening and closing may be effective for noise reduction in a binary image, eliminating small internal holes. It should be noted that all images acquired with imaging devices, including digital cameras and CCD cameras, are susceptible to geometric distortion, which is a discrepancy on an image between the actual and the ideal image coordinates, and is caused mainly by the properties of the lens. As a result, there are many types of distortion. It is effective to image orthogonal grids such as a section of paper for identifying the degree and type of distortion caused by the imaging device. Geometric correction of the image includes affine, conformal, and projective transformation.

Although an 8-bit or 24-bit color image (8 bits for each color) is normally used in digital image analysis with a personal computer, some CCD cameras can output images with a number of gray levels greater than 8 bits. A higher number of gray levels involves more information on light intensity and affords a more detailed analysis of light intensity, while requiring more computing time and a greater file size. The number of gray levels required depends on the purpose of the analysis. A reduction in the number of gray levels may suppress the effects of noise in imaging. The number of gray levels should be reduced to a level matching the purpose of the analysis.

The spatial resolution of 640 × 480 pixels corresponding to Video Graphics Array (VGA) aspects has historically often been used for digital analysis, but now we can acquire an image more than several thousand pixels in both width and height. High-resolution imaging has an advantage in macroscopic imaging, particularly for plant cell culture (Ibaraki, 2006). Imaging a whole culture at high resolution yields information on cell and cell clusters. However, analysis of high-spatial-resolution images requires more time, and reduction of the spatial resolution has the merits of effectively reducing not only the computing time but also noise.

The properties of a camera should also be considered. Given that the relationship between input and output of a digital camera is generally not linear, a gamma correction is needed in order to obtain the linear relationship between them. The gamma value mainly depends on the electrical properties of the camera and should be predetermined for each camera. In addition, the linearity of the relationship of pixel values to input is often limited to a certain range of pixel values. Particularly, in the region of low and high pixel values, linearity may not be observed. Therefore, the conditions in which linearity is observed should be confirmed, particularly when pixel values are used to estimate light intensity entering the camera (Ibaraki et al., 2012).

To enhance the visual appearance of an image, a grayscale image is converted into an 8-bit or fewer color image using a lookup table that lists the pixel value corresponding to each color. This method is often applied in fluorescence imaging-based ion mapping, and a color bar representing the lookup table as a bar chart should be added to in the images. Histogram stretching and tone curve adjustment are also effective ways to enhance the visual appearance of the image and can be easily performed with commercially available software. However, these methods change the pixel values, and therefore should not be used for analyses based on the pixel values.

Logarithmic transformation of pixel values is often effective for an image under transmitted lighting because the relationship between the optical density and transmitted light intensity is not linear, and it follows a logarithmic relationship.

2.2.4 Extraction of objects of interest

The pixels corresponding to objects of interest are selected for further analysis. This process is sometimes referred to as segmentation or thresholding. One of the popular methods for selecting pixels to be analyzed is thresholding with a fixed gray-level value (threshold value). For determination of the threshold value, a histogram of pixel values, which is a frequency distribution of numbers of pixels with the same pixel values, is often used. In the histogram, an object consisting of pixels with similar pixel values is expressed as a distribution with a peak, so that pixels of two different objects can be distinguished by setting the threshold value at the valley between two peaks. For color images, the threshold value can be set for each color component individually, or in the image converted to grayscale based on the values derived from the color components.

Several automatic methods for the determination of threshold value, such as discriminant analysis based method (Otsu, 1979) have been developed. The details of automatic segmentation methods are described in Chapter 1 of this book. Methods using machine learning techniques such as support vector machine (Yu et al., 2011) or neural networks (Fu and Chi, 1996) have also been reported. Robust and multipurpose methods for thresholding, however, have not yet been developed. Tajima and Kato (2011) compared 16 automatic thresholding algorithms for rice root images and observed that the accuracy of root length estimation varied with the algorithms. It should be noted that the effectiveness of each thresholding method may depend on the chromatic and structural characteristics of the objects (plants) and imaging conditions. Therefore, it is very important to acquire an image in which target pixels can be distinguished from the background.

2.3 Color analysis

Color is one of the main characteristics used in image analysis for plants, given that plant cells have various kinds of pigments, which are the source of their color. In practical cultivation, leaf color, which is normally visualized with the naked eye, has been used for evaluating the plant status in order to support management practices. For example, some plant diseases can be detected by color degradation in parts of a leaf, and fertilization timing can be based on leaf color information for crops such as rice. Foliar color has always been of great interest and value to resource managers and scientists as a visual indicator of plant health (Murakami et al., 2005).

Normally, an RGB color coordinate system is used in digital image analysis using a personal computer, although the JPEG color images acquired by commercially available imaging devices such as digital still and video cameras use a YCbCr color format, which allows performing lousy subsampling to reduce the file size.

For color analysis, RGB data are often converted into a color appearance system such as a hue-saturation-intensity (HSI; occasionally referred to as HLS) color model or hue-saturation-value (HSV; occasionally referred to as HSB) color model because the color appearance system is more suitable for expressing human sense impressions. A formula for converting RGB into HSI or HSV values has been proposed, and a function for converting an RGB color image into an HSI or HSV image is provided in most commercially available software. Extraction of leaves from an image is often performed using the image converted into an HSI image (Bardsley and Ngugi, 2013; Möller et al., 2007). HSI color components have also been used for the estimation of pigment production in hairy root culture (Berzin et al., 1999).

In recent studies, R, G, and B values have sometimes been directly used or by converting into the component ratios, which are referred to as *r*, *g*, and *b*, respectively, for color analysis, combined with principal component analysis (PCA) or nonlinear identification methods such as neural networks (Prasad and Dutta Gupta, 2008; Dutta Gupta et al., 2013). Flower color could be analyzed using RGB values and the derived values (Keyser et al., 2013).

Plant leaves contain many types of pigments, among which chlorophyll is the richest and most important. Greenness of a leaf depends on chlorophyll content and provides key information for the diagnosis of plant physiological status, including nitrogen or water status. Yadav et al. (2010) estimated the leaf chlorophyll content of micropropagated potato plantlets using *rgb* values. Leaf greenness index from *g* values was calculated for the comparison of leaf color in bedding plants (Parsons et al., 2009). Wang et al. (2008) used a ratio of R to G for estimation of leaf chlorophyll content.

In color analysis, the most important point is to keep the imaging conditions constant. The RGB values in an image depend on the spectral properties of the light source and the imaging device. Therefore, a color standard or a color chart should be imaged together with the target object for proper color analysis, particularly under natural sunlight conditions, in which imaging conditions vary with time.

2.4 *Shape analysis*

In plants, organs such as leaves, flowers, and fruits have different shapes, and shape features can be used for identification or selection of the target organs. Given that development processes involve morphological changes, shape analysis also provides valuable information for classification of developmental stages. Moreover, morphological information of leaves is used with color information to identify plant species and to select weeds in the crop canopy (Golzarian and Frick, 2011).

Simple shape analysis is accomplished by extracting geometrical features of the target area. These include length (major and minor axes, perimeter, etc.), area, centroid, moment, and indices derived from combinations of these features, including aspect ratio, circularity ($4\pi \times$ area/perimeter^2), compactness (perimeter^2/area), and symmetry. Elliptic Fourier descriptors along the contour from the centroid are often used for morphological analysis of biological objects. Features extracted from the Fourier descriptors have been used for morphological classification of somatic embryos (Uozumi et al., 1993), for analysis of leaf shape variations (Iwata et al., 2002; Keyser et al., 2013), and for description of root morphology (Lootens et al., 2007). Sets of these geometrical parameters have often been used as inputs to determine a model describing the morphological feature of interest by statistical analyses such as regression analysis, PCA, and discriminant analysis, or by nonlinear identification systems such as artificial intelligence using support vector machines or neural networks.

Skeleton analysis is also used for shape analysis. Leaf morphology could be analyzed by skeleton analysis (Wilder et al., 2011). Somatic embryos of carrots were evaluated morphologically using skeleton images extracted by a thinning process (Kurata et al., 1993). Midlines of root or stem can be used for shape change analysis (Spalding and Miller, 2013). Midline length and the distribution of local curvature along the midline can provide a useful morphological description of a plant root or stem (Silk, 1984).

Template matching technique is one of the pattern recognition methods and can be used for shape analysis, particularly for detection of an object with desirable morphological features. In template matching, similarity score, the degree of the matching, is evaluated by the sum of squared differences (SSD), sum of absolute differences (SAD), or normalized correlation coefficient (NCC). Although template matching is a robust way to select objects with the required properties (local features) related not only to shape but also to color and texture, it requires more computing time and is susceptible to rotation and size of the template. Scale-invariant feature transform (SIFT) has been proposed as a way to extract local features independent of rotation and size (Lowe, 1999).

2.5 *Particle analysis*

Particle analysis is a procedure that recognizes closed areas as particles and calculates features related to pixel value (color information) and geometric features for each particle. The analysis yields frequency distributions for the features. It is effective for images in which target objects are assembled, including microscopic images of an assembly of organelles and cells or macroscopic images of plants or plant canopy in which multiple leaves, fruits, and flowers are present.

Particle analysis is also used for the detection of lesions in a leaf infected by plant disease. The size distribution or numbers of lesions are good indices for evaluating the disease level (severity), normally assessed by visual inspection.

Success in particle analysis depends on thresholding, particularly of small particles. The area and shape of a particle consisting of a small number of pixels are generally strongly affected by thresholding. In detection of lesions on a leaf, a small lesion is subject to thresholding processes (Wijekoon et al., 2008). In addition, small particles are subject to noise. These problems can be avoided by increasing spatial resolution of the image by using a camera of high resolution (greater number of pixels) or limiting the field of view in imaging (close-up imaging).

2.6 *Growth analysis*

Images can be used for estimation of dimension (size) of a target object or measuring line length or projected area as number of pixels. Differences between size features of an object, such as shoot or root estimated from images acquired at different times, are attributable to growth of the object, including elongation or expansion. Machine vision, in which digital images are automatically acquired, is applicable to the measurement of plant growth (Spalding and Miller, 2013). It permits not only morphometrics, which is the study of geometric features in growth, but also kinematics, which is the study of the internal material processes that create the geometry (Spalding and Miller, 2013). Time-lapse images of a plant organ are also used for analysis of growth-dependent oscillation, called circumnutation (Iwabuchi and Hirafuji, 2002).

Time-lapse imaging has also been applied for root growth and development (French et al., 2009; Lobet et al., 2011). Root growth pattern and complexity were used for phenotyping (Iyer-Pascuzzi et al., 2010; Zhong et al., 2009). There have been many reports on root growth patterns using images. Fractal dimension has often been used for analyzing root complexity (Tatsumi et al., 1989; Walk et al., 2004), and recently alternative methods have been proposed (Zhong et al., 2009; Iyer-Pascuzzi et al., 2010).

Nondestructive acquisition of size information, such as of leaf area, for a whole plant permit the analysis of growth rate, normally a destructive analysis. Relative growth rate (RGR) is commonly used for growth analysis and is based on invasive measurement of dry weight. In contrast, relative leaf growth rate (RLGR) can be estimated nondestructively from images. RLGR of *Arabidopsis thaliana* was estimated automatically and used for phenotyping (Arvidsson et al., 2011). Normally, a projected area can be measured from an image. If a linear relationship between projected area and actual leaf area is observed, RLGR can be estimated simply by image analysis.

In forestry application, digital hemispherical photography, which captures an image with a fish-eye lens from below a canopy, is often used to estimate the leaf area index (LAI). Liu and Pattey (2010) showed the effectiveness of digital photography for agricultural crops using a rectilinear lens at the top of the canopy.

2.7 *Texture analysis*

Texture analysis does not require selection of pixels of a target object in an image. Instead, it analyzes whole regions of the image and extracts characteristic features related to the texture of the image, i.e., the macroscopic pattern of light intensity. Texture analysis can characterize individual objects in a macroscopic image in which individual objects are not clearly identified (Shono et al., 1994). Mean gray level, variance, range (the difference between maximum and minimum values of gray level), and other statistical features derived from a gray-level histogram, including skewness and kurtosis, are used as simple texture features for classification and segmentation of images based on texture, although these texture features may not involve information on spatial distribution (Ibaraki, 2006).

Tuceryan and Jain (1998) divided texture analysis methods into four categories: statistical, geometrical, model based, and signal processing. Of these categories, histogram-derived features, the gray-level run lengths method (Galloway, 1975), and the spatial gray-level dependence method (SGDM) are classified as statistical methods, and two-dimensional (2D) frequency transformation is classified as a signal processing method. Two-dimensional frequency transformation has been widely used for image analysis. It can derive the power spectrum image (frequency-domain image), which expresses periodic features in the image texture. In the gray-level run lengths method (Galloway, 1975), features are extracted from the matrix that yield a set of probabilities that a particular-length line consisting of pixels with the same gray level will occur at a distinct orientation. It is useful for analysis of band pattern texture. Texture features extracted using SGDM, developed by Haralick et al. (1973), have often been used for texture analysis of biological objects. In SGDM, a co-occurrence matrix is determined and 14 texture features are calculated from the matrix. Color co-occurrence matrices derived from image matrices for each color attribute—intensity, hue, and saturation—have also been used in texture analysis (Shearer and Holmes, 1990).

Geometrical methods consider texture to be composed of texture primitives, describing the primitives and the rules that govern their spatial organization (Ojala and Pietikäinen, 2003). Model-based methods hypothesize the underlying texture process, constructing a parametric generative model that could have created the observed intensity distribution (Ojala and Pietikäinen, 2003).

In remote sensing, texture analysis has been extensively used for classification of land use or plant species identification (Ibaraki, 2006). In proximal remote sensing for plant canopies, applications of texture analysis have been reported. Shearer and Holmes (1990) identified plant species using color co-occurrence matrices. Shono et al. (1995) compared the effectiveness of several methods for texture analysis, including the gray-level run lengths, SGDM, and power spectrum methods, on estimation of the species composition in a pasture. Shono et al. (1995) analyzed leaf orientation by texture features extracted by the power spectrum method. Murase et al. (1994) quantified plant growth by analyzing texture features using a neural network.

Texture features have also been used as an input of PCA for classification to separate wheat from weeds (Golzarian and Frick, 2011). Texture features could also be used as promising markers for identifying calcium-deficient lettuce plants (Story et al., 2010).

2.8 *Emerging applications and future perspectives*

Three-dimensional image analysis is expected to be used for analyzing growth or structure of plants in botanical research, given that plants have complex structures. Recently, several applications have been reported for 3D image analysis for plants (see Chapter 9 in this book). Use of 3D images is promising for plant image analysis, although difficulties in image acquisition remain to be resolved.

At present, visualization of invisible objects or phenomena is a key process in biological research, and image analysis is expected to play an important role in this process. Use of electromagnetic waves other than visible, or hyperspectral imaging, is promising for the analysis of physiological and functional properties of plants. Fluorescence imaging has already become an essential tool for cell biology using microscopy, and development of a user-friendly system for acquiring fluorescence images at the macroscopic level is expected to contribute to progress in plant stress biology.

In recent years, advances in imaging devices have been remarkable, and high-quality images can now be easily acquired. However, progress in the development of software for image analysis is limited. Multipurpose software packages are expensive and must be customized for the user's purpose, requiring considerable knowledge. NIH Image© or Scion Image© is the most popular software for multipurpose image analysis. Several manuals describe image analyses for a special purpose using this software (e.g., Robinson et al., 2009). Key features to be considered in the development of software for image analysis are flexibility and the possibility of automation.

Use of an image database is another promising aspect for practical application of image analysis. Cloud computing in which data can be accumulated on an Internet server is now widely distributed and easily utilized. Without any special device or knowledge (hardware and software), a researcher or grower can save many images and easily browse them. There are now available cloud systems for agriculture (e.g., Akisai, Fujitsu; myAgri, NEC). Although a principal issue remains how to extract useful information from a mass of images, the use of cloud computing enhances the utility of image analysis for plants in both research and practical applications.

References

Arvidsson, S., Pérez-Rodríguez, P., and Mueller-Roeber, B., A growth phenotyping pipeline for *Arabidopsis thaliana* integrating image analysis and rosette area modeling for robust quantification of genotype effects, *New Phytol.*, 191, 895–907, 2011.

Bardsley, S.J., and Ngugi, H.K., Reliability and accuracy of visual methods to quantify severity of foliar bacterial spot symptoms on peach and nectarine, *Plant Pathol.*, 62, 40–474, 2013.

Berzin, I., Mills D., and Merchuk, J.C., A non-destructive method for secondary metabolite determination in hairy root cultures, *J. Chem. Eng. Jpn.*, 32, 229–234, 1999.

Corkidi, G., Balderas-Ruíz, K.A., Taboada, B., Serrano-Carreón, L., and Galindo, E., Assessing mango anthracnose using a new three-dimensional image-analysis technique to quantify lesions on fruit, *Plant Pathol.*, 55, 250–257, 2006.

Cui, D., Zhang, Q., Li, M., Hartman, G.L., and Zhao, Y., Image processing methods for quantitatively detecting soybean rust from multispectral images, *Biosystem. Eng.*, 107, 186–193, 2010.

Dutta Gupta, S., Ibaraki, Y., and Pattanayak, A.K., Development of a digital image analysis method for real-time estimation of chlorophyll content in micropropagated potato plants, *Plant Biotechnol. Rep.*, 7, 91–97, 2013.

French, A., Ubeda-Tomás, S., Holman, T.J., Bennett, M.J., and Pridmore, T., High-throughput quantification of root growth using a novel image-analysis tool, *Plant Physiol.*, 150, 1784–1795, 2009.

Fu, H., and Chi, Z., Combined thresholding and neural network approach for vein pattern extraction from leaf images, *IEE Proc. Vision Image Signal Process.*, 153(6), 881–892, 2006.

Galloway, M.M., Texture analysis using gray level run lengths, *Comput. Graphic. Image Process.*, 4, 172–179, 1975.

Golzarian, M.R., and Frick, R.A., Classification of images of wheat, ryegrass and brome grass species at early growth stages using principal component analysis, *Plant Methods*, 7, 28, 2011.

Haralick, R.M., Shanmugam, K., and Dinstein, I., Textural features for imaging classification, *IEEE Trans. Syst. Man Cybernet.*, SMC-3, 610–621, 1973.

Ibaraki, Y., Evaluation of plant suspension cultures by texture analysis, in *Plant tissue culture engineering*, ed. S. Dutta Gupta and Y. Ibaraki, 69–79, Springer, Berlin, 2006.

Ibaraki, Y., and Matsumura, K., Non-destructive evaluation of the photosynthetic capacity of PSII in micropropagated plants, *J. Agric. Meteorol.*, 60, 1073–1076, 2005.

Ibaraki, Y., Matsumura, K., and Dutta Gupta, S., Low-cost photochemical reflectance index measurements of micropropagated plantlets using image analysis, *Comput. Electron. Agric.*, 71, 170–175, 2010.

Ibaraki, Y., Yano, Y., Okuhara, H., and Tazuru, M., Estimation of light intensity distribution on a canopy surface from reflection images, *Environ. Control Biol.*, 50, 117–126, 2012.

Iwabuchi, K., and Hirafuji, M., Potential use of time-lapse images: determination of circumnutational movement to assess plant vigor, presented at *Proceedings of World Congress of Computers in Agriculture and Natural Resources*, 2002.

Iwata, H., Nesumi, H., Ninomiya, S., Takano, Y., and Ukai, Y., Diallel analysis of leaf shape variations of citrus varieties based on elliptic Fourier descriptors, *Breeding Sci.*, 52, 89–94, 2002.

Iyer-Pascuzzi, A.S., Symonova, O., Mileyko, Y., Hao, Y., Belcher, H., Harer, J., Weitz, J.S., and Benfey, P.N., Imaging and analysis platform for automatic phenotyping and trait ranking of plant root systems, *Plant Physiol.*, 152, 1148–1157, 2010.

Jones, H.G., Stoll, M., Santos, T., de Sousa, C., Chaves, M.M., and Grant, O.M., Use of infrared thermography for monitoring stomatal closure in the field: application to grapevine, *J. Exp. Bot.*, 53, 2249–2260, 2002.

Keyser, E., Lootens, P., Van Bockstaele, E., and De Rick, J., Image analysis for QTL mapping of flower color and leaf characteristics in pot azalea (*Rhododendron simsii* hybrids), *Euphytica*, 189, 445–460, 2013.

Kurata, K., Komine, M., Liyanage, H.K., and Ibaraki, Y., A thinning-based algorithm for evaluation of somatic embryos, *Trans. ASAE*, 36, 1485–1489, 1993.

Lenk, S., Chaerle, L., Pfündel, E.E., Langsdorf, G., Hagenbeek, D., Lichtenthaler, H.K., Van Der Straeten, D., and Buschmann, C., Multispectral fluorescence and reflectance imaging at the leaf level and its possible applications, *J. Exp. Bot.*, 8, 807–814, 2007.

Liu, J., and Pattey, E., Retrieval of leaf area index from top-of-canopy digital photography over agricultural crops, *Agric. Forest Meteorol.*, 150, 1485–1490, 2010.

Lobet, G., Pagès L., and Draye, X., A novel image-analysis toolbox enabling quantitative analysis of root system architecture, *Plant Physiol.*, 157, 29–39, 2011.

Lootens, P., Van Waes, J., and Carlier, L., Description of the morphology of roots of *Chicorium intybus* L. *partim* by means of image analysis: comparison of elliptic Fourier descriptors and classical parameters, *Comput. Electron. Agric.*, 58, 164–173, 2007.

Lowe, D.G., Object recognition from local scale-invariant features, in *Proceedings of IEEE International Conference on Computer Vision (ICCV)*, 1999, pp. 1150–1157.

Möller, M., Alchanatis, V., Cohen, Y., Meron, M., Tsipris, J., Naor, A., Ostrovsky, V., Sprintsin, M., and Cohen, S., Use of thermal and visible imagery for estimating crop water status of irrigated grapevine, *J. Exp. Bot.*, 58, 827–838, 2007.

Murakami, P.F., Hitchcock, M.R., van den Berg, A.K., and Schaberg, P.G., *An instructional guide for computer-based leaf color analysis*, General Technical Report NE-327, 2005. http://www.nrs.fs.fed.us/pubs/gtr/ne_gtr327.pdf (accessed July, 2013).

Murase, H., Honami, N., and Nishiura, Y., A neural network estimation technique for plant water status using textural features of pictorial data of plant canopy, *Acta Hort.*, 339, 255–262, 1994.

Novák, K., Determination of symbiotic nodule occupancy in the model *Vicia tetrasperma* using a fluorescence scanner, *Ann. Bot.*, 107, 709–715, 2011.

Ojala, T., and Pietikäinen, M., Texture analysis, in *CVonline: the evolving, distributed, non-proprietary, on-line compendium of computer vision*, ed. R.B. Fisher. http://homepages.inf.ed.ac.uk/rbf/CVonline/LOCAL_COPIES/OJALA1/texclas.htm (accessed July, 2013).

Omasa, K., and Takayama, K., Simultaneous measurement of stomatal conductance, non-photochemical quenching, and photochemical yield of photosystem II in intact leaves by thermal and chlorophyll fluorescence imaging, *Plant Cell Physiol.*, 44, 1290–1300, 2003.

Otsu, N., A threshold selection method from gray-level histograms, *IEEE Trans. Syst. Man Cybern*, SMC-9-1, 62–66, 1979.

Parsons, N.R., Edmondson, R.N., and Song, Y., Image analysis and statistical modeling for measurement and quality assessment of ornamental horticulture crops in glasshouses, *Biosys. Eng.*, 104, 161–168, 2009.

Prasad, V.S.S., and Dutta Gupta, S. Photometric clustering of regenerated plants of gladiolus by neural networks and its biological validation, *Comput. Electron. Agric.*, 60, 8–17, 2008.

Robinson, S.C., Laks, P.E., and Turnquist, E.J., A method for digital color analysis of spalted wood using scion image software, *Material*, 2, 62–75, 2009.

Shearer, S.A., and Holmes, R.G., Plant identification using colour co-occurrence matrixes, *Trans. ASAE*, 38, 2037–2044, 1990.

Shono, H., A new method of image measurement of leaf tip angle based on textural feature and a study of its availability (in Japanese with English abstract), *Environ. Control Biol.*, 33, 197–207, 1995.

Shono, H., Okada, M., and Higuchi, S., Texture analysis of photographic images from close distance: an application to estimate species composition in a mixed pasture field (in Japanese with English abstract), *J. Agric. Meteorol.*, 49, 227–235, 1994.

Silk, W.K., Quantitative descriptions of development, *Annu Rev. Plant Physiol.*, 35, 479–518, 1984.

Spalding, E.P., and Miller, N.D., Image analysis is driving a renaissance in growth measurement, *Curr. Opin. Plant Biol.*, 16, 100–104, 2013.

Stephan, D., Slabber, C., George, G., Ninov, V., Francis, K.P., and Burger, J.T., Visualization of plant viral suppressor silencing activity in intact leaf lamina by quantitative fluorescent imaging, *Plant Methods*, 7, 25, 2011.

Story, D., Kacira, M., Kubota, C., Akoglu, A., and An, L., Lettuce calcium deficiency detection with machine vision computed plant features in controlled environments, *Comput. Electron. Agric.*, 74, 238–243, 2010.

Tajima, R., and Kato, J., Comparison of threshold algorithms for automatic image processing of rice roots using freeware ImageJ, *Field Crops Res.*, 121, 460–463, 2011.

Tatsumi, J., Yamauchi, A., and Kono, Y., Fractal analysis of root systems, *Ann. Bot.*, 64, 499–503, 1989.

Tuceryan, M., and Jain, A.K., Texture analysis, in *The handbook of pattern recognition and computer vision*, ed. C.H. Chen, L.F. Pau, and P.S.P. Wang, 207–248, World Scientific Publishing Co., Hackensack, NJ, 1998.

Uozumi, N., Yoshino, T., Shiotani, S., Suehara, K.I., Arai, F., Fukada, T., and Kobayashi, T., Application of image analysis with neural network for plant somatic embryo culture, *J. Ferment. Bioeng.*, 76, 505–509, 1993.

Usha, K., and Singh, B., Potential applications of remote sensing in horticulture—a review, *Scientia Hortic.*, 153, 71–83, 2013.

Walk, T.C., Van Erp, E., and Lynch, J.P., Modelling applicability of fractal analysis to efficiency of soil exploration by roots, *Ann. Bot.*, 94, 119–28, 2004.

Wang, F., Yamamoto, H., and Ibaraki, Y., Measuring leaf necrosis and chlorosis of bamboo induced by typhoon 0613 with RGB image analysis, *J. Forestry Res.*, 19, 225–230, 2008.

Wijekoon, C.P., Goodwin P.H., and Hsiang, T., Quantifying fungal infection of plant leaves by digital image analysis using Scion Image software, *J. Microbiol. Method.*, 74, 94–101, 2008.

Wilder, J., Feldman, J., and Singh, M., Super ordinate shape classification using natural shape statistics, *Cognition*, 119, 325–340, 2011.

Yadav, S.P., Ibaraki, Y., and Dutta Gupta, S., Estimation of the chlorophyll content of micropropagated potato plants using RGB based image analysis, *Plant Cell Tiss. Org. Cult.*, 100, 183–188, 2010.

Yu, Z., Wong, H.S., and Wen, G., A modified support vector machine and its application to image segmentation, *Image Vision Comput.*, 29, 29–40, 2011.

Zhong, D., Novais, J., Grift, T.E., Bohn, M., and Han, J., Maize root complexity analysis using a support vector machine method, *Comput. Electron. Agric.*, 69, 46–50, 2009.

chapter three

Applications of RGB color imaging in plants

S. Dutta Gupta, Yasuomi Ibaraki, and P. Trivedi

Contents

3.1 Introduction

The application of imaging techniques to agriculture and plant science has previously been confined to images captured through the techniques of remote sensing, involving aircraft or satellites, which were then processed and analyzed using computers. With new technological advancements in image capture and data processing, imaging techniques have brought solutions to various practical problems in the fields of biology, medicine, and agriculture. Various types of imaging techniques such as thermal imaging, fluorescence imaging, hyperspectral imaging, and photometric feature-based imaging have contributed significantly to the advancement of various plant phenotyping aspects. One of them, the photometric feature or RGB-based imaging is worth mentioning because of its dependency on color variation of different biological samples. In recent years, considerable progress in the applications of RGB-based imaging has been witnessed in the various fields of agriculture and plant science. It has replaced the human vision system to assess the quality of different food articles, recognition of weeds and diseases in large agricultural fields, and determination of nutrient status of plants. Specifically, RGB-based image analysis has been applied in agriculture for weed identification (Hemming and Rath, 2000), weed and crop mapping (Tillet et al., 2001), weed and crop discrimination (Aitkenhead et al., 2003; Rasekhi et al., 2010), quantification of turf grass color (Karcher and Rechardson, 2003), quantitative analysis of a specially variable physiological process across a leaf surface (Aldea et al., 2006), weed recognition (Ahmad et al., 2006), seed color test for identification of commercial seed traits (Dana and Ivo, 2008), nodulation in soybean (Gwata et al., 2004), and noninvasive estimation of nitrogen (Mercado-Luna, 2010). In plant tissue culture, RGB-based image analysis has been confined to the identification and estimation of shoot length (Honda et al., 1997), secondary metabolite determination in hairy root cultures (Berzin et al., 1999), clustering of regenerated plants into groups (Mahendra et al., 2004; Prasad and Dutta Gupta, 2008), and estimation of chlorophyll content in micropropagated plants (Yadav et al., 2010; Dutta Gupta et al., 2013). The fundamental concepts of the RGB color model and its application for the determination of nitrogen status in plants have been detailed in Chapter 4 of this book. In this review, we describe briefly the applications of RGB imaging in various fields of agriculture and illustrate its potential in plant tissue culture systems.

3.2 *In vivo applications of RGB imaging in plants*

Color is used as an important parameter for measuring plant health, quality of biological products, and agricultural processes in general (Alchanatis et al., 1993; Cliff et al., 2002). Some significant physiological changes in the biological sample may be depicted by change in color. A simple technique of RGB imaging can track down the changes in color and can directly help in monitoring the status of plant developmental stage and health. The following section narrates the application of the RGB color model in various fields of agriculture.

3.2.1 *Grain quality determination*

Test weight, varietal purity, soundness, vitreousness, and maximum limit of foreign material are the five factors, depending on which grains are basically graded worldwide. The latter four factors require visual observation by trained personnel. However, grading decisions are subjective and influenced by the individual experience of inspectors despite training. Digital image analysis uses digital images to measure the size of individual grains, and extract different features and shape-related information from the images, thus offering an objective and quantitative method for estimation of morphological parameters (Poblaciones et al.).

Paliwal et al. (1999) developed a machine vision system (MSV) that can distinguish among the kernels of Canada Western Red Spring (CWRS) wheat, Canada Western Amber Durum (CWAD) wheat, barley, oats, and rye using a kernel signature. For each kernel, three attributes, viz., length, shape function (Fourier descriptors in polar coordinates), and color, were extracted that were collectively called the kernel signature. Using this kernel signature data, a training set was formed. In the training set, each grain type was represented by a kernel signature consisting of three attributes: length of the kernel, shape of the kernel (in terms of Fourier descriptors), and color (histograms in R, G, and B components). The image was first read in PBM (Portable Bit Map) format, and the morphological features, namely, length and shape function, were extracted and the image was converted into binary form. The third attribute, color, was determined by reading the color image, segmenting the kernels, and then plotting three color histograms for the R, G, and B components for each kernel. For each grain type, average histograms were obtained for all the kernels in the training set. For any unidentified kernel, the algorithm first calculated these attributes for the kernel and then compared them with the corresponding values of each grain type in the training set using three different distance functions, one each for each attribute. The classification accuracies obtained for CWRS wheat, CWAD wheat, barley, oats, and rye were 100, 94, 93, 99, and 95%, respectively. Shape was found to be the best feature for classification, followed by color and length of the kernels.

A hybrid method for cereal grain identification was developed by Mehrez et al. (2006). In this method, a statistical pattern recognition method (using morphological and color features) was combined with a fuzzy logic decision-making method to classify three grain types: hard wheat, tender wheat, and barley. The statistical pattern recognition method operated directly on distribution intervals of the morphological and color parameters. The recognition rate with this method was weak, with an average of 69.74%. Therefore, a fuzzy logic-based decision-making method was introduced to preclude the overlaps of distribution curves of grains. This method gave an improved average recognition rate of 83.42%. When both the methods were combined (statistical pattern recognition method and fuzzy logic-based method) to develop a hybrid method, an average recognition rate of 91.49% was obtained.

3.2.2 *Inspection and grading of fruits*

Computer vision systems due to their qualities such as rapidness, consistency, and objective assessment have been used increasingly in the food and agricultural industry for inspection and evaluation purposes. Advances in hardware and software for digital image processing have motivated the development of machine vision systems to evaluate the quality of diverse and processed foods.

It is a common belief that a good quality sweet orange is reddish in color, medium sized, with a height-to-width ratio of less than 1, and having a glossy surface. However, these recognition criteria were difficult to follow in their grading, as they varied from people to people and across locations. This prompted Kondo et al. (2000) to introduce a machine vision and neural network system to evaluate the quality of the Iyokan orange and automate the orange classification operation. Two sets of parameters were included in the study:

1. Color component ratio (R/G), weight, height–width ratio (H/W), and degree of roughness (last two parameters were manually observed)
2. R/G, weight, Feret's diameter ratio, and a texture feature (all parameters were extracted from images)

These parameters were investigated and their correlation with sugar content and pH was determined. The correlation coefficient between measured sugar content values and predicted sugar content values was 0.79 when parameter set 1 was used, while it was 0.84 when parameter set 2 was used. The correlation coefficients between measured pH values and predicted pH values were 0.83 and 0.68 when parameter sets 1 and 2 were used, respectively.

Leemans et al. (1998) introduced a machine vision system to automatically detect the defects in Golden Delicious apples. A novel algorithm that was able to detect defects, using the color information, was presented by them. The algorithm comprised three steps to detect both well-contrasted and diffuse defects. The first was a coarse defect segmentation based on a statistical comparison between the color of an individual pixel and the global color of the fruit by making use of the Mahalanobis distances. Different algorithm combinations were tested using a set of 80 apples, including fruits of different qualities and damaged fruits. The defects observed were bitter pit, fungi attack, growth defects, bruising, punches, insect holes, russeting and scab. The algorithm used gave satisfactory results with well-contrasted defects. Two approaches were used by them to enhance the detection process: global approach and local approach. The first enhancement, consisting of the global approach, was well suited for diffuse defects, but increased the part of healthy tissues segmented as defects. The second enhancement, consisting of a local approach, was able to correct these problems.

Grading the oil palm fruit manually by human graders is prone to error, as the method is subjective and inconsistent. The degree of ripeness affects the quality and quantity of oil that can be extracted. May and Amaran (2011), using the RGB color model, developed an automated grading system to distinguish underripe, ripe, and overripe classes of oil palm fruit. The grading system comprises a charge-coupled device (CCD) camera and a computer to acquire and analyze the images. An Olympus E-520 digital camera was used to acquire the images, and the noise was removed by the background subtraction method. The color features of the oil palm fruit were analyzed based on the RGB model, and finally, grading was done by a fuzzy inference system using MATLAB®. The system has been able to distinguish the three different types of oil palm fruit with 86.67% classification accuracy.

3.2.3 Weed identification

Weeds are the unwanted plants and compete with crops for water, soil nutrients, light, and space in a field, reducing crop yields. They impose an increase in the expenditure on labor and equipment, render harvesting difficult, and reduce the quality and marketability of agricultural produce. Identification of weeds and their removal from the field is problematic and ultimately results in arbitrary application of herbicides causing crop damage and environmental pollution. Hence, the use of machine vision systems has been proposed through which weeds can be identified, counted, and mapped in fields for planning future herbicide applications (Woebbecke et al., 1995). This would lead to the possibility of localized spraying of herbicides, and thus reduce chemical waste, crop damage, and environmental pollution.

A machine vision system was developed by Hemming and Rath (2001) in order to identify and locate weeds in crops using digital image analysis. Two different crops, cabbage (*Brassica oleracea* L.) and carrots (*Daucus carota* L.), were studied in both greenhouse and open field conditions: In the experiment, each image captured was converted to R, B, G parameters, but the separation of the green plants and the brown soil was found to be insufficient by applying only the RGB color image. Subsequently, each image was transformed from the RGB to the hue, saturation, and intensity (HSI) color space. The best parameters used for classification for cabbage were area/circumference in almost all cases and area and hue in more than 90% of the cases. The experiment achieved average identification accuracy of 88% for cabbage and 72% for carrots.

Yang et al. (2000) captured field images of corn with the most common weeds found in cornfield. The weeds are velvetleaf (*Abutilon theophrasti*), quack grass (*Agropyron repens*), lamb's-quarters (*Chenopodium album*), and yellow nut sedge (*Cyperus esculentus*). The field images were then cropped to 100 × 100 pixels, so as to leave only one object in each image: either a corn plant or a group of weeds. The greenness method based on a comparison of the intensities of the three primary colors (i.e., red, green, and blue) on a pixel-by-pixel basis was applied to all of the images, and finally, information extracted from the images was subjected to artificial neural network (ANN) modeling in order to create weed coverage and weed patchiness maps. These maps were used as input in a fuzzy logic system to determine herbicide application rates for each spot in the field. The success rates of identification ranged from 70 to 90% for corn and 68 to 100% for *Abutilon theophrasti*, and 88 to 100% for corn and 66 to 74% for *Chenopodium album*. When all the weeds were taken in one class, the success rate ranged from 52 to 66% of corn and 76 to 85% weeds. The results of this work illustrated the potential benefits of using image processing methods, ANNs, and fuzzy logic to develop herbicide application maps for precision farming. In order to develop a method for automated localization of weeds for precision spraying of herbicides in corn, Yang et al. (2003) combined image processing, ANNs, and fuzzy logic and constructed the herbicide application maps. The images were taken by a digital camera, recognized by ANNs, and application maps were developed using a fuzzy logic algorithm.

Discriminant analysis was performed by Jafari et al. (2006) to establish the relationship between the three main color components, R, G, and B, of weeds and sugar beet classes. Information was extracted from 300 digital images of sugar beet plants and 7 types of common sugar beet weeds at different normal lighting conditions (from sunny to cloudy sky, from morning to afternoon) and fed to the discriminant analysis procedure. As the major problem in color segmentation schemes was variation in light intensity, which directly affects R, G, B components, two separate algorithms–one for parts in light and the second for parts in shadow–were adopted. Luminance value was also calculated separately for parts in the shadow and parts in the light using the equation

$$L = R + G + B/3$$

where L is pixel luminance, and R, G, and B are red, green, and blue components, respectively. By comparing with this luminance value, it was determined whether the photograph was taken in light or dark. Discriminant analysis (DA) is a technique used to build a predictive model of group membership based on observed characteristics of each case. Discriminant functions (DFs) are the linear combinations of the standardized independent variables, which yield the biggest mean differences between the groups. A stepwise discriminant analysis was performed to determine each pixel membership to eight groups of plants, including sugar beet and seven types of weeds, based on their independent variables R, G, and B. Comparing the results from the photographs taken in light and shadow, the correct classification rate (CCR) was noted to be more for parts in light than parts in shadow. A classification rate of 88.5% was achieved for all the weeds except *Portulaca* in sunlight and 88.1% for the weeds except *Convolvulus* in shadow. Such classification rates were best suited for spot spraying with selective herbicides.

3.2.4 Disease detection in plants

Efforts to reduce crop damage due to plant diseases using agrochemicals in the past decades proved to be futile. Early detection and classification of diseases are essential for selective application of pesticides. Traditional approaches by naked eye observation method by the farmers and experts who have the ability to detect subtle changes in color are laborious, time-consuming, and impractical for large fields. The application of image processing techniques has the potential for fast and accurate detection of plant diseases (Cui et al., 2009; Kai et al., 2011).

Identification of diseased and normal citrus leaves under laboratory conditions was made with a color co-occurrence method employing hue, saturation, and intensity (HSI) color features in conjunction with statistical classification algorithms (Pydipati et al., 2006). Four different classes of citrus leaves were selected for this study. The diseased leaf samples investigated were greasy spot (*Mycosphaerella citri*), melanose (*Diaporthe citri*), scab (*Elsinoe fawsettii*), and normal citrus leaf. In this method, the image was transformed from a red, green, blue (RGB) color space to a hue, saturation, and intensity (HSI) color model. Once the RGB image was transformed into HSI pixel maps, each pixel map was used to generate a color co-occurrence matrix (CCM), resulting in three CCMs. The spatial gray-level dependency matrices (SGDMs) were then generated for each color pixel map of the image, one each for hue, saturation, and intensity. From the SGDM matrices, texture statistics were generated for each image using the three color feature CCMs. Discriminant analysis was then used to evaluate the potential classification accuracies. Classification accuracies achieved were above 81% in all the data models using the intensity feature, and above 95.8% for hue and saturation features.

For disease spot detection in rice, RGB images were converted into the HSI model, and the threshold method was applied on the H component for segmenting infected areas from plant leaves (Phadikar and Sil, 2008). Cui et al. (2009) introduced a fast manual threshold-setting method based on the HSI color model to segment the disease spot of soybean rust. In maize, an RGB image was converted into YCbCr color space to detect the disease spot (Kai et al., 2011). A comparison on the effect of CIELAB, HIS, and YCbCr color space was made on the monocot and dicot plant leaves for disease spot detection (Chaudhary et al., 2013). In this work the Otsu threshold method was used for segmentation, and an algorithm that is independent of background noise, plant type, and disease spot color was developed.

3.2.5 Determination of leaf water potential

Zakaluk and Sri Ranjan (2007, 2008) examined the use of ANN modeling of RGB images, captured by a 5-megapixel digital camera to predict the leaf water potential of potato plants (*Solanum tuberosum* L.) grown in a greenhouse as well as in the field. The greenhouse experiment was conducted with cv. Russet Burbank, while the field study examined cv. Sangre. Images were acquired over different soil nitrate and water content levels, and were calibrated radiometrically. Principal component analysis (PCA) was applied to transform the RGB images, six image transformations, and nine vegetation indices. An inverse linear relationship was observed between soil nitrogen and leaf reflection in the green image band. ANN modeling of RGB images indicates its potential in predicting the leaf water potential of potato plants.

3.2.6 Estimation of leaf chlorophyll content and nitrogen status

In recent years, several image analysis methods have been developed to monitor various parameters of plant health status using the RGB color space model. In most of the studies, digital cameras were used to acquire leaf images, which were then analyzed to examine the relationship between the R, G, and B values and chlorophyll content and nitrogen status of plants. The development of low-cost digital cameras with charged-couples device (CCD) arrays to capture images offers an advantage of a low-cost real-time monitoring process over an optical sensor-based SPAD (Soil Plant Analysis Development) meter.

To determine the foliar chlorophyll status in wheat, Kawashima and Nakatani (1998) used the function (R – B)/(R + B). In contrast, G/(R + G + B) gave good results in broccoli (Suzuki et al., 1999), while the ratio R/(R + G + B) was found to predict chlorophyll content in cabbage (Cai et al., 2006). Pagola et al. (2009) applied PCA to digital leaf images of barley and, using RGB components, obtained an index of greenness that was correlated with SPAD value and yield. For estimation of leaf chlorophyll content and phenotyping soybean nodulation, the SPAD readings were compared with features extracted from RGB images (Volmann et al., 2011). The leaf chlorophyll content measured with the SPAD meter was significantly correlated to the green color of the RGB model.

The potential of color image analysis for nitrogen determination on tomato seedlings was investigated in a field experiment by Mercado-Luna et al. (2010). Color images were captured with a digital camera, processed in MATLAB, and various RGB parameters were determined. The color parameters correlated better with the status of plant nitrogen than the SPAD values. Red and blue colors appeared to be more accurate predictors of nitrogen status. The study indicates the potential of color image analysis as an accurate and quick way of nitrogen estimation for early detection of nitrogen deficiency in tomato seedlings. For estimation of nitrogen in sugar beet leaf, different combinations of R, G, and B functions were analyzed by linear regression and neural network. Compared to regression analysis, the neural network showed better performance (Moghaddam et al., 2010).

Color image analysis was adopted to investigate the suitable index of the diagnosis of nitrogen status of pepper plants (Yuzhu et al., 2011). Digital images of pepper coverage were acquired by digital camera at the flowering and fruiting stages. The images were processed to determine the averages of R, G, and B colors. Regression analysis for inorganic nitrogen in soil and total nitrogen in plants and SPAD readings were performed, and the results show negative relations between the G/(R + G + B) ratio of coverage image and the indices of pepper nitrogen status. An automatic nitrogen estimation method based on leaf color was developed in sugarcane by Auearunyawat et al. (2012). The parameters R, G, B, G/B, G/R, and R/B significantly correlated with nitrogen concentration in the sugarcane leaves.

A new leaf color-based algorithm to determine foliar chlorophyll contents of lettuce, broccoli, and tomato was developed by Ali et al. (2012). The proposed algorithm maps the normalized value of G, with respect to R and B, using a logarithmic sigmoid transfer function. The developed optileaf algorithm produced superior correlations with the true value of chlorophyll content measured with the SPAD meter and spectrophotometer. Further, a consistently better performance was achieved with this algorithm over a wide range of species than the other image processing-based methods.

3.3 Applications of RGB imaging in plant tissue culture systems

Plant tissue culture technique is one of the key components of plant biotechnology due to its application in micropropagation, secondary metabolite, and transgenic plant production. Methods revealing real-time phenotyping of regenerated plants in terms of plant growth status, photosynthetic abilities, and hyperhydric status of in vitro grown plants have an impact on quality transplant production. Image sensing-assisted analysis of cell, tissue, and organ culture in a controlled environment may help in optimization of culture conditions and mechanization of the process in order to achieve efficient micropropagation, and high-value production of natural compounds. In this context, digital color (RGB) image analysis is increasingly becoming popular with its potential in phenotyping plant culture systems noninvasively.

3.3.1 *Nondestructive evaluation of in vitro stored plants*

Among the various methods of ex situ plant conservation, seed banking is the most efficient for many species. However, many crops produce recalcitrant or short-lived seeds, and in the case of clonal crops, seeds are not the best material to conserve in view of their genetic heterogeneity and unknown worth. In such cases, the following approaches of germplasm conservation may be applied:

1. Freeze preservation
2. Slow growth cultures
3. Desiccated somatic embryos/artificial seeds

The main advantages of in vitro culture-assisted conservation include the requirement of less storage space as well as the maintenance of germplasm in a pest- and pathogen-free environment. Cold-stored in vitro plants show a decline in quality and health over time, and it is difficult to monitor their growth aseptically by visual examination. Thus, a system that can detect the growth status of plants is essential to reduce the loss during storage.

Aynalem et al. (2006) proposed an image analysis technique to evaluate the health of plantlets during slow growth storage and compared the results of the digital image system to those of standard visual evaluations. Digital images of the first two nodes of each plantlet were evaluated for red, green, blue, modified normalized differences of vegetation index (MNDVI), green/red ratio, intensity, hue, and saturation. Four *Pyrus communis* L. cultivars (Luscious, Bartlett-Swiss, Belle Lucrative, and Louise Bonne de Jersey) were selected for the study based on responses previously determined through visual assessment. Color images were represented as 24-byte images with red, blue, and green bands taken by a digital camera. The mean values for blue, green, and red in each image were calculated using Microsoft Excel. The RGB values were converted to hue, intensity, and saturation (HIS) and MNDVI (R – G/R + G) for additional analysis. The health of each genotype, as indicated by the visual rating scores of all four genotypes, declined at different rates over the 15 months of the monitoring period. At 6 months, visual ratings had declined steadily for *P. communis* cv. Luscious and Bartlett-Swiss, while Belle Lucrative and Louise Bonne de Jersey ratings did not show significant declines until 9 months. Correlations between visual ratings and G/R and MNDVI values were significant for all the cultivars. Regression analysis indicated that the MNDVI and G/R ratios changed significantly over the 15 months. Intensity, hue, and saturation values were not consistently significant and did not correlate with the visual ratings. The study demonstrated the potential of image analysis in evaluating the health condition of in vitro stored tissue-cultured plantlets with a feasibility of developing an automated evaluation system.

3.3.2 *Analysis of pigment-producing cell cultures*

Secondary metabolites often play a major role in the adaptation of plants to their environment and plant defense against herbivory. These molecules also represent an important source of bioactive pharmaceuticals and natural colorants. Natural anthocyanin pigments from plant cells can be used as a substitute for the synthetic food colorants. Anthocyanin extraction from fresh fruits yields an irregular, unpredictable product, which is limited by seasonal availability of raw materials and geographic barriers. In vitro plant cell culture systems have been extensively utilized with the objective of improving the production of secondary compounds. However, these secondary metabolites are produced only by a fraction of the cells. A selection of high-producing cell lines from the heterogeneous culture may contribute to an enhancement of productivity. The economic feasibility of this process will hinge on isolating elite high-producing cell lines, which requires an expert to identify those cells. Obviously, detection by humans will be prone to error and time-consuming, and warrants application of imaging techniques.

A machine vision system was introduced by Smith et al. (1995) with two objectives:

1. To identify the visual characteristics of pigment-producing cell cultures using color machine vision
2. To compare the performance of machine vision analysis with the data obtained by conventional means

Macroscopic images of callus colonies and microscopic images from suspension cultures of anthocyanin-producing *Ajuga pyramidalis* cells were analyzed by machine vision to ascertain culture productivity. Callus mass was accurately estimated using data from simultaneous capture of top and side view images. Several combinations of these image areas were used to develop mathematical models. The models summing top and side view combinations had high coefficients of determination and estimation errors less than 10%. In suspension cultures, no single RGB plane could provide adequate segmentation to separate pigmented cells from other objects like nonpigmented cells, pigmented aggregates, nonpigmented aggregates, and the background. But conversion of RGB data to the HSI (hue-saturation-intensity) coordinate system permitted clear separation. Hue (H) was quite distinct for pigmented plant cells and aggregates. The saturation characteristics of the HSI representation provided information on the degree of pigmentation of the cells and are as follows:

If ($S > 0.0$ and $S < 0.10$), pigmentation = low.
If ($S > 0.10$ and $S < 0.20$), pigmentation = medium.
If ($S > 0.20$ and $S < 1.0$), pigmentation = high.

Miyanaga et al. (2000) introduced an improved version of the machine vision system to evaluate anthocyanin accumulation in individual cells of strawberry cell cultures. In their study, cell aggregates were macerated by enzymatic treatments. Images of the enzyme-treated cell suspension were obtained by using a microscope and a charge-coupled device (CCD) camera. The image was digitized through a video-capture board and the brightness data for each of the red (R), green (G), and blue (B) components were stored in a personal computer. The color characteristics of anthocyanins produced by strawberry cell culture showed a higher value of red component (R) than the green (G) or blue (B) component. The difference between R and G, compared to that of the (R-G) value, was defined as an index for pigment accumulation. A significant correlation (0.98) was observed between the average anthocyanin's content by extraction method and the average (R-G) value.

3.3.3 *Estimation of shoot length of regenerated rice callus*

Commercial micropropagation generally involves two steps:

1. Shoot regeneration and their further proliferation into plantlets, preferably in a bioreactor
2. Acclimatization and ex vitro transfer of plantlets

Plantlets with long shoots of high photosynthetic ability may survive better under ex vitro conditions. The development of an automatic system for selection of such competent shoots using image analysis would be helpful in achieving the desired goal of micropropagation.

A fuzzy neural network (FNN) system of image analysis was reported by Honda et al. (1997) to estimate the shoot length of regenerated rice callus. The original images were stored as data sets of R, G, and B with brightness level of 256. A model using multiple regression analysis and FNN was developed to extract the shoot region from the image data. The shoot length was estimated from the original image of regenerated rice

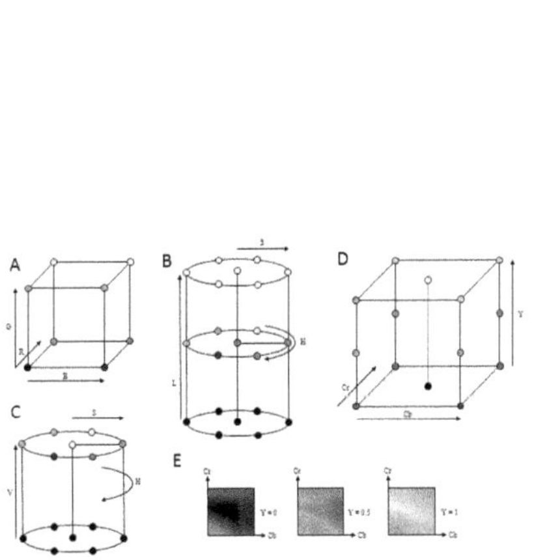

Figure 1.4 Diagrammatic representations of common color spaces. (a) The RGB cube. (b) The HSL cylinder. (c) The HSV cylinder. (d) The YCbCr cube. (e) A selection of planes taken from the YCbCr cube, demonstrating how color changes with Y, Cb, and Cr.

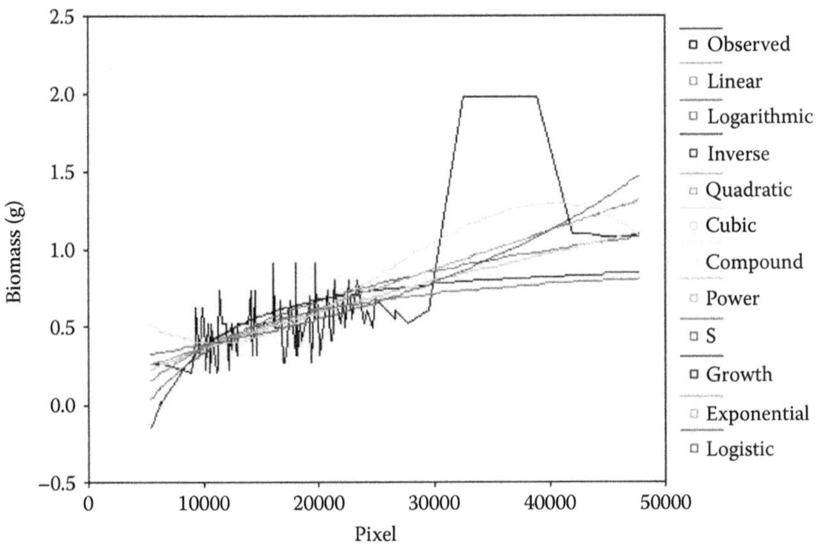

Figure 6.2 Simulation on the relationship between pixel and biomass of *O. violaceus* plantlets in vitro.

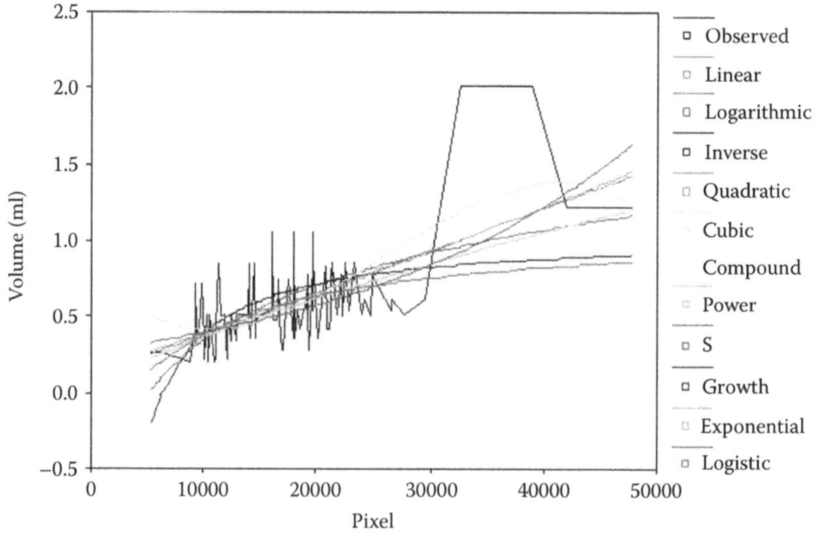

Figure 6.3 Simulation on the relationship between pixel and volume of *O. violaceus* plantlets in vitro.

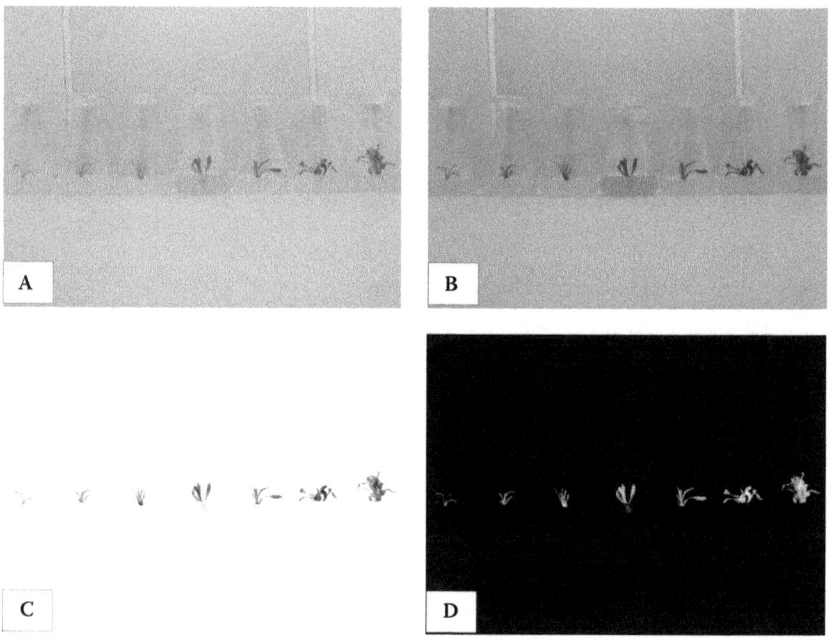

Figure 6.9 Steps in the image analysis procedure: (a) original picture, (b) grayscale image, (c) image after brightness and contrast of (b) adjusted, and (d) image after the color (c) reversed.

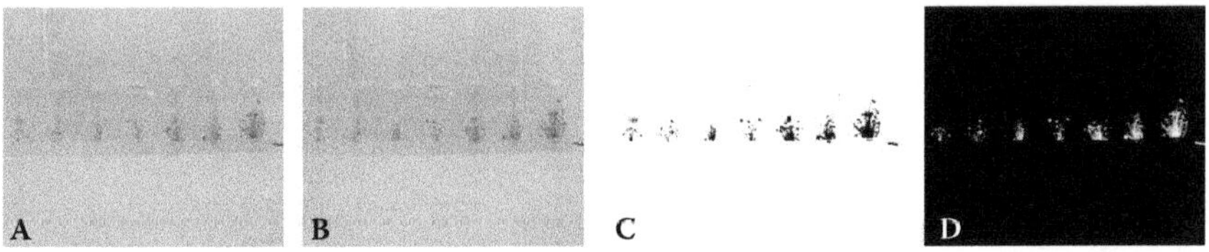

Figure 6.12 Steps in the image treatment procedure: (a) original picture, (b) gray-scale image, (c) image after brightness and contrast of (b) adjusted, and (d) image after color of (c) reversed.

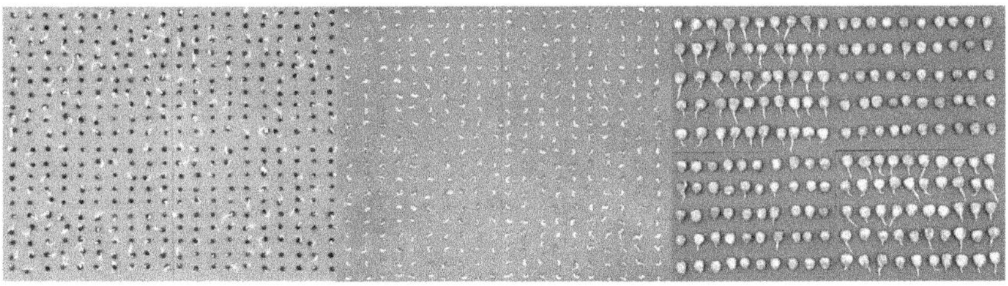

Figure 7.3 Example of images obtained at 20°C. From left to right, oilseed rape 48 h after sowing, *M. truncatula* 24 h after sowing, and maize 72 h after sowing.

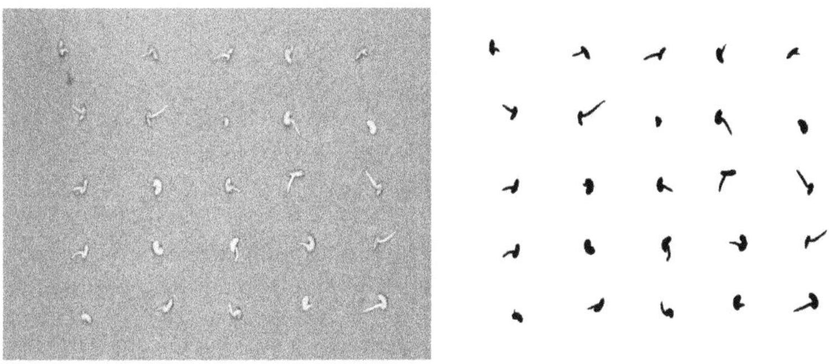

Figure 7.4 Images obtained with a video camera and after segmentation in the red channel. (*M. truncatula* 30 h after sowing at 20°C.)

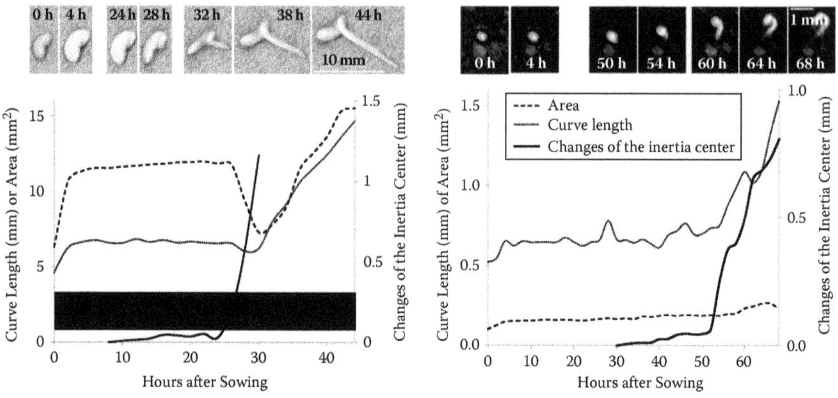

(a) *Medicago truncatula* (b) *Arabidopsis thaliana*

Figure 7.5 Sequence of images and main image analysis parameters extracted from sowing to radicle elongation stages of germination for (a) *M. truncatula* and (b) *A. thaliana.*

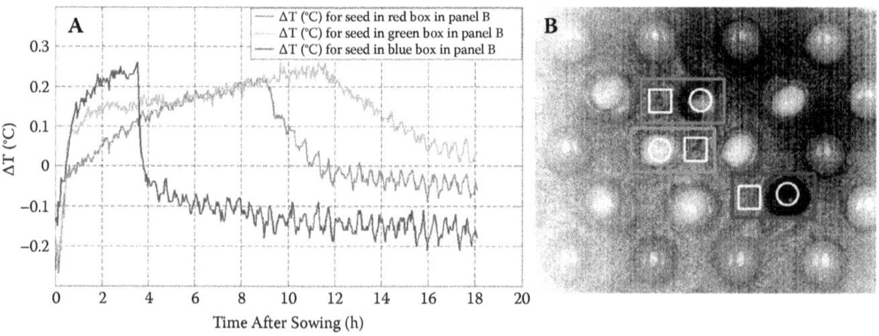

Figure 8.2 Evolution of pea seeds' temperature during imbibition. (a) Evolution of the relative seed temperature ΔT after sowing for three different seeds. (b) Pea seeds sowed on a blotter. Studied seeds are identified with red, green, and blue boxes. Growth chamber, blotter, and water temperatures are fixed at 24°C. The relative seed temperature ΔT ($T_{seed} - T_{env.}$) is defined as the difference between mean temperature T_{seed} of individual seed (white circles) and mean temperature of close environment $T_{env.}$ of seed (white squares).

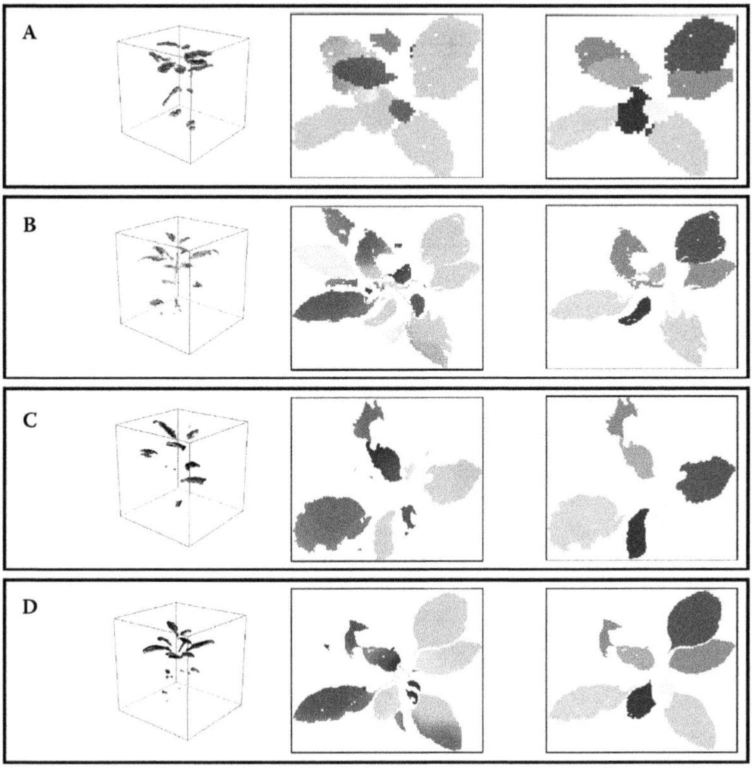

Figure 9.5 Comparison of four depth cameras with the plant 1 of Figure 9.4. Each figure line is composed of the point cloud (left), the depth image (center), and the segmented image (right). In the image segmented with the algorithm of Chéné et al. (2012), only leaves presented in all depth images are represented with the same color for each depth camera. (a) Depth camera is a Microsoft Kinect that uses a pseudorandom dot grid as the structured lighting pattern. (b) Depth camera is a laser scanner that uses triangulation. (c) Depth camera is a hybrid depth camera that uses 2D structured lighting and stereovision (structured lighting pattern allows correspondences between two cameras). (d) Depth camera is a 2D structured lighting depth camera that uses a binary stripe sequential projection as the structured lighting pattern.

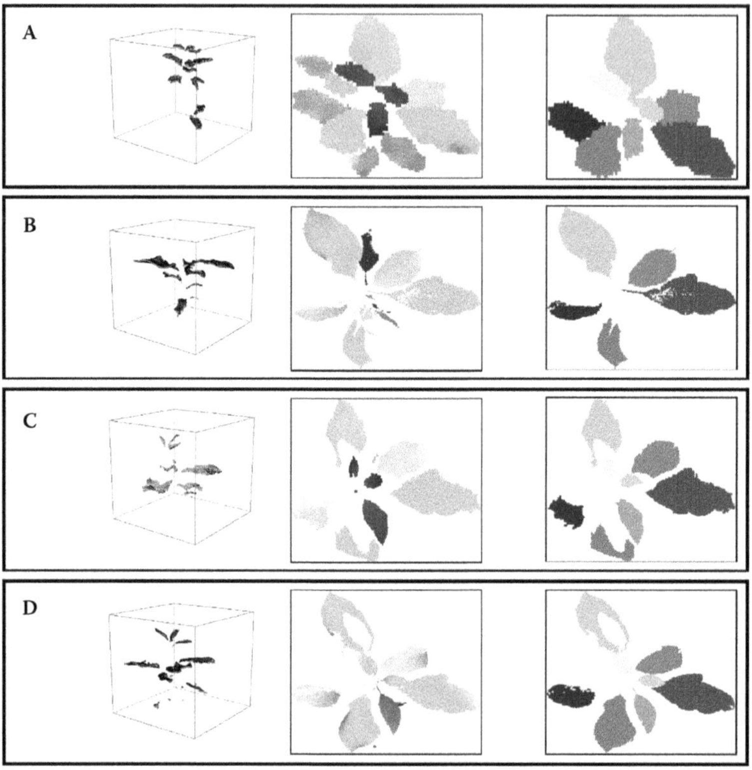

Figure 9.6 Same as in Figure 9.5 with plant 2 of Figure 9.4.

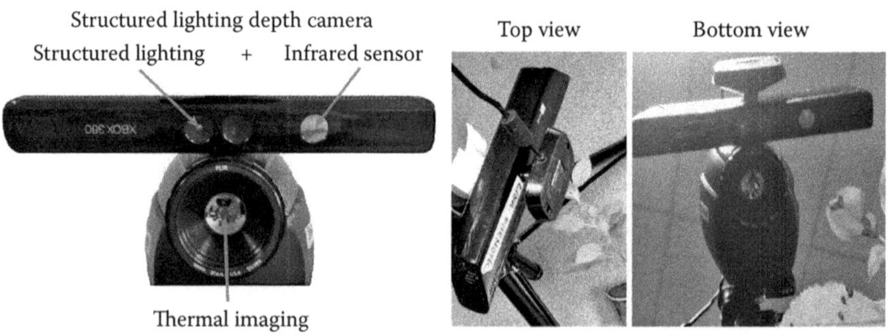

Figure 9.9 Left: The anatomofunctional imaging system in this chapter is composed of a depth camera using the 2D structured lighting (binary dot pseudorandom grid) method and a thermal imaging. Right: An acquisition on a plant is done by one shot taken in plant top view.

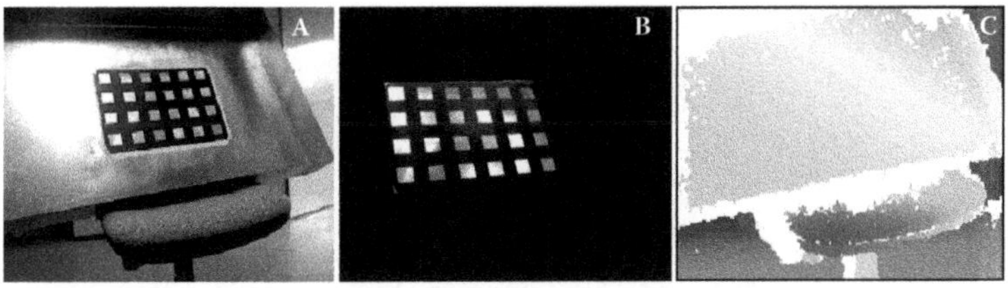

Figure 9.10 A calibration scene. (a) The RGB image of the calibration chessboard on the metallic plane. (b) The thermal image of the calibration chessboard on the metallic plan. (c) The disparity map of the calibration chessboard plan.

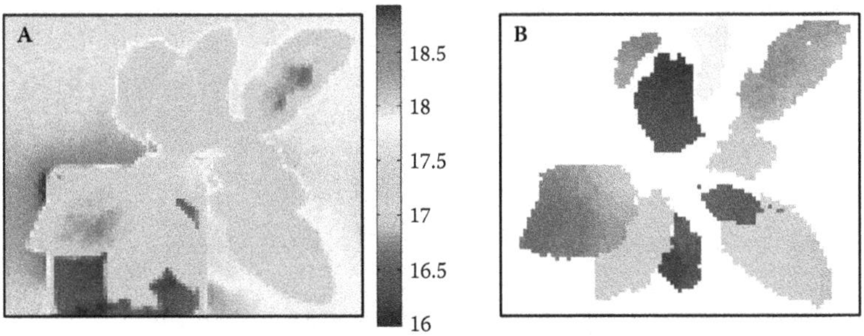

Figure 9.11 An anatomofunctional imaging composed of depth camera and a functional imaging allows us to have two registered images: the depth map and the functional image. In this chapter, the anatomofunctional imaging system is composed of (a) a thermal imaging providing a thermal image and (b) a depth camera providing a depth map.

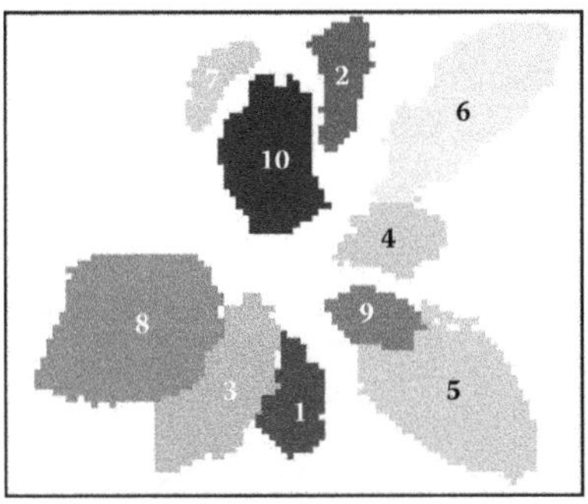

Figure 9.12 Plant leaves are segmented by application of the depth segmentation algorithm of Chéné et al. (2012). Like the depth map, the segmented image is registered with the thermal image. In this image, each leaf is associated with a number. These numbers allow us to create a mask for each individual leaf.

Figure 9.13 The edge of each leaf mask is applied on the thermal image. Colors of these edges are equal to the color of the leaf-segmented image of Figure 9.12. In this case only leaf 6 is hosting the pathogen.

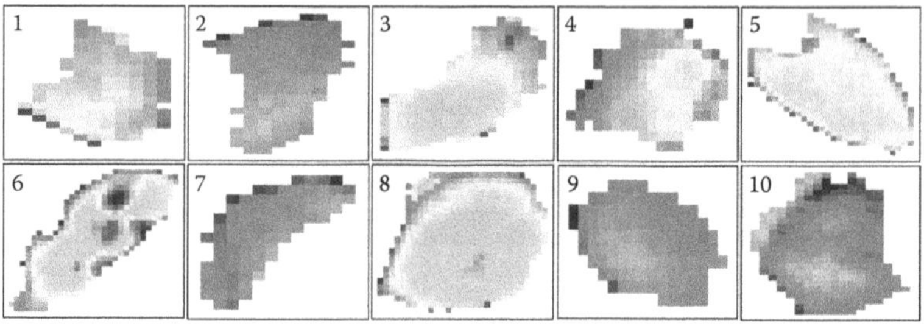

Figure 9.14 Thermal images of each individual leaf segmented from the plant. Numbers correspond to numbers in Figure 9.12. The image resolution is the functional image resolution.

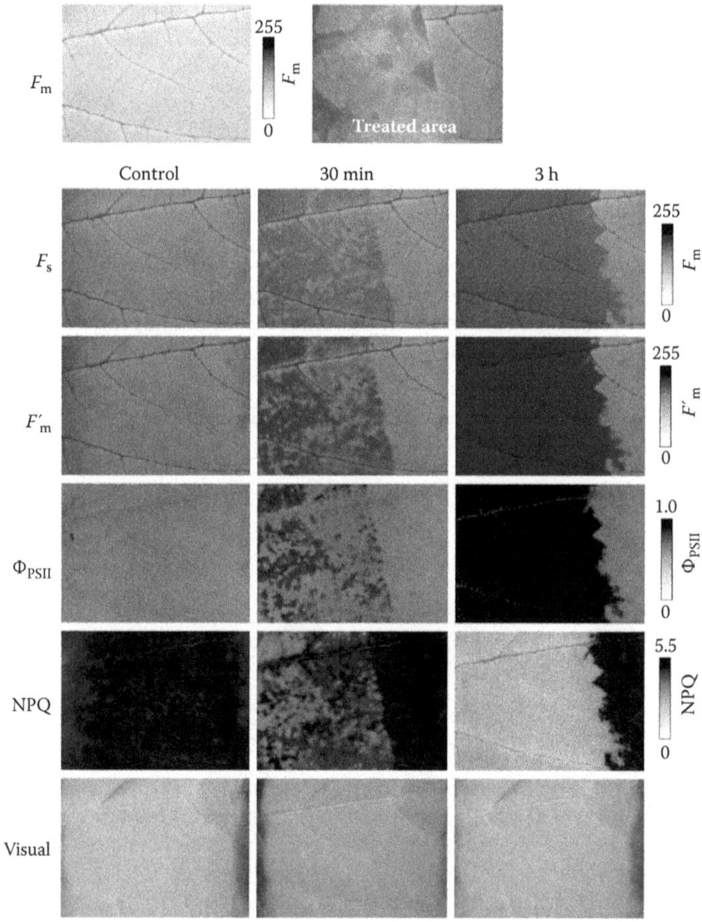

Figure 10.7 Effects of Basta treatment on a *Phaseolus vulgaris* L. leaf. The fluorescence image titled "Treated area" represents the Basta treatment. Chlorophyll fluorescence intensity images (F_s and F'_m) were captured before (control), 30 min, and 3 h after treatment, and Φ_{PSII} and NPQ images were calculated each time.

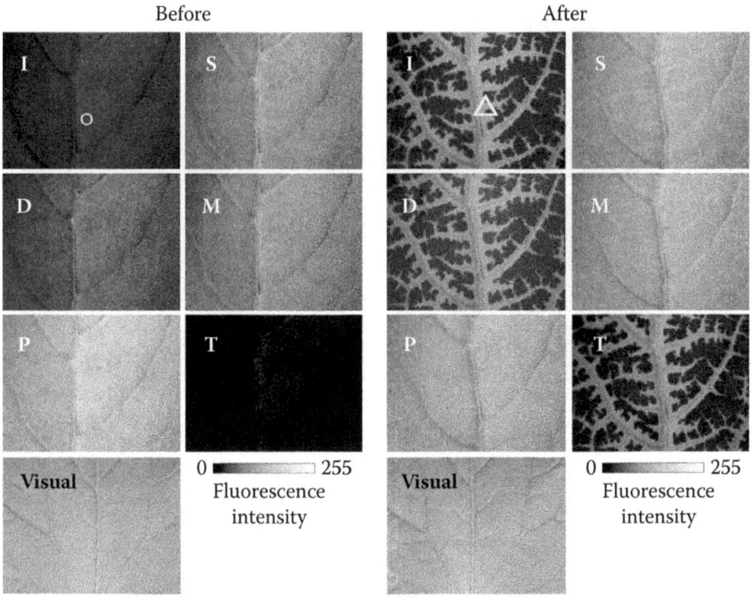

Figure 10.10 Changes in chlorophyll fluorescence induction images of a kidney bean (*Phaseolus vulgaris* L.) leaf before and after herbicide treatment. The petiole was soaked in 1/1000 diluted solution of Nekosogi-ace (Rainbow Chemicals) for 20 min. The symbols ○ and Δ indicate the leaf areas used for the chlorophyll fluorescence induction analysis shown in Figure 10.11.

Figure 11.1 An example of thermal images of rose canopy in a greenhouse. The values in the figure indicate photosynthetic photon flux density (PPFD) measured with a quantum sensor.

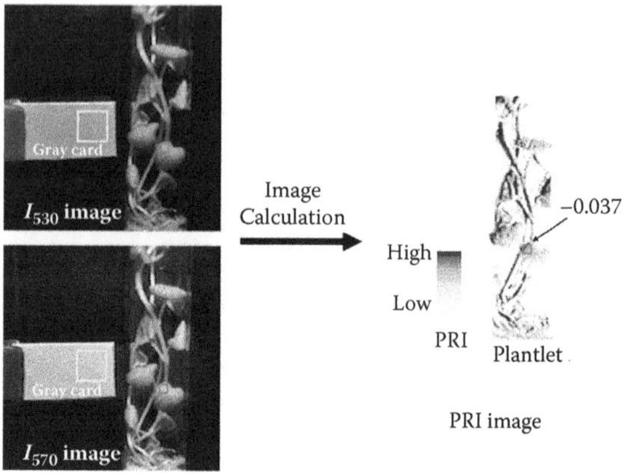

Figure 11.3 An example of acquired images (I530, I570) and the PRI image of a potato plantlet. (Reprinted from Ibaraki et al., *Comput. Electron. Agric.*, 71, 170–175, 2010. With permission of Elsevier.)

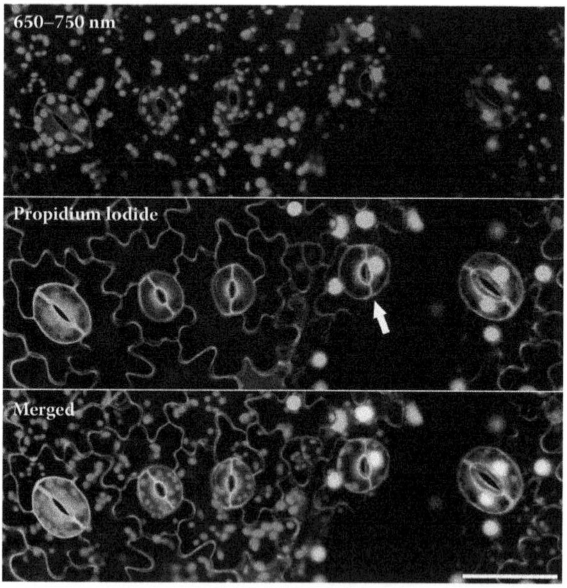

Figure 13.3 Use of propidium iodide (PI) to monitor cell damage. PI (10 μg/ml) is infiltrated into a tobacco leaf using a syringe applicator. The region in the vicinity of the syringe injection (right half of the image) shows damaged cells with nuclear PI labeling. The left half of the middle panel shows intact leaf pavement and guard cells with PI-labeled cell walls. The arrow indicates a transition zone stoma with two guard cells, one of which is damaged, while the other one is intact. Red-colored fluorescence (mostly chlorophyll emission between 650 and 750nm) and PI (green pseudocolored, 550–650 nm emission) images are merged at the lowermost panel. Excitation wavelength = 488 nm. Scale bar = 50 μm.

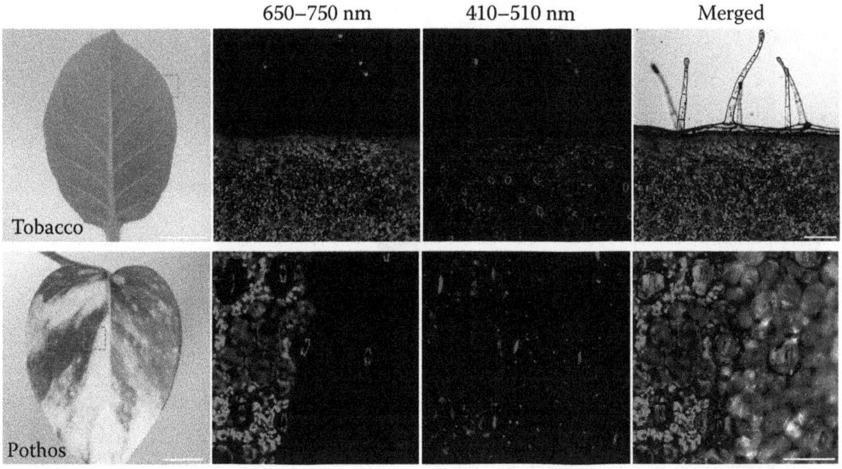

Figure 13.4 Red (650–750 nm emission excited by 488 nm) and blue (410–510 nm emission excited by 405 nm) autofluorescence on leaf border of tobacco with trichomes (*Nicotiana tabacum,* upper panel) and variegated leaf of golden pothos (*Epipremnum aureum,* lower panel). Border region sampled for microscopy imaging is indicated with a red square on the tobacco leaf photo. The dotted red square on the pothos leaf photo indicates a variegated leaf area with green (chlorophyll-rich) and white (etiolated) sectors. Microscopy images from a variegated area are shown as close-up images of the pothos leaf. Last panels show merged images of red, blue, and bright field images. Scale bars = 1 cm (leaf macro-photos) and 100 μm (microscopy images).

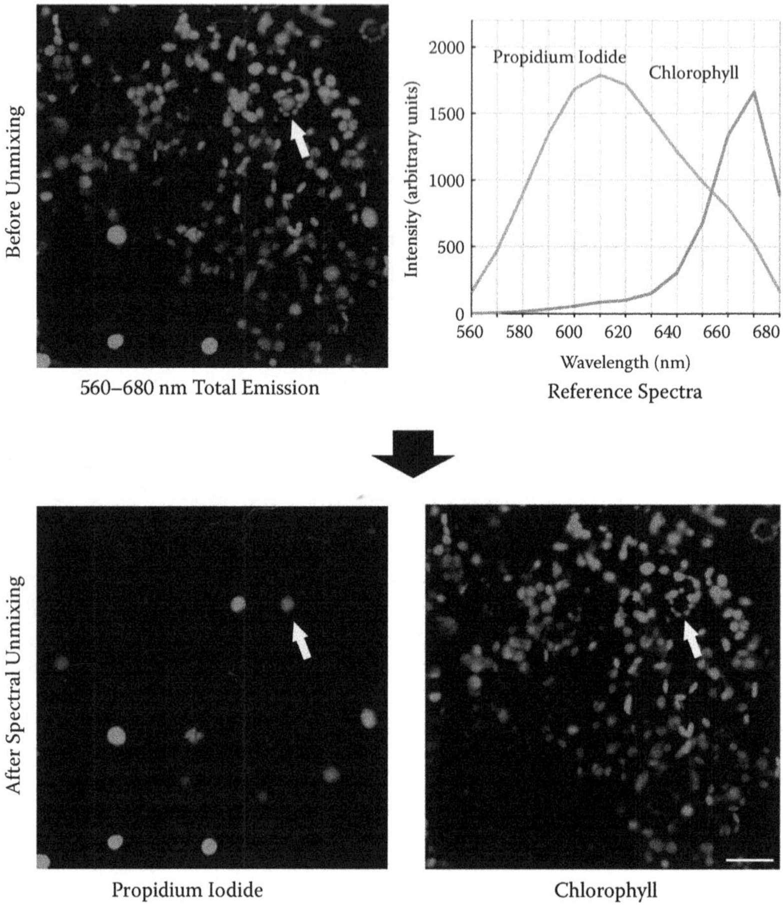

Figure 13.5 Spectral imaging and computational unmixing of propidium iodide (PI) and chlorophyll fluorescence. Upper panel shows intensity projection over lambda axis (excitation = 488 nm; emission = 560–680 nm, 10 nm/step, 20 nm resolution) of a PI-infiltrated tobacco leaf with chlorophyll autofluorescence. A labeled nucleus and a well-isolated chloroplast are used to capture in vivo reference spectra for PI (orange) and chlorophyll (red), respectively (graph, upper panel). A mixed image is unmixed into PI and chlorophyll channels using the reference spectra and the spectral unmixing software of Olympus Fluoview (lower panel). Arrows indicate clear separation of a PI-labeled nucleus surrounded by chloroplasts. All images are pseudocolored. Scale bar = 20 μm.

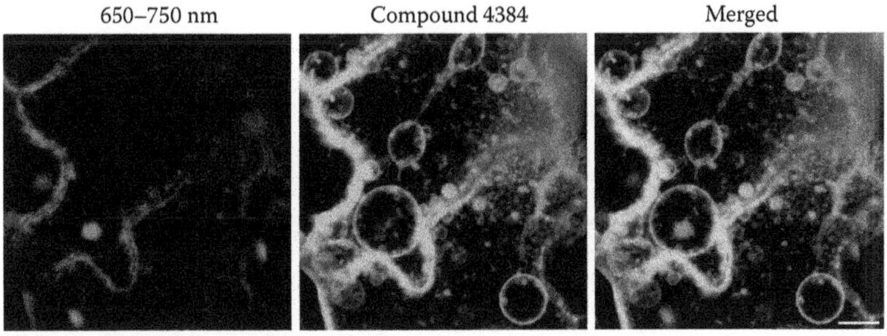

Figure 13.6 Tobacco leaves infiltrated with a 1,8-naphthalimide derivative (compound-4384). Last panel shows the merged images of red emission (excitation = 488 nm, emission = 650–750 nm, mostly chlorophyll autofluorescence) and green pseudocolored chemical fluorescence (excitation = 405 nm, emission = 500–590 nm). Scale bar = 10 μm.

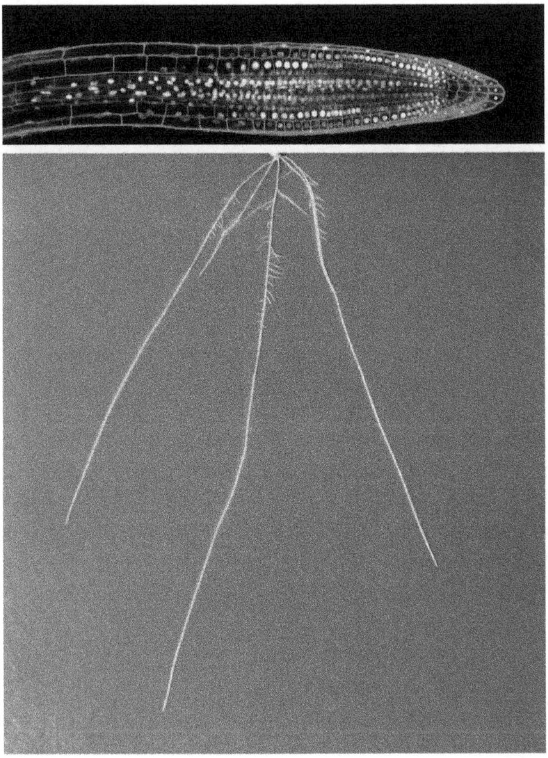

Figure 14.1 Example images of roots at the cellular (top) and whole architecture (bottom) scales. The top image is a confocal image of an *Arabidopsis thaliana* root tip with fluorescent markers on cell walls and nuclei. The bottom image is a wheat root system growing on a filter paper system, imaged with a digital camera. (Images courtesy of, respectively, Dr. Antoine Larrieu and Mr. Jonathan Atkinson, University of Nottingham.)

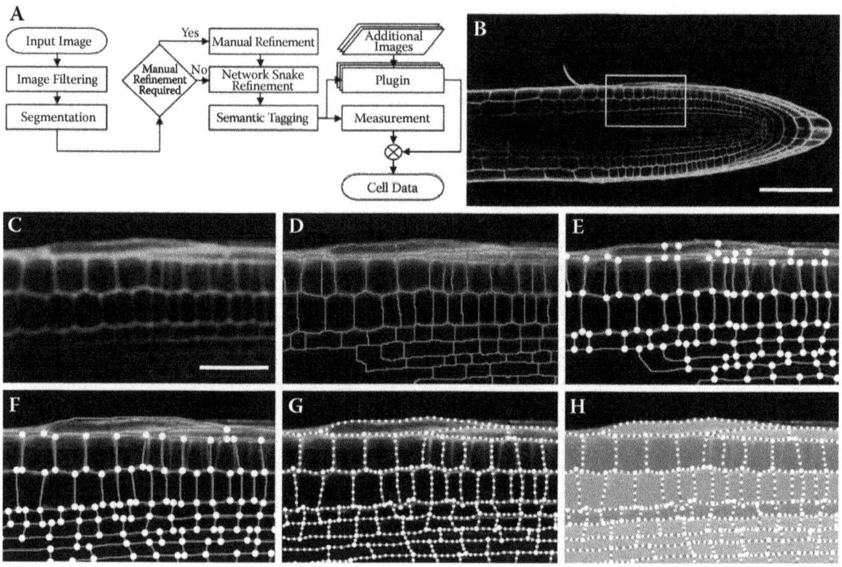

Figure 14.4 The CellSeT processing pipeline. (a) The steps involved in confocal image analysis using CellSeT. (b) A typical input CLSM image. Cell walls have been stained with propidium iodide (here displayed green) to reveal their cellular organization. (c) A zoomed portion of the input image after filtering. (d) Two-level watershed segmentation is used to initially distinguish cells. In cases where cell walls are poorly defined, this may lead to under- or oversegmentation. (e) The refinement graph structure generated using the segmented image. (f) Manual refinement of the graph structure; oversegmentation in the epidermal cells has been removed, and undersegmentation in the stele cells has been corrected. (g) CellSeT uses a network snake algorithm to refine the results of manual segmentation. (h) Cells can be semantically tagged by the user to provide additional information during the output stage. Bar in (b) = 100 μm, bar in (c) = 30 μm. (From Pound et al., *Plant Cell*, 24, 1353–1361, 2012. Copyright © American Society of Plant Biologists.)

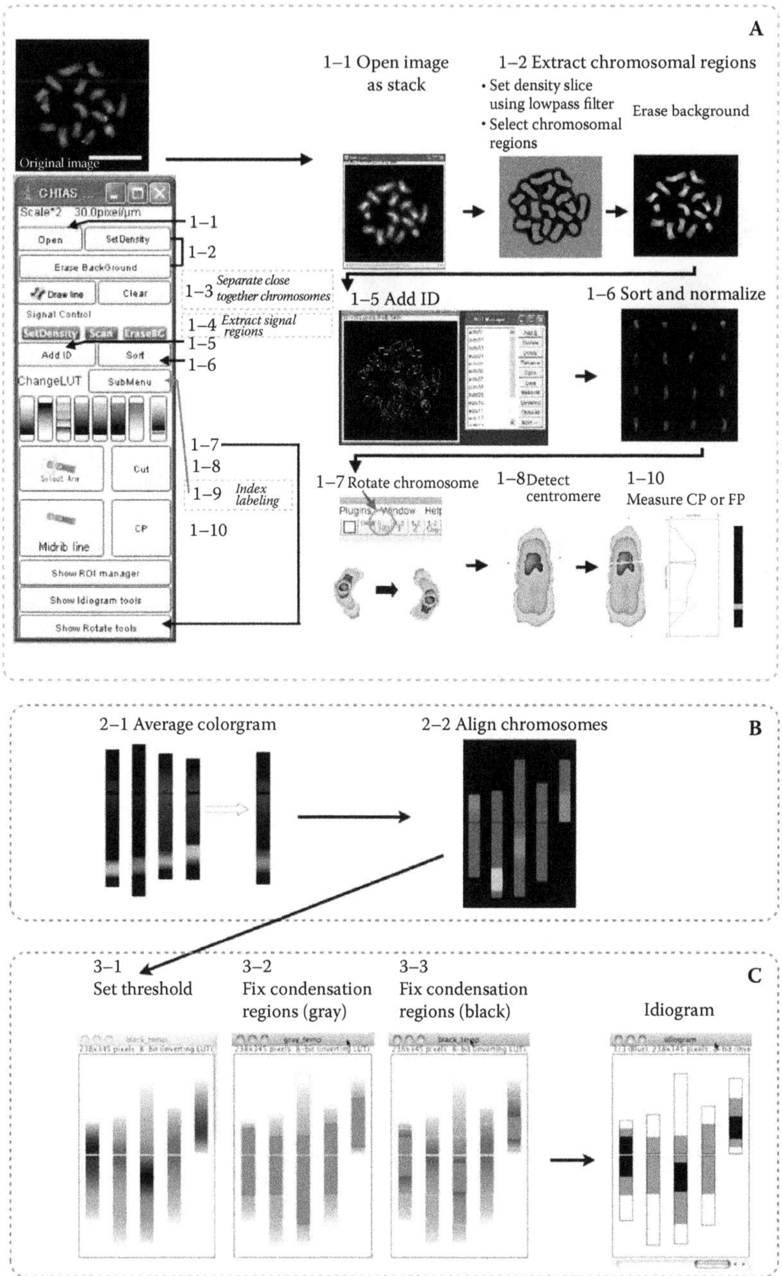

Figure 15.1 Operation diagram of Chromosome Image Analyzing System (CHIAS) IV showing the three-step analysis of the chromosome images of the red clover obtained at the prometaphase. (a) Step 1: Measurement of condensation pattern (CP) or fluorescent pattern (FP) of the chromosomes. (b) Step 2: Create the standard profile from the CP or FP of each chromosome. (c) Step 3: Create the ideogram based on the standard profile.

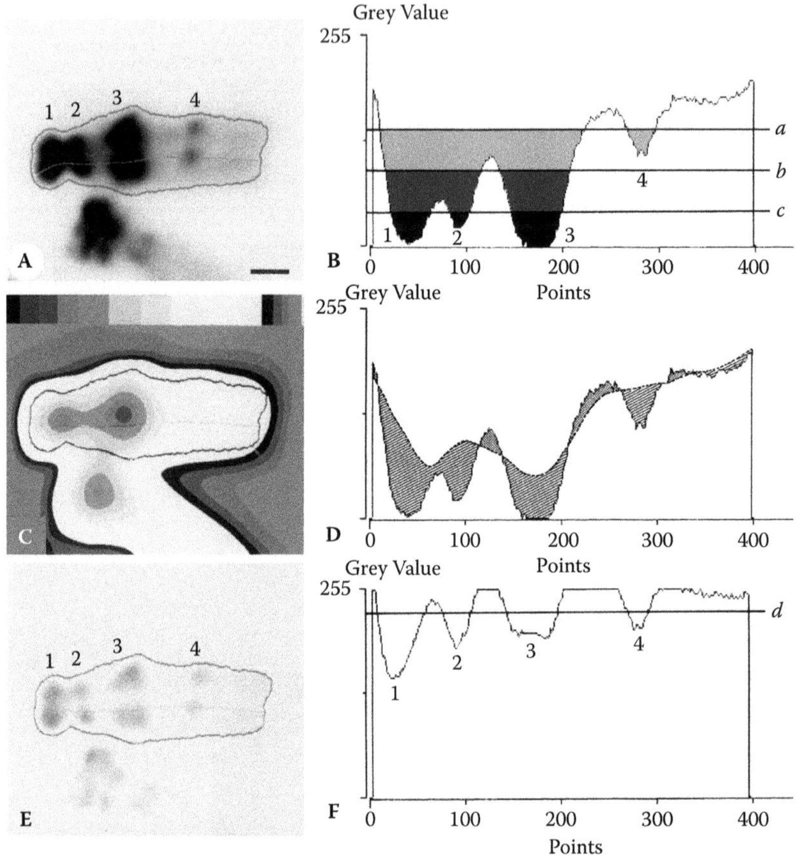

Figure 15.3 Identifying C-banded *Crepis* chromosomes using simulated human vision. (a) Digital image of a C-banded *Crepis* chromosome (2C). The chromosome contour line and the chromatid mid-rib line were superimposed with an overlay line. The human eye can easily recognize four bands. Bar = 1 μm. (b) Density profile of the gray values along the mid-rib of the chromatid. The x- and y-axes of the graph represent the number of pixels and the gray values, respectively. High values correspond to light gray values. Lines a, b, and c indicate gray values of 140, 90, and 40, respectively. Along lines a and b, bands 1 and 2 are combined as a single large band, and along lines b and c, band 4 is not recognized as a band. (c) Pseudocolor representation of the low-pass filtered image. (d) Difference between the gray values of the original image and the values for the low-pass filtered image. Two representative density profiles at the mid-rib of the chromatid were used in the calculation. (e) Standardized gray image generated by subtracting the low-pass filtered image from the original image. (f) Density profile at the mid-rib of the standardized image. In this processed image, all four bands were successfully picked up and separated along line d. (Modified from Fukui and Kamisugi, *Chromosome Res.*, 3, 79–86, 1995. With kind permission from Springer Science and Business Media.)

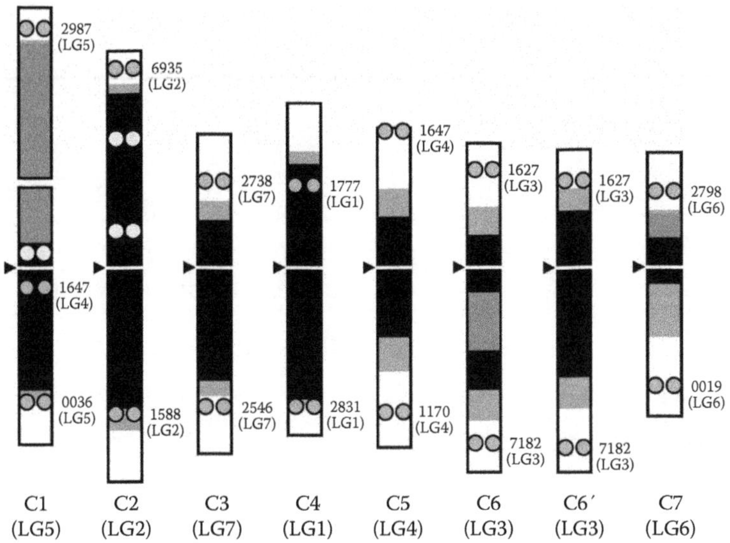

Figure 15.6 Chromosome map of red clover. The loci of then BAC clones that harbor linkage group-specific microsatellite markers are indicated by green circles; 26S rDNA loci are indicted by red boxes; and 5S rDNA loci are indicated by orange circles. Arrowheads indicate the positions of the centromeres. Reproduced from Kataoka et al., *Cytogen. Genome Res.*, 137, 60–69, 2012; with kind permission from Karger Publishers.

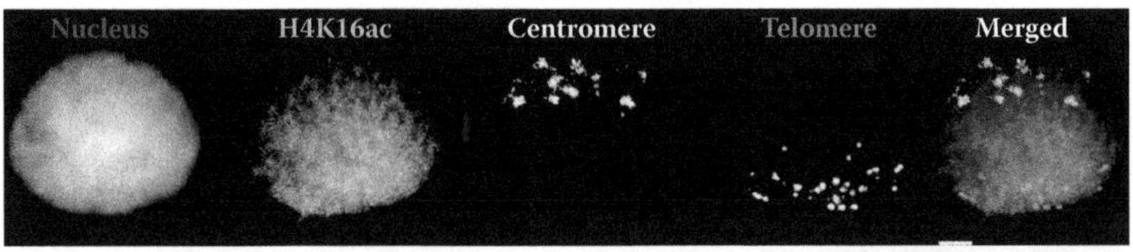

Figure 15.10 Immuno-FISH images of barley nucleus. Each panel is a stacked image from optical slices obtained with a confocal microscope. The image was reconstructed using the 3D viewer plug-in of ImageJ. Each 2D image was processed using a median filter, normalization, and thresholding. Reproduced from Wako et al., *Plant Mol. Biol.*, 49, 645–653, 2002; with kind permission from Springer Science and Business Media.

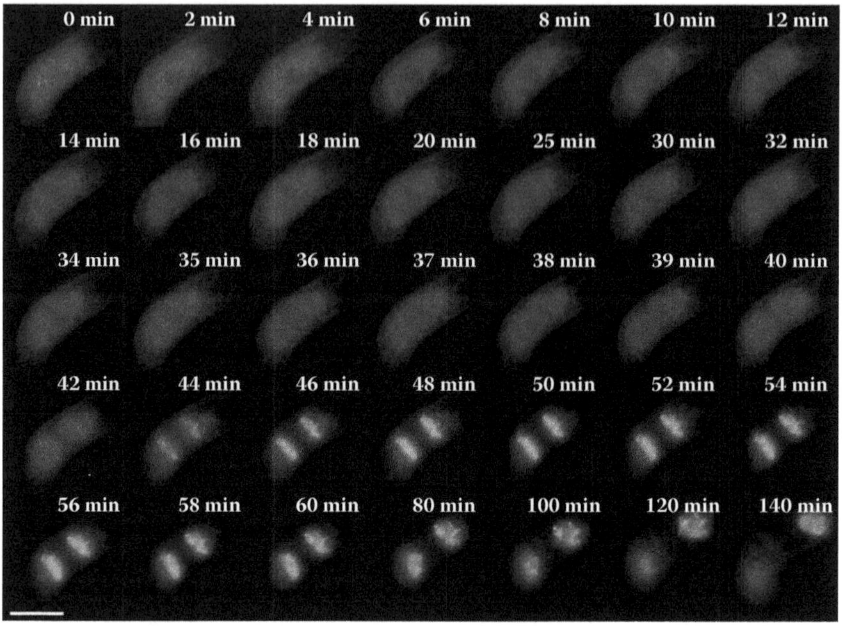

Figure 15.11 Time-lapse image of tobacco chromosomes (blue) during mitosis transformed with GFP-fused AtLHP1 (green). Bar = 30 μm.

(a) RGB

(b) Thermal 24.1-31.1 degrees Celcius

(c) Thermal 23.5-37.0 degrees Celcius

(d) RGB Outdoor

(e) Thermal 21.8-25.9 degrees Celcius

(f) Thermal 21.8-25.8 degrees Celcius

Figure 16.2 (b, c) Thermal images of fruit and foliage brought from ambient (hot) conditions into a cool (20°C) environment (a) with RGB image for reference. (e, f) Thermal images of fruit on tree collected in late afternoon (ambient temperatures decreasing) (d) with RGB image for reference.

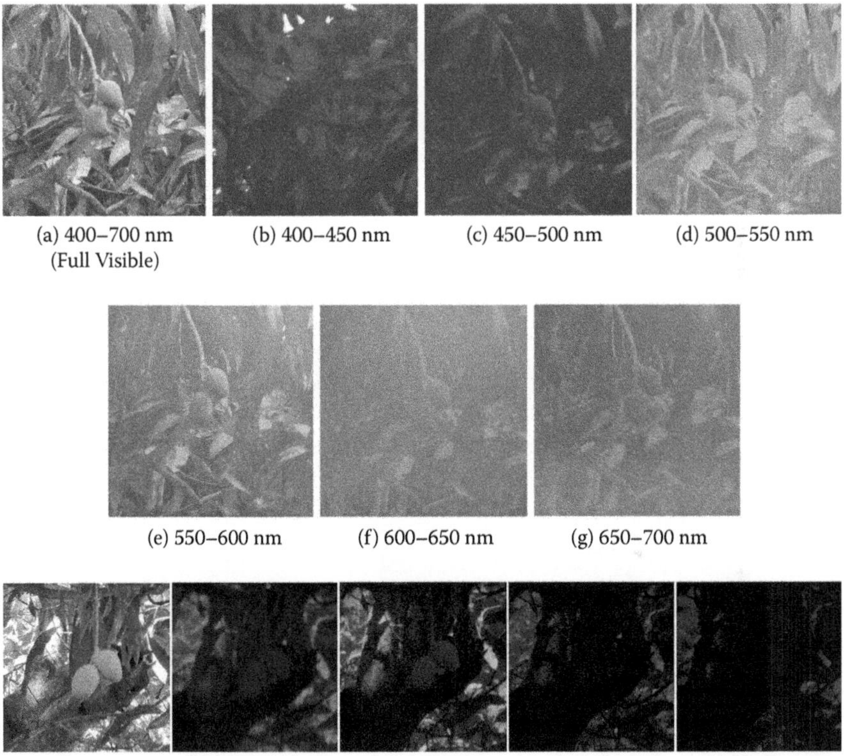

(a) 400–700 nm (Full Visible) (b) 400–450 nm (c) 450–500 nm (d) 500–550 nm

(e) 550–600 nm (f) 600–650 nm (g) 650–700 nm

(h) Full Visible (i) 700–750 nm (j) 750–800 nm (k) 800–850 nm (l) 850–900 nm

Figure 16.5 Image of mango tree (fruit and foliage) acquired using a CCD (RGB) camera with filters: (a) 400–700 nm, (b) 400–450 nm, (c) 450–500 nm, (d) 500–550 nm, (e) 550–600 nm, (f) 600–650 nm, (g) 650–700 nm, (h) 400–700 nm, (i) 700–750 nm, (j) 750–800 nm, (k) 800–850 nm, (l) 850–900 nm.

callus using two different algorithms. The data obtained from image analysis were correlated with the actual shoot length. The image analysis was able to predict the shoot length with an average error of 1.3 mm.

3.3.4 Selection of viable somatic embryos

Somatic embryogenesis is an asexual form of plant propagation in nature that mimics many of the events of sexual reproduction. Somatic embryogenesis can be induced by two modes: directly on explants and indirectly from callus. The most important advantage of cloning by somatic embryogenesis is that the embryogenic tissue can be cryopreserved without changing its genetic makeup and without loss of juvenility. During manual harvest, somatic embryos are picked by a skilled worker with the use of a stereomicroscope. This procedure is cumbersome and has the limitations of human vision as well as skill. Over time, this can result in inconsistency in the harvest of conversion-competent embryos. Development of an efficient machine vision classifier may provide an alternative to identify conversion-competent embryos that are virtually impossible with human vision.

Padmanabhan et al. (1998) developed a machine vision system to separate competent and noncompetent classes of somatic embryos in sweet potato. Size, shape, and color-related measurements were used to classify the somatic embryos following discriminant analysis. Somatic embryos at the torpedo and the cotyledonary stage were first characterized by machine vision and then transferred to a conversion medium. Images of 378 embryos were captured by a color video camera and saved in the computer before the embryos were transferred to a conversion medium. The color-related features, such as hue, intensity, and saturation, were extracted using a color image-processing algorithm. The embryo size- and shape-related features from the images were extracted using 32 radii length or distances from the embryo centroid to 32 perimeter points at constant angular increments. A total of 303 embryos were subjected to a 20-day conversion study following computer vision analysis and were scored for root and shoot conversions. An embryo was scored competent or noncompetent in terms of its subsequent conversion into a plantlet. Three color descriptors (hue, saturation, and intensity of the centroid, cotyledon, and radicular region of the embryo), along with 7 size descriptors and 10 shape descriptors, were subjected to Fisher discriminant analysis. The analysis revealed that there were differences in the morphological parameters between competent and noncompetent classes of somatic embryos. About 60% of the competent and 60–75% of the noncompetent embryos could be correctly classified as competent and noncompetent, respectively, in terms of their subsequent conversion into plantlets. This resulted in a substantial improvement in regeneration efficiency (about 15%), when compared to the regeneration efficiency realized through manual harvest of somatic embryos, which was about 44%.

3.3.5 Photometric clustering of regenerated plants

The main goal of commercial micropropagation is to produce genetically identical and healthy plants. However, one of the major drawbacks in commercialization of micropropagation is the poor survival of regenerated plants upon ex vitro transfer. The intrinsic quality of the plantlets largely regulates the success of acclimatization. Variations in the physiological status of the regenerated plants and their rooting ability, hyperhydric status, and adaptability to ex vitro conditions all together contribute to the survival rate. Development of an automatic decision-making system reflecting the variations of in vitro regenerated plants is necessary to achieve success in micropropagation. The decision making may be made in the form of clustering and grouping of regenerated plants based on their intrinsic properties. Since the phenotyping of regenerated plants is difficult to manifest by human visual evaluation, machine vision-coupled neural network-based clustering might be an attractive option.

Mahendra et al. (2004) proposed a machine vision system to test whether regenerated plants can be sorted into groups based on their photometric behavior using an image analysis system coupled with a neural network algorithm. The leaves of gladiolus were scanned in a deskjet scanner and the trichromatic features (R, G, and B) were extracted from the images using Adobe Photoshop 7.0. The leaves were then sorted out into groups using an adaptive resonance theory (ART2)-based algorithm. ART2 is a clustering algorithm that clusters a given set of input patterns into some groups in an unsupervised manner. Properties of ART2 depend upon the vigilance parameter (VP) that defines the minimum similarity between patterns in one cluster. The effect of VP was evaluated with increasing magnitude from 0.59 to 0.99. The number of generated groups increased from 1 to 2 with the VP range over 0.985. A significant difference in the grouping pattern was observed with VP values of 0.985 and 0.99. Leaves having maximum similarity in terms of inherent pixel properties were clustered in a single group. The study demonstrated that the leaf photometric property could provide a classifying feature with which the variations among the regenerated plants can be projected.

To refine the clustering approach and to ascertain its biological significance in terms of indexing the in vitro corm development potential of gladiolus, Prasad and Dutta Gupta (2008) adopted a fuzzy ART-assisted photometric clustering. The clustering result was compared with ART2, self-organizing maps (SOMs), hierarchical clustering (HC), and k-means algorithm (KM). Biological validation of such clustering was assessed by the ability of in vitro corm induction of the respective groups. Photometric features were extracted from the digitized leaf images as described by Mahendra et al. (2004). The data were then normalized to a value lying between 0 and 1. The training set data of leaf images were subjected to fuzzy ART clustering analysis with VP ranging from 0.1 to 1.0 at an increment of 0.01. The number of groups was found to be seven at vigilance parameter 0.91, compared to other clustering methods that produced only two groups. A significant difference in corm induction potential between the groups was observed only with ART2, where corm induction was 36.8% in group 0, while in group 1 it was 69.4%. Fuzzy ART-assisted clustering patterns are not conducive in segregating the potential corm-producing shoots. ART2 aided in the sorting of regenerated plants using RGB features that appeared to be more promising for selecting a group of plants capable of corm development than the other clustering methods.

3.3.6 Estimation of chlorophyll content in micropropagated potato plants

Leaf color is proposed to be an indicator of the plant nutrient and health status, which is closely related to the amount and proportion of chlorophyll content. Leaf color analysis based on the determination of chlorophyll content using an optical sensor-based chlorophyll meter, particularly the SPAD meter, has become popular in the last decade as an alternative to the destructive spectrophotometric measurement of chlorophyll content (Wu et al., 2007; Uddlinng et al., 2007; Pagola et al., 2009; Hu et al., 2013). However, culture growth in an enclosed vessel under aseptic conditions makes the SPAD chlorophyll meter inaccessible to an in vitro system.

Yadav et al. (2010) developed a method for rapid and noninvasive determination of chlorophyll content of leaves of micropropagated potato plants using RGB-based image analysis. In this method, the leaves of regenerated plants were scanned with the help of a scanner. The color histogram was obtained from the image using Adobe Photoshop v. CS image processing software. From the histogram, the mean bright of primary colors red (R), green (G), and blue (B) was recorded and the spectral parameters hue, saturation, and luminosity were estimated from the RGB values. Mean brightness ratio (rgb) was also calculated. The brightness values and their ratios were modeled as linear correlation functions for chlorophyll content. The model equations are $Y = aR + bG + cB$ and $y = ar + bg + cb$, respectively, for RGB and rgb, where Y and y are the predicted chlorophyll contents of the regenerated plants; R, G, B and r, g, b, are, respectively, the mean brightness value and mean brightness ratio of each primary color; and a, b, and c are model parameters. The model parameters a, b, and c were determined by using the matrices from the RGB and rgb models. The models were solved with the help of MATLAB v. 7.1 software. The matrices are

$$[\ a \quad b \quad c\ ]^T = [A_{RGB}^T \bullet A_{RGB}]^{-1} \bullet A_{RGB}^T \bullet Y'$$

and

$$[\ a' \quad b' \quad c'\ ]^T = [A_{rgb}^T \bullet A_{rgb}]^{-1} \bullet A_{rgb}^T \bullet Y'$$

respectively, for RGB and rgb. *ARGB* represented the mean brightness values of the primary colors, *Argb* represented the mean brightness ratio of the primary colors, and vector *Y* represented the chlorophyll content determined by the chlorophyll meter. A significant correlation was observed between the model-predicted chlorophyll content and the chlorophyll content measured by SPAD meter. The parameters luminosity and saturation were found to be negatively correlated with the chlorophyll content. An RGB-based image analysis system as described in this work demonstrates its potential for noninvasive estimation of chlorophyll content and subsequent analysis of plant health status. In order to make the system amenable for real-time estimation of chlorophyll content, Dutta Gupta et al. (2013) introduced a digital image analysis method. The component steps of digital image analysis are presented in Figure 3.1. Leaf images were acquired using a simple digital still camera. The RGB color image was saved as a RAW format file (3648 × 2736 pixels). RGB features were extracted from the image and correlated with the SPAD values. Simple correlation studies failed to reveal any correlation between the mean brightness parameters and the actual chlorophyll content. A MATLAB-based linear RGB/rgb model as described earlier and a feedforward backpropagation neural network model were applied to provide a relationship between input variables (photometric features) and output variables (chlorophyll content). For neural network modeling, Easy-NN Plus v. 7.0 was used. A significant correlation between the predicted and actual chlorophyll content was obtained with mean brightness ratio (rgb) in both the models. The correlation between the predicted and actual chlorophyll contents as derived by linear and ANN modeling using the mean brightness ratio is shown in Figure 3.2a and b. Considering the level of correlation between the model-predicted and SPAD-derived chlorophyll content, the developed digital image analysis system is capable of real-time estimation of chlorophyll content during micropropagation. An RGB-based color sensor has also been developed to monitor and control plant growth in a modified plant micropropagation system (Seelye et al., 2011). However, our study suggests a better predictability of the rgb model than RGB for estimation of chlorophyll content during in vitro culture.

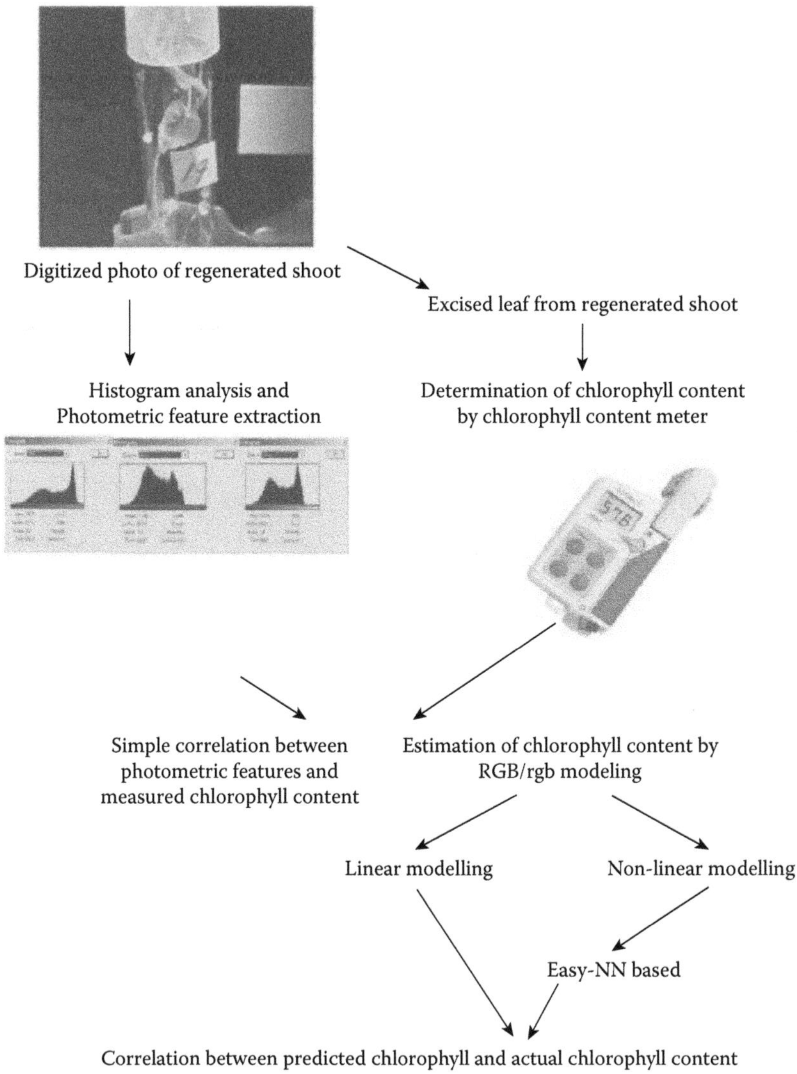

Figure 3.1 The component steps of digital image analysis system to study the correlation of RGB features as well as model-predicted chlorophyll content to that of actual chlorophyll content measured with SPAD meter. (Reproduced from Dutta Gupta et al., *Plant Biotechnol. Rep.*, 7, 91–97, 2013. With permission from Korean Society for Plant Biotechnology and Springer.)

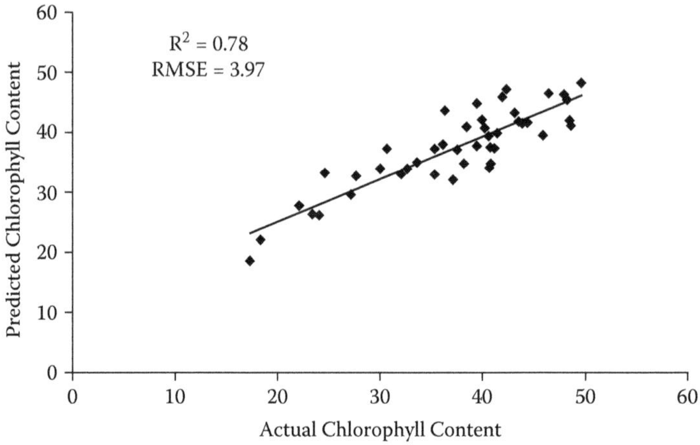

(a)

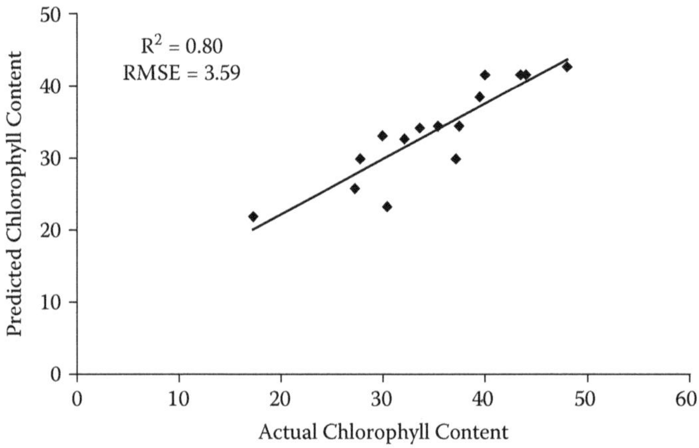

(b)

Figure 3.2 The correlation of chlorophyll content estimated by (a) linear and (b) Easy-NN-based rgb models to that of SPAD-measured chlorophyll content. (Reproduced from Dutta Gupta et al., *Plant Biotechnol. Rep.*, 7, 91–97, 2013. With permission from Korean Society for Plant Biotechnology and Springer.)

3.4 Conclusions

The present chapter presents an overview of the recent developments and applications of RGB image analysis in plant science. Image processing is recognized as the core of machine vision and has been used in different engineering and biological fields. In recent years there has been exciting developments in imaging approaches to the study of various plant phenotyping aspects. The nondestructive nature of this technique and its potential to replace the low-resolution, ambiguous human vision system with a high-throughput vision system makes it an ideal approach for assessing the quality of food articles, recognition of weeds, detection of diseases, and determination of nutrient/health status of plants. Even though a large number of efficient algorithms have been developed, it still requires further improvement in the level of accuracy with a user-friendly interface.

References

Ahmad, I., Muhamin, A., and Naeem Islam, M., Real-time specific weed recognition system using histogram analysis, *Proc. World Acad. Sci. Eng. Technol.*, 16, 145–148, 2006.

Aitkenhead, M.J., Dalgetty, I.A., Mullins, C.E., McDonald, A.J.S., and St. Rachan, N.J.C., Weed and crop discrimination using image analysis and artificial intelligence methods, *Comput. Electron. Agric.*, 39, 157–171, 2003.

Alchanatis, V., Pelege, K., and Ziv, M., Classification of tissue culture segments by color machine vision, *J. Agric. Eng. Res.*, 55, 299–311, 1993.

Aldea, M., Frank, T.D., and Delucia, E.H., A method for quantitative analysis for spatially variable physiological processes across leaf surfaces, *Photosynth. Res.*, 90, 161–172, 2006.

Ali, M.M., Al-Ani, A., Eamus, D., and Tan, D.K.Y., A new image processing based technique to determine chlorophyll in plants, *Am.-Eurasian J. Agric. Environ. Sci.*, 12, 1323–1328, 2012.

Auearunyawat, P., Kasetkasem, T., Wongmaneeroj, A., Nishihara, A., and Keinprasit, R., An automatic nitrogen estimation method in sugarcane leaves using image processing techniques, in *International Conference on Agricultural, Environment and Biological Sciences (ICAEBS 2012)*, 2012, pp. 26–27.

Aynalem, H.M., Righetti, T.L., and Reed, B.M., Non-destructive evaluation of in vitro-stored plants: a comparison of visual and image analysis, *In Vitro Cell. Dev. Biol.-Plant*, 42, 562–567, 2006.

Berzin, I., Mills, D., and Merchuk, J.C., A non-destructive method for secondary metabolite determination in hairy root cultures, *J. Chem. Eng. Jpn.*, 32, 229–234, 1999.

Cai, H., Haixin, C., Weitang, S., and G, Lihong, G., Preliminary study on photosynthetic pigment content and color feature of cucumber initial blooms, *Trans. CSAE*, 22, 34–38, 2006.

Chaudhary, N., Singh, S., Agrawal, S.B., and Agrawal, M., Assessment of six Indian cultivars of mung bean against ozone by using foliar injury index and changes in carbon assimilation, gas exchange, chlorophyll fluorescence and photosynthetic pigments, *Environ. Monit. Assess.*, 185, 7793–7807, 2013.

Cliff, M., Sanford, K., Wismer, W., and Hampson, C., Use of digital images for evaluation of factors responsible for visual preference of apples by consumers, *Hortic. Sci.*, 37, 1127–1131, 2002.

Cui, D., Minzan, L., and Zhang, Q., Development of an optical sensor for crop leaf chlorophyll content detection, *Comput. Electron. Agric.*, 69, 171–176, 2009.

Dana, W., and Ivo, W., Computer image analysis of seed shape and seed color of flax cultivar description, *Comput. Electron. Agric.*, 61, 126–135, 2008.

Dutta Gupta, S., Ibaraki, Y., and Pattanayak, A., Development of a digital image analysis method for real-time estimation of chlorophyll content in micropropagated potato plants, *Plant Biotechnol. Rep.*, 7, 91–97, 2013.

Gwata, E.T., Wofford, D.S., Pfahler, P.L., and Boote, K.J., Genetics of promiscuous nodulation in soybean: nodule dry weight and leaf color score, *J. Hered.*, 95, 154–157, 2004.

Hemming, J., and Rath, T., Computer-vision based weed identification under field condition using controlled lighting, *J. Agric. Eng. Res.*, 78, 233–243, 2001.

Honda, H., Takikawa, N., Noguchi, H., Hanai, T., and Kobayashi, T., Image analysis associated with a fuzzy neural network and estimation of shoot length of regenerated rice callus, *J. Ferment. Bioeng.*, 84, 342–347, 1997.

Hu, H., Zhang, J., Sun, X., and Zhang, X., Estimation of leaf chlorophyll content of rice using image color analysis, *Can. J. Remote Sens.*, 39, 185–190, 2013.

Jafari, A., Mohtasebi, S.S., Jahromi, H.E., and Omid, M., Weed detection in sugar beet fields using machine vision, *Int. J. Agric. Biol.*, 8, 602–605, 2006.

Kai, S., Zhikun, L., Hang, S., and Chunhong, G., A research of maize disease image recognition of corn based on BP networks, in *IEEE Third International Conference on Measuring Technology and Mechatronics Automation*, 2011, pp. 246–249.

Karcher, D.E., and Rechardson, M.D., Quantifying turf grass color using digital image analysis, *Crop Sci.*, 43, 943–951, 2003.

Kawashima, S., and Nakatani, M., An algorithm for estimating chlorophyll content in leaves using a video camera, *Ann. Bot.*, 81, 49–54, 1998.

Kondo, N., Ahmed, U., Monta, M., and Murase, H., Machine vision based quality evaluation of Iyokan orange fruit using neural networks, *Comput. Electron. Agric.*, 29, 135–147, 2000.

Leemans, V., Magein, H., and Destain, M.F., Defects segmentation on "Golden Delicious" by using machine color vision, *Comput. Electron. Agric.*, 20, 117–130, 1998.

Mahendra, Prasad, V.S.S., and Dutta Gupta, S., Trichromatic sorting of in vitro regenerated plants of gladiolus using adaptive resonance theory, *Curr. Sci.*, 87, 348–353, 2004.

May, Z., and Amaran, M.H., Automated ripeness assessment of oil palm fruit using RGB and fuzzy logic technique, presented at *Proceedings of the 13th WSEAS International Conference on Mathematical and Computational Methods in Science and Engineering*, 2011.

Mehrez, A., Ali, D., and Mohamed, A., Hybrid method for cereal grain identification using morphological and color features, in *13th IEEE International Conference on Electronics, Circuits and Systems, ICECS '06*, 2006, pp. 870–873.

Mercado-Luna, A., Rico-García, E., Lara-Herrera, A., Soto-Zarazúa, G., Ocampo-Velázquez, R., Guevara-González, R., Herrera-Ruiz, G., and Torres-Pacheco, I., Nitrogen determination on tomato (*Lycopersicon esculentum* Mill.) seedlings by color image analysis (RGB), *Afr. J. Biotechnol.*, 9, 5326–5332, 2010.

Miyanaga, K., Seki, M., and Furusaki, S., Analysis of pigment accumulation heterogeneity in plant cell population by image-processing system, *Biotechnol. Bioeng.*, 67, 493–497, 2000.

Moghaddam, P.A., Derafshi, M.H., and Shayesteh, M., A new method in assessing sugar beet leaf nitrogen status through color image processing and artificial neural network, *J. Food Agric. Environ.*, 8, 485–489, 2010.

Padmanabhan, K., Cantillife, D.J., Harrell, R.C., and Harrison, J., Computer vision analysis of somatic embryos of sweet potato [*Ipomea batatus* (L.) Lam.] for assessing their ability to convert to plants, *Plant Cell Rep.*, 17, 681–684, 1998.

Pagola, M., Ortiz, R., Irigoyen, I., Bustince, H., Barrenechea, E., Aparicio-Tejo, P., Lamsfus, C., and Lasa, B., New method to assess barley nitrogen nutrition status based on image color analysis, comparison with SPAD-502, *Comput. Electron. Agric.*, 65, 213–218, 2009.

Paliwal, J., Shashidhar, N.S., and Jayas, D.S., Grain color identification using kernel signature, *Trans. ASAE*, 42, 1921–1924, 1999.

Phadikar, S., and Sil, J., Rice disease identification using pattern recognition techniques, in *Proceedings of 11th International Conference on Computer and Information Technology (ICCIT)*, Khulna, Bangladesh, 2008, pp. 420–423.

Poblaciones, M.J., Lopez-Bellido, L., and Lopez-Bellido, R.J., Field estimation of technological bread making quality in wheat, *Field Crops Res.*, 112, 253–259, 2009.

Prasad, V.S.S., and Dutta Gupta, S., Photometric clustering of regenerated plants of gladiolus by neural network and its biological validation, *Comput. Electron. Agric.*, 60, 8–17, 2008.

Pydipati, R., Burks, T.F., and Lee, W.S., Identification of citrus disease using color texture features and discriminant analysis, *Comput. Electron. Agric.*, 52, 49–59, 2006.

Rasekhi, R., Asadi, V., and Jafari, A., Weeds and corn classification by image processing and neural network techniques, *Int. J. Nat. Eng. Sci.*, 4, 41–46, 2010.

Seelye, M., Sen Gupta, G., Bailey, D., and Seelye, J., *Low cost color sensors for monitoring plant growth in a laboratory*, IEEE, New York, 2011.

Smith, M.A.L., Reid, J.F., Hansen, A.C., Li, Z., and Madhavi, D.L., Non-destructive machine vision analysis of pigment-producing cell cultures, *J. Biotechnol.*, 40, 1–11, 1995.

Suzuki, T., Murase, H., and Honamin, N., Non-destructive growth measurement cabbage pug seedlings population by image information, *J. Agric. Mech. Assoc.*, 61, 45–51, 1999.

Tillet, N.P., Hague, T., and Miles, S.J., A field assessment of a potential method for weed and crop mapping geometry, *Comput. Electron. Agric.*, 32, 229–246, 2001.

Uddlinng, J., Gelang-Alfredsson, J., Piiki, K., and Pleijel, H., Evaluating the relationship between leaf chlorophyll concentration and SPAD-502 chlorophyll meter reading, *Photosynth. Res.*, 91, 37–46, 2007.

Vollmanna, J., Waltera, H., Satoa, T., and Schweigerb, P., Digital image analysis and chlorophyll metering for phenotyping the effects of nodulation in soybean, *Comput. Electron. Agric.*, 75, 190–195, 2011.

Woebbecke, D.M., Meyer, G.E., Mortensen, D.A., and Von Bargen, K., Shape features for identifying young weeds using image analysis, *Trans. ASAE*, 38, 271–281, 1995.

Wu, J., Wang, D., Rosen, C.J., and Bauer, M.E., Comparison of petiole nitrate concentrations, SPAD chlorophyll readings, and Quick-Bird satellite imagery in detecting nitrogen status of potato canopies, *Field Crops Res.*, 101, 96–103, 2007.

Yadav, S.P., Ibaraki, Y., and Dutta Gupta, S., Estimation of the chlorophyll content of micropropagated potato plants using RGB based image analysis, *Plant Cell Tiss. Org. Cult.*, 100, 183–188, 2010.

Yang, C.C., Prashaer. S.O., Landry, J.A., Perret, J., and Ramaswamy, H.S., Recognition of weeds with image processing and their use with fuzzy logic for precision farming, *Can. Agric. Eng.*, 42, 195–200, 2000.

Yang, C.C., Prasher, S.O., Landry, J.A., and Ramaswamy, H.S., Development of a herbicide application map using artificial neural networks and fuzzy logic, *Agric. Syst.*, 76, 561–574, 2003.

Yuzhu, H., Xiaomei, W., and Shuyao, S., Nitrogen determination in pepper (*Capsicum frutescens* L.) plants by color image analysis (RGB), *Afr. J. Biotechnol.*, 10, 17737–17741, 2011.

Zakaluk, R., and Sri Ranjan, R., Artificial neural network modeling of leaf water potential for potatoes using RGB digital images: a green house study, *Potato Res.*, 49, 255–272, 2007.

Zakaluk, R., and Sri Ranjan, R., Predicting the leaf water potential of potato plants using RGB reflectance, *Can. Biosyst. Eng.*, 50, 7.1–7.12, 2008.

chapter four

RGB imaging for the determination of the nitrogen content in plants

Gloria Flor Mata-Donjuan, Adán Mercado-Luna, and Enrique Rico-García

Contents

4.1 Introduction

Precision agriculture is a recent concept aimed at increasing land productivity through technology. It involves factors such as soil characterization, crop management, and others. The nutrimental deficiencies of the plants are used as a scouting diagnosis to determine the nutrient stage of the crops. Both field and laboratory diagnoses can occur, and they each have their drawbacks. A field diagnosis is based on observations made after the symptoms are visible, when the plant has already suffered a nutrimental disorder. A laboratory diagnosis gives reliable results for most nutrients, but it is time-consuming and requires an available laboratory, which can also be problematic. Thus, when the laboratory results are available, the nutrimental stage of the plant has often already changed. No method exists that determines the nutrimental stage of a plant with precision and speed at a low cost. Some researchers have developed plant nutritional meters based on the interaction between light and the plant's foliar structure. It is known that when plants suffer stress due to a lack of water or other nutrient, the foliar structure inside the leaf changes, altering its reflectance. These reflectance changes can be measured and have been correlated with the plant's chlorophyll and nitrogen (N) levels. The goal of this chapter is to describe the basis of plant reflectance and its correlation with the nitrogen status of plants. A red, green, and blue (RGB) color model is the most frequently used in these types of studies. In this color model, the R and B bands are the best indicators of a plant's nitrogen status. Greenness indices that are based on the RGB model are another way to accurately estimate the N content. However, some difficulties are associated with using light reflectance to determine the N content of plants. The biochemical components, pigments, proteins, carbohydrates, oils, water, leaf morphology, and leaf architecture of the plant influence its tendency to absorb, transmit, or reflect radiation. Additionally, not all plants have the same reflectance, and the same plant can even change its reflectance as a response to its development stage, leaf orientation, color soil effect, atmospheric conditions, and other factors. Thus, an improved hue, luminance, and saturation (IHLS) color model has been proposed to overcome some of the aforementioned issues. This method is based on the RGB model and is less susceptible to light changes. It uses RGB values that have not been normalized, so no information is lost. Combined with the IHLS model, digital images can be used to determine the plant's nitrogen status, yielding a promising technique that is inexpensive, reliable, noninvasive, and quick.

4.2 *The importance of nitrogen in plants*

Among the 16 essential elements required for a plant's growth and development, N is considered the most important because it is found in such large quantities. Nitrogen comprises 1 to 3% of a plant's dried weight depending on the species and the phenological stage. N serves both structural and osmotic functions in a plant. Its structural functions are specific and relate to the synthesis of essential molecules required for growth: nucleic acids, amino acids, proteins, chlorophyll, and alkaloids.

Its osmotic function is associated with the osmotic regulatory process and the effects of ion nitrate and other reduced forms of N, which reduce the hydric potential of vacuoles (Mcintyre, 1997).

Nitrogen participates directly in the synthesis of amino acids, proteins, nucleic acids, and other cellular components that are needed for plant growth and development (Sánchez et al., 2009). It is also found in the chlorophyll molecule and participates in plant metabolism (Alcantar et al., 2007). N reaches plant roots through the process of mass flow, in which it is transported via transpiratory flow through the soil solution by a hydric gradient.

The primary N source for plants is ammonia and nitrate that is generated by soil organic matter. For N to be assimilated by plants, it must be absorbed as NO_3^-, which is then reduced to NO_2^- and converted to NH_4^+. Then, that NH_4^+ is incorporated into organic compounds (Sánchez et al., 2009).

These ions reach the open spaces in the outer cellular walls of root cells and then penetrate their membranes to reach the vegetal cells. Nitrate absorption takes place against an electrochemical gradient; both the roots and the nitrate ions carry negative charges, and the nitrate concentration is larger in root cells than in the soil around them. This transfer against a gradient implies that metabolic energy (ATP) is being spent. Ammonia absorption, on the other hand, takes place through a passive mechanism that does not require energy (Novoa and Loomis, 1981). Once absorbed, nitrate is reduced by the energy gained from photosynthesis. Ammonia does not need to be reduced; instead, it is rapidly converted into an amine or amide because in its natural state, it is toxic to plants. In low light, the absorption and reduction of N and the fixation and reduction of carbon can fight each other for the available energy (Novoa and Loomis, 1981).

4.3 Nitrogen deficiency symptoms

Nitrogen deficiency affects plant metabolism because of the diverse functions of this nutrient. The first symptom to develop is the end of a cell's biochemical reactions. Detection only occurs when a great number of these reactions are not taking place, long after the first alteration has happened. Then, normal reactions diminish, and alterations, transformations, and function losses are present in cells, tissue, and organs (Sánchez et al., 2007).

A deficiency of any nutrient can result in marked changes in the assimilation and metabolic activities of other nutrients. N deficiency decreases the thylakoid quantity, carboxylation activity, photochemical efficiency, chlorophyll content, and stomatal conductance of leaves, and these effects are ultimately reflected in the plant's biomass production.

Typical symptoms of nutrimental deficiencies include one of three characteristics: abnormal appearance within a single plant, as indicated by asymmetric organs forming on opposite sides of the plant, differing nutrient mobility in young and old leaves, and appearance differences occurring between separate plants (Gárate and Bonilla, 2003; Sánchez et al., 2007).

A plant with a nitrogen deficiency has a reduced growth rate, as evidenced by smaller leaves that are evenly yellow in color instead of light green; this yellowing will also be observed on old leaves as their chlorophyll levels diminish (CPHA, 2008). Leaf necrosis can occur under strong deficiencies.

Foliar symptoms of nitrogen deficiency have been classified into five types: (1) interveinal chlorosis, (2) necrosis on leaf edges, (3) inhibition of new growth, which could ultimately cause the death of leaves as well as the main and side shoots, (4) anthocyanin accumulation, which results in an intense red coloration, and (5) no visible symptom or bleak green coloring, which primarily occurs if the diagnosis is made during symptom development. For example, nitrogen deficiency can cause chlorosis on the underside of the plant leaves, and when the deficiency is at an advanced stage, it can provoke death, causing the leaves to fall (Barker and Pílbeam, 2007).

4.4 Visible spectrum region

The visible spectrum is the small portion of the electromagnetic spectrum that oscillates between 0.4 μm (a shorter wavelength with a larger frequency) and 0.7 μm (a longer wavelength with a smaller frequency). What we perceive as an object's color is a result of the surface of that object reflecting a portion of the white light spectrum and absorbing the rest. The human eye can only perceive electromagnetic radiation wavelengths that fall in the visible spectrum. Reflected light from objects excites eye cells called cones and rods in the retina, which are sensible to red, blue, and green light (Gonzalez and Woods, 2002).

4.5 The effect of electromagnetic energy incidence on objects

Three phenomena are present when electromagnetic radiation hits an object's surface. Different portions of the energy are either reflected, absorbed, or transmitted by the objects, as shown in Equation 4.1:

$$EI(\lambda) = ER(\lambda) + EA(\lambda) + ET(\lambda) \tag{4.1}$$

where EI denotes applied energy and ER, EA, and ET denote reflected energy, absorbed energy, and transmitted energy, respectively.

Every object in nature has unique distributions for its reflected, transmitted, and absorbed radiation. This spectrum characteristic can be used to distinguish one object from others or to obtain data about the object's form, size, and other physical and chemical properties. The particular electromagnetic properties of each material produce a specific distribution of its reflectance in different wavelengths, generating a unique pattern known as a spectrum sign. Spectrum signs differ according to the wavelength of electromagnetic energy (Soria et al., 1998).

In the visible electromagnetic spectrum, variations in reflectance result in the visual effect called color (Hecht, 2000). Color is not an intrinsic property of objects. An object's perceived color depends on the light incidence properties of the object and on the visual system perceiving it. Our visual system determines the final chromatic sensation perceived by our brain, which is caused by the light waves transmitted or reflected by the object that has been introduced to the eye (Ortiz-Zamora, 2002). For example, we call an object blue when it reflects primarily the blue portion of the electromagnetic spectrum, and we say and object is green when it reflects the green portion of the spectrum.

4.6 The relationship between physiology and spectral reflectance in plants

The fundamental principle of spectrometry is that radiation undergoes observable and measurable changes when interacting with substances, and these changes depend on the chemical composition of the object's physical and structural attributes (Nicolai et al., 2007; Armenta et al., 2007).

Many authors affirm that light applied to objects causes molecular vibrations, which absorb one type, or wavelength, of energy and reflect others. The portion of the spectrum reflected allows researchers to estimate the quantities of certain molecules in the object or determine the functional groups (strong bonds between light atoms) present, such as C-H, O-H, N-H, and S-H (Roggo et al., 2007; He et al., 2007; Xie et al., 2007; Christy, 2008).

The largest effect, caused by a physical or structural component of the object, is energy dispersion, which is affected by particle size, optic effects, or crystallized shapes in the sample (Roggo et al., 2007). The primary elements causing radiation dispersion in fruits and vegetables are cell walls and suspended particles, such as starch; these structures induce abrupt changes in refraction indices. Chloroplasts and mitochondria also induce dispersion by presenting refraction indices on their surfaces that are different from those of the particles surrounding them (Nicolai et al., 2007).

Chloroplasts contain 70% of the total N found in leaves (Madakadze and Madakadze, 1999). Chlorophyll is the primary element responsible for a plant's reflectance and transmittance of visible radiation (Ritchie, 2003). Thus, light reflectance can be correlated with the N concentration of a plant's foliage (Costa et al., 2001). The incidence of visible spectrum reflectance from leaves fluctuates between a maximum reflectance near 550 nm and a minimum near 660 nm; changes in the chlorophyll concentration affect the visible spectrum in this range and can be used to identify chlorosis associated with N deficiencies (Carter and Spiering, 2002).

The interaction between plant foliar molecules and solar radiation causes reflectance in both the visible spectrum (VIS) and the infrared range (IR). Biochemical components such as pigments, proteins, carbohydrates, oils, and water, along with leaf morphology (mesophyll) and architecture, influence the plant's tendency to absorb, transmit, and reflect solar radiation of short wavelengths from 280 to 2800 nm (Curran et al., 2001). The short wavelength radiation that is absorbed by plants is controlled by the molecular interactions inside the vegetative tissue, where the electrons absorb different wavelengths according to molecules' structures and chemical bonds (Jones, 1997). Chlorophyll, the most important photosynthetic pigment in plants, controls visible light absorbance at wavelengths of 430, 460, 640, and 660 nm by transitioning its electrons. Therefore, chlorophyll effectively absorbs the blue (400–500 nm) and red (600–700 nm) spectrums, reflecting green (Curran, 1989).

4.7 *Spectrum reflectance under nitrogen deficiency conditions*

Many studies have focused on the effects of nitrogen on the light spectrum reflected by leaves; N is considered the most important element in a plant because it is found in the largest quantities, from 1 to 3% of a leaf's dry weight. Spectrum reflectance by a plant's foliage is a precise indicator of the nutrimental state of the plant (De la Cruz et al., 2011). The reflection of red and green radiation by plants is strongly influenced by chlorophyll's light absorption (Costa et al., 2001).

To investigate the relationship of photosynthesis to foliar nitrogen, the influence of nitrogen on photosynthesis was divided into soluble proteins and complex proteins, i.e., pigments contained in chloroplast thylakoids. This division of soluble and complex proteins represents the dark and light phases of photosynthesis, respectively. In thylakoids, only 4 moles of nitrogen are spent per mole of chlorophyll, while in the complex pigment protein, 40 moles of N are used per mole of chlorophyll (Santos et al., 2008). Thus, leaf color intensity is directly related to the amount of chlorophyll and nitrogen in leaves (Mercado-Luna et al., 2010). Stress caused by low nitrogen can be measured in the reflectance between 700 and 740 nm (Osborne et al., 2002; Noh et al., 2006).

Due to its structural characteristics, foliar mesophyll tissue is a good reflector of radiance energy, including wavelengths corresponding to the near infrared spectrum. This energy passes through the palisade parenchyma, which has chloroplasts capable of absorbing blue and red light and reflecting green. If any type of stress is present, the normal relationships between the tissues are disrupted and the plant begins to lose vigor; the mesophyll collapses, resulting in a reduction of near-infrared reflectance by the leaves (Sá et al., 1999).

One of the limitations in estimating foliar nitrogen using spectrometry is the presence of water in fresh foliage; water masks the absorption characteristics of biochemical compounds, primarily in the mid-infrared range. Other effects that complicate analysis include leaf orientation, the soil bottom effect, and atmospheric interference (Mutanga and Skidmore, 2007).

4.8 RGB color space theory

A color space (also called a color model or color system) is a mathematical system in which each point of the space represents a color. The representation of color is numeric, and two, three, or more variables can be used with the system (Gonzalez and Woods, 2002). There are several color space models in use today, including the RGB (red, green, and blue) color space model.

The RGB color space is represented by the first plane of a tridimensional Cartesian system; the value of each red, blue, and green color is represented by a range of normalized values between 0 and 1. The X-axis represents the color green, the Y-axis represents blue, and the Z-axis represents red (Figure 4.1).

RGB color space theory is based on the way the human eye perceives colors, which is called trichromatic. Two types of light receptors exist in our retinas: rods and cones. Each one of these receptor types allows humans to sense the longitude of a wave in the electromagnetic spectrum

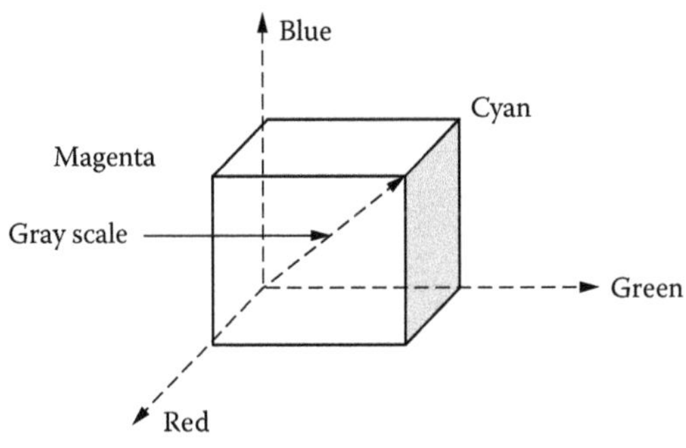

Figure 4.1 RGB color model.

(Pratt, 2007). One group senses the wave longitude corresponding to green, a second group senses the wave longitude corresponding to blue, and a third group senses the wave longitude corresponding to red. This forms the RGB color model, made of three bands representing red, green, and blue.

4.9 Applying the RGB model to nitrogen estimation in plants

Several recent proposals for nitrogen estimation use color models and image processing. This section presents some of the techniques for N and chlorophyll estimation that use the RGB color space. Notably, R and B are the best indicators of N and chlorophyll contents.

4.9.1 Yuzhu index

Yuzhu et al. (2011) evaluated the indices of correlation between the RGB components and the N content of pepper plants. They found that the index represented by Equation 4.2 estimated the level of N in pepper plants with a correlation factor of $R^2 = 0.76$.

$$\frac{G}{R+G+B} \tag{4.2}$$

4.9.2 Pagola index

Another method proposed for N estimation is to determine the greenness indices of images taken from conventional cameras. Pagola et al. (2008) proposed Equation 4.3 as a greenness index for barley that is based on RGB components.

$$Ipca = 0.7582|R-B| - 0.1168|R-G| + 0.6414|G-B| \tag{4.3}$$

This index was applied to two varieties of barley (Scarlet and Pewter), and their values were compared with the results of SPAD measuring. The researchers found that their results were as reliable as, and in some cases better than, the results of SPAD. This index was also applied to cultured barley in different stages of development. At the beginning and end of the culture, the index's correlation with the N content was low-N demand is low in early growth stages, and N is absorbed by the grains in the latter stages. Thus, the best correlations were obtained in crops between 53 and 67 days after planting, when an R^2 range of 0.89 to 0.92 was obtained.

4.9.3 Estimating the chlorophyll content in leaves using the Kawashima and Nakatani indices

The nitrogen and chlorophyll contents of plants are directly related. A proposed method to estimate N is to determine a plant's chlorophyll levels. One tested method involved processing video camera images of rye and wheat crops. An index based on R and B normalized components (Equation 4.4) was used:

$$Index = \frac{R - B}{R + B} \tag{4.4}$$

The values were normalized to remove any noise caused by meteorological conditions such as radiation, relative humidity, or temperature. Another condition that can modify the correlation between the chlorophyll content and the value obtained from the index is the angle of the leaf. In this study, Kawashima and Nakatani (1998) obtained a correlation value of $R^2 = 0.8$.

4.10 Nitrogen estimation based on histograms of RGB model components

The method described in this section uses image processing to determine the RGB components and account for each component in the image. Once the image has been obtained, it is segmented, and the values of each pixel of the foliage are extracted. These values are transferred into the RGB color model, so each pixel has three numeric values representing R, G, and B. The values obtained for each component are between 0 and 255. A histogram of each color is then created (Equation 4.5):

$$\begin{aligned} Tr &= \sum_{i=0}^{i<255} (fr)(i) \\ Tg &= \sum_{i=0}^{i<255} (fg)(i) \\ Tb &= \sum_{i=0}^{i<255} (fb)(i) \end{aligned} \tag{4.5}$$

where fr = frequency of each red value, fg = frequency of each green value, fb = frequency of each blue value, i = ith component value, and Tb, Tr, Tg = total pixels in each component.

It is known that the G band does not give any reliable information about the N or chlorophyll contents in leaves. However, the R and B bands do correlate with the N and chlorophyll levels. Measurements made of tomato seedlings using this method had a negative correlation for the R band equal to $R^2 = 0.91$ (Mercado-Luna et al., 2010), indicating that as the N levels decreased, the amount of red increased. If the nitrogen deficiency is great, the chlorophyll content and the G band are reduced. This increase in red and decrease in green makes the leaves appear yellow in color, representing the physiological absence of chlorophyll. The B band also showed a negative correlation for N estimation, $R^2 = 0.89$.

It is important to note that the RGB color space is sensitive to light variations. For this reason, it is necessary to control the light intensity to maintain the same conditions and prevent a distortion effect. The values of the RGB components have a positive correlation with light intensity.

4.11 *Applying the IHLS color space to nitrogen estimation in plants*

To improve upon the results obtained using the RGB color space, the IHLS (improved hue, luminance, and saturation) color space was proposed for N estimation. This method uses the true RGB values without normalization, thus preventing information loss. The unnormalized data are less sensible to light variations (Mata-Donjuan et al., 2012). This color space is represented in polar coordinates. IHLS is a quantitative model, whereas the RGB model is qualitative. Therefore, the IHLS model is able to undergo image processing (Hanbury and Serra, 2003; Angulo and Serra, 2005). The model is represented by Equations 4.6 to 4.8:

$$l = 0.0213r + 0.715g + 0.072b \tag{4.6}$$

$$s = \max(r, g, b) - \min(r, g, b) \tag{4.7}$$

$$h = \begin{cases} \dfrac{g-b}{\max(r,g,b)-\min(r,g,b)}; & \text{if } r = \max(r,g,b) \\[2ex] \dfrac{b-r}{\max(r,g,b)-\min(r,g,b)}+2; & \text{if } g = \max(r,g,b) \\[2ex] \dfrac{r-g}{\max(r,g,b)-\min(r,g,b)}+4; & \text{if } b = \max(r,g,b) \end{cases} \tag{4.8}$$

where l = luminance, s = saturation, h = hue, g = green component in the RGB space, b = blue component in the RGB space, and r = red component in the RGB space.

The ranges of values taken by each component in this color space are 0–255 for luminance and saturation and 0–360° for hue.

This model has been used to quantify the color of tomato and lettuce plants to estimate N levels. The hue and saturation values have also been evaluated under different light conditions. These two components are the values that carry the chromatic information. Mata-Donjuan et al. (2012) determined that the hue component specifically correlates with the N content in leaves. This component takes angular values between 0 and 360°, where 0° corresponds to red, 120° to green, and 240° to blue. It is noteworthy that the saturation component must have values above 20 to ensure correct measuring.

Mata-Donjuan (2010) evaluated the N levels of tomato seedlings and found this model to correlate positively and linearly, with $R^2 = 0.86$. The hue values obtained were between 50 and 90°. To study the effect of light, an experiment on green lettuce was carried out using conditions of both natural light under a greenhouse and artificial light. The correlations obtained between hue (H) and N levels in the plants were $R^2 = 0.879$ and $R^2 = 0.87$, respectively; the small variation indicates that this method can adequately adjust to differing light levels. The same images, taken in natural light under a greenhouse or in artificial light, were analyzed using the RGB model. In those two different light conditions, correlation values of $R^2 = 0.8$ and $R^2 = 0.71$ were found for the R bands, and values of $R^2 = 0.87$ and $R^2 = 0.86$ were found for the B bands, respectively. These data show that the RGB model is more susceptible to light variations than the IHLS model.

4.12 Measuring nitrogen content using digital images

Based on the studies described above, researchers have developed novel methodologies to estimate the N and chlorophyll levels in plants (Mercado-Luna et al., 2010; Pagola et al., 2008; Yadav et al., 2010; Dutta Gupta et al., 2013). Some studies claim that determining N and chlorophyll levels by digital image processing is better than using the SPAD chlorophyll meter. Several factors must be taken into consideration when measuring N or chlorophyll levels with image processing.

The chemical composition and texture of the leaves can modify the light reflected by the leaves. Every crop has different chemical compositions and physiology; even the same plant will change according to its development stage, altering the light that is being transmitted, reflected, or absorbed.

The use of the RGB model requires a controlled light condition to avoid measurement distortion. Standardization is also necessary for the proper use of digital image processing because different devices, methods of measurement, and light conditions have been used in various studies.

Because this methodology is easy to implement using only a digital camera and a computer, standardization will allow future researchers to generate a useful database and reduce uncertainty when correlating different sets of measurements.

This new technology makes the simple and inexpensive measurement of N levels in plants a reality. Ultimately, digital image processing can be used to estimate the status of not only nitrogen, but any nutrient in a plant (Wiwart et al., 2009).

References

Alcantar, G. G., Trejo-Téllez, L. I., Fernández, P. L., and Rodríguez-Mendoza, M., Elementos esenciales, in *Nutrición de cultivos*, ed. G. G. Alcantar and L. I. Trejo-Téllez, 7–48, Editorial Mundi-Prensa, México, 2007.

Angulo, J., and Serra, J., Image color segmentation using bivariate histograms in luminance/saturation/hue polar color spaces, *Comp. Sist.*, 8, 1–19, 2005.

Armenta, S., Garrigues, S., and De la Guardia, M., Partial least squares-near infrared determination of pesticides in commercial formulations, *Vib. Spectrosc.*, 44, 273–278, 2007.

Barker, A., and Pílbeam, D. J., *Handbook of plant nutrition*, Taylor and Francis, Boca Raton, FL, 2007.

California Plant Health Association (CPHA), *Manual de fertilizantes para cultivos de alto rendimiento*, Editorial Limusa, México, 2008.

Carter, G. A., and Spiering, B. A., Optical properties of intact leaves for estimating chlorophyll concentration, *J. Environ. Qual.*, 31, 1424–1432, 2002.

Christy, C. D., Real-time measurement of soil attributes using on-the-go near infrared reflectance spectroscopy, *Comput. Electron. Agric.*, 61, 10–19, 2008.

Costa, C., Dwyer, L. M., Dutilleul, P., Stewart, D. W., Ma, B. L., and Smith, J. D., Interrelationships of applied nitrogen, SPAD and yield of leafy and non-leafy maize genotypes, *J. Plant Nutr.*, 24, 1173–1194, 2001.

Curran, P. J., Remote sensing of foliar chemistry, *Remote Sens. Environ.*, 30, 271–278, 1989.

Curran, P. J., Dungan, J. L., and Peterson D. L., Estimating the foliar biochemical concentration of leaves reflectance spectrometry testing the Kokaly and Clark methodologies, *Remote Sens. Environ.*, 76, 349–359, 2001.

De la Cruz-Durán, J., Sánchez-García, P., Galvis-Spínola, A., Carrillo-Salazar, J., and Carrillo-Salazar, J., Índices espectrales en pimiento para el diagnóstico nutrimental de nitrógeno, *Terra Latinoam.*, 29, 259–265, 2011.

Dutta Gupta, S., Ibaraki, Y., and Pattanayak, A. K., Development of a digital image method for real-time estimation of chlorophyll content in micropropagated potato plants, *Plant Biotechnol. Rep.*, 7, 91–97, 2013.

Gárate, A. E., and Bonilla, I., Nutrición mineral y producción vegetal, in *Fundamentos de fisiología vegetal*, ed. J. Azcón-Bieto and M. Talón, 113–130, McGraw Hill, Spain, 2003.

Gonzalez, R. C., and Woods, R. E., *Digital image processing*, 2nd ed., Prentice Hall, 2002.

Hanbury, A., and Serra J., Colour image analysis in 3D-polar coordinates, presented at DAGM, Magdeburg, Germany, 2003.

He, Y., Huang, M., Garcia, A., Hernández, A., and Song, H., Prediction of soil macronutrients content using near infrared spectroscopy, *Comput. Electron. Agric.*, 58, 144–153, 2007.

Hecht, E., *Óptica*, Addison Wesley, Reading, MA, 2000.

Jones, M. J., *Organic chemistry*, W. W. Norton and Company, New York, 1997.

Kawashima, S., and Nakatani, M., An algorithm for estimating chlorophyll content in leaves using a video camera, *Ann. Bot.*, 81, 49–54, 1998.

Madakadze, I. C., and Madakadze R. M., Field evaluation of the chlorophyll meter to predict yield and nitrogen concentration of switchgrass, *J. Plant Nutr.*, 22, 1001–1010, 1999.

Mata-Donjuan, G. F., Estimación del contenido de nitrógeno en cultivos bajo invernadero en diferentes condiciones de iluminación usando el espacio de color LSM, Master diss., Universidad Autónoma de Querétaro, 2010.

Mata-Donjuan, G. F., Mercado-Luna, A., Rico-García, E., and Herrera-Ruiz, G., Use of improved hue, luminance and saturation (IHLS) color space in the estimation of nitrogen on tomato seedlings (*Lycopersicon esculentum*), *Sci. Res. Essays*, 7, 2343–2349, 2012.

Mcintyre, G. I., The role of nitrate in the osmotic and nutritional control of plant development, *Aust. J. Plant Physiol.*, 68, 107–112, 1997.

Mercado-Luna, A., Rico-García, E., Lara-Herrera, A., et al. Nitrogen determination on tomato (*Lycopersicon esculentum* Mill.) seedlings by color image analysis (RGB), *Afr. J. Biotechnol.*, 9(33), 5326–5332, 2010.

Mutanga, O., and Skidmore, A. K., Red edge shift and biochemical content in grass canopies, *ISPRS J. Photogrammetry Remote Sensing*, 62, 34–42, 2007.

Nicolai, B. M., Beullens, K., Bobelyn, E., Peirs, A., Saeys, W., Theron, K. I., and Lammertyna, J., Nondestructive measurement of fruit and vegetable quality by means of NIR spectroscopy: a review, *Postharvest Biol. Technol.*, 46, 99–118, 2007.

Noh, H., Zhang, Q., Shin, B., Han, S., and Feng, L. A., Neural network model of maize crop nitrogen stress assessment for a multi spectral imaging sensor, *Biosyst. Eng.*, 4, 477–485, 2006.

Novoa, R., and Loomis, R. S., Nitrogen and plant production, *Plant Soil*, 58, 177–204, 1981.

Ortiz-Zamora, F. G., Procesamiento morfológico de imágenes en color. Aplicación a la reconstrucción geodésica, PhD diss., Universidad de Alicante, 2002.

Osborne, S. L., Schepers, J. S., Francis, D. D., and Schlemmer, M. R., Detection of phosphorus and nitrogen deficiencies in corn using spectral radiance measurements, *Agron. J.*, 6, 1215–1221, 2002.

Pagola, M., Ortiz, R., Irigoyen, I., et al., New method to assess barley nitrogen nutrition status based on image color analysis comparison with SPAD-502, *Comput. Electron. Agric.*, 65, 213–218, 2008.

Pratt, W., *Digital image processing*, 4th ed., Wiley-Interscience, NY, 2007.

Ritchie, L. G., *Use of ground-based canopy reflectance to determine ground cover, nitrogen and water status and final yield in wheat*, Utah State University Press, 2003.

Roggo, Y., Chalus, P., Maurer, L., Lema-Martinez, C., Edmond, A., and Jent, N., A review of near infrared spectroscopy and chemometrics in pharmaceutical technologies, *J. Pharm. Biomed. Anal.*, 44, 683–700, 2007.

Sá, I. B., Antonio, R., and Almorox, J. A., Aplicación de sensores remotos en la detección y evaluación de plagas y enfermedades en la vegetación, presented at VIII Congreso Nacional de Teledetección, Albacete, España, 1999.

Sánchez, E., Ávila-Quezada, G., Gardea, A. A., Muñoz, E., Ruiz, J. M., and Romero, L., Nitrogen metabolism in roots and leaves of green bean plants exposed to different phosphorus doses, 2009, http://www.scielo.org.ar/scielo.php?script=sci_arttext&pid=S185156572009000100003&lng=es (accessed December 18, 2012).

Sánchez, G. P., Molinos, S. C., Alcantar G. G., and Sandoval M. V., Diagnóstico nutrimental en plantas, in *Nutricion de cultivos*, ed. G. G. Alcántar and L. I. Trejo-Téllez, 201–248, Mundi-Prensa, México, 2007.

Santos, M. A., Rosados, M. J. R., Quinteiro, M. F. I., Ortiz, V. R., Pose, S. L., Rodríguez, D. C., and Lliteras, M. T. F., *Clorofila en una población de laurusnobilis del parque nacional de las islas atlánticas*, Galicia, *Cuad. Soc. Esp. Ciem. For.*, 25, 61–66, 2008.

Soria, R. J., Ortiz, S. C. A., Islas G. F., and Volke, H. V., *Sensores remotos. Principios y aplicaciones en la evaluación de los recursos naturales, Experiencias en México*, Sociedad Mexicana de la Ciencia del Suelo, Chapingo, México, 1998.

Wiwart, M., Fordonski, G., Zuk-Golaszewska, K., and Suchowisska, E., Early diagnostics of macronutrient deficiencies in three legume species by color image analysis, *Comput. Electron. Agric.*, 65, 125–132, 2009.

Yadav, S. P., Ibaraki, Y., and Dutta Gupta, S., Estimation of the chlorophyll content of micropropagated potato plants using RGB based image analysis, *Plant Cell Tiss. Org. Cult.*, 100, 183–188, 2010.

Xie, L., Ying, Y., and Ying, T., Combination and comparison of chemometrics methods for identification of transgenic tomatoes using visible and near infrared diffuse transmittance technique, *J. Food Eng.*, 82, 395–401, 2007.

Yuzhu, H., Xiaomei, W., and Shuyao, S., Nitrogen determination in pepper (*Capsicum frutescens* L.) plants by color image analysis (RGB), *Afr. J. Biotechnol.*, 77, 17737–17741, 2011.

chapter five

Sterile dynamic measurement of the in vitro nitrogen use efficiency of plantlets

Yanyou Wu and Kaiyan Zhang

Contents

5.1 Introduction

Near-infrared spectroscopy is a rapid and nondestructive method that enables sterile sample analysis. It does not require extensive sample pretreatment using chemical reagents. This quick and efficient modern analysis technique, developed in the late 1980s, is considered a green detection technology. Near-infrared spectroscopy has wide applications in petrochemical (Liu et al., 2002; Gao et al., 2005), food (Uddin et al., 2002; Han et al., 2003; Uddin and Okazaki, 2004), agricultural (Kong et al., 2004; Fan et al., 2006; Y.X. Li et al., 2006a; Lu et al., 2007), tobacco (Jiang et al., 2006), tea (Lin and Niu, 2008), and medical (Chen et al., 2004; Yue et al., 2006; Luypaert et al., 2007) industries. The International Association for Cereal Science and Technology and the American Association of Cereal Chemists has classified near-infrared quantitative analysis as a standard method. This technique is also a standard test method for food analysis worldwide (Nicolai et al., 2007; Shiroma and Rodriguez-Saona, 2009; Bobelyn et al., 2010).

Researchers worldwide gradually began using spectral methods to determine and analyze the nitrogen content in different environments. Pan (2007) used remote sensing to monitor wheat growth and estimated nitrogen accumulation in soil. Zhang and Wang (2008) used fresh leaf spectra and estimated nitrogen nutrition conditions. Tang et al. (2003) used hyperspectral imaging and a chlorophyll meter for the rapid determination of the nitrogen nutrition status of barley. Xue et al. (2003) used the rice canopy reflectance spectrum to monitor the nitrogen nutrition conditions of colony leaves. Song et al. (2004) used data obtained by aviation imaging spectrometry to study the effect of soil nitrogen fertilization and variable rate fertilization on winter wheat growth. Bronson et al. (2003) used spectral information to determine the nitrogen conditions of cotton during irrigation. Nevertheless, the nitrogen content of a semisolid medium (e.g., tissue culture) needs to be determined using near-infrared spectral analysis. The present study aims to use visible and near-infrared spectroscopy for analyzing the nitrogen content of a tissue culture medium to estimate the growth conditions of plantlets cultured in vitro. The results obtained may serve as a reference for industrial plantlet production.

Plant tissue culture is performed in a sterile, partially closed environment. During this culture, a long-term sterile condition is maintained when plantlets are transferred to a conical glass flask containing the culture medium. A nondestructive and sterile method is required for determining the amount of nitrogen consumed by the plantlets and the conditions for their growth. Nondestructive, online monitoring of the nitrogen content of the culture medium can be performed using near-infrared spectroscopy.

5.2 Traditional nitrogen measurement methods

5.2.1 Importance of nitrogen in plant growth and development

Nitrogen is a life element (Huang et al., 2010) because it is an essential nutrient for plant growth. Nitrogen considerably influences plant yield and quality, and excess or lack of nitrogen greatly affects plant growth and development. Appropriate nitrogen availability is necessary for plants to absorb phosphorus, potassium, calcium, and magnesium (Chio et al., 2001; Zheng et al., 2005). Nitrogen in plants comprises organic nitrogen, nitrogen in the form of nitrates, and organic nitrogen contained in plant proteins.

As a basic physiological process of plants, nitrogen metabolism is important for the geochemical cycle. The main mechanism of nitrogen assimilation in plants is the reduction of nitrate to ammonia. This mechanism is directly involved in amino acid synthesis and transformation. During nitrogen assimilation, some key enzymes are involved in catalysis and regulation—nitrate reductase, glutamine synthetase, glutamate synthase, and asparagine transaminase. After proteins are synthesized using amino acids, they are modified, classified, transported, stored, and incorporated into plant tissues. Nitrogen and carbon metabolisms in plants occur in harmony. Both of these processes are fundamental to plant life activities (Wang et al., 2009).

Nitrogen deficiency directly affects the biosynthesis of biomolecules such as amino acids, proteins, and nucleic acids. It decreases plant photosynthesis, thereby reducing crop yield (Peng et al., 2006). Accurate, rapid, and economical methods for determining the plant nitrogen status are important for the adequate scientific use of nitrogen and for the sustainable development of agricultural production.

Considering the physiological differences in plants, different plantlet types grown in vitro present with different optimum nitrogen concentrations. Under different culture conditions, plants have different photosynthetic capacities, whereas under identical culture conditions, plants have different growth rates, nitrogen use efficiencies, and growth cycles. In large-scale cultivation, there is no scientific basis to determine the growth and development of plantlets in vitro by the appearance of plants and the volume of consumed medium. Therefore, delaying transplantation beyond the optimal transplantation time may cause economic losses. In addition, biomass of plantlets grown in vitro will not substantially increase with time, if they slowly metabolize nitrogen. Moreover, decreased nitrogen metabolism efficiency causes a pseudomorphism of nitrogen use, resulting in slow and poor plantlet growth in vitro. This condition delays plantlet cultivation and is thus unfavorable for plantlet transplantation.

Considering the above-mentioned factors, the nitrogen content of the culture medium during cultivation must be determined for investigating nitrogen consumption from the medium at different times during plantlet growth. The observations of nitrogen consumption can be related to the growth cycle of plantlets in vitro. The identification of these relationships will help in studying the physiological properties of plantlets, determining the optimal inoculation period for plantlets, and deciding whether the residual nitrogen in the growth medium can be recycled to reduce culture costs. These findings can be used as the basis for industrial plantlet production.

5.2.2 Traditional nitrogen estimation methods

5.2.2.1 Kjeldahl digestion

The basic principle of Kjeldahl digestion comprises heating a test material with sulfuric acid. Carbon from carbohydrates is converted into carbon dioxide. Nitrogen and hydrogen are converted to ammonia, which then reacts with sulfuric acid to form ammonium sulfate. In the present study, we added a catalyst to accelerate organic matter decomposition and improve the boiling point of the digestive solution. Distillation, performed after digestion, involved the addition of a base to the digestive solution in an azotometer to break down ammonium sulfate into ammonium. Ammonia and water vapor were distilled using a boric acid absorption solution and absorbed by the boric acid. During titration, the boric acid absorption solution was titrated using calibrated hydrochloric acid. Subsequently, the nitrogen content of the digested sample was calculated on the basis of the amount of hydrochloric acid added (Huang et al., 1998). The Kjeldahl method is commonly used because it is simple, can provide stable determination results, and has higher accuracy than other methods available for nitrogen content determination. However, this method involves many complex steps, the results are sensitive to human error, and the analysis efficiency is low. Because the sample processing time is 60 min, this method is not suitable for mass sample analysis. It is also not suitable for treating samples containing high amounts of fixed nitrogen and nitrate.

5.2.2.2 Chemiluminescence

The chemiluminescence method involves the emission of light because of chemical reactions, in which the molecular components (reactants, products, and intermediates) absorb the released energy, shift from the ground state into the excited state, and then return to the ground state. During this process, energy is released in the form of radiation and chemical luminescence is produced. This method, which is based on the relationship between the molecular luminous intensity and the content of analyte, has many advantages, such as high sensitivity and wide linear range. Moreover, it only requires simple instruments and equipment.

The essential difference between chemiluminescence and other luminescence analysis methods is the energy absorbed when a system produces luminescence (radiation). The system producing chemical luminescence must involve radiation and chemical reactions in a single step to provide sufficient energy for light production. Because the reaction does not need an external light source, the advantages of chemiluminescence are that it can reduce or eliminate scattering, prevent the interference of background photosynthetic stray light, reduce the influence of noise, and improve the signal-to-noise ratio (Li et al., 2006).

5.2.2.3 *Double-wavelength method*

The double-wavelength method is used for identifying nitrogen-based compounds. Nitrate and nitrite have a strong ultraviolet absorption at 203 and 230 nm, respectively. The double-wavelength coefficient ratio method is used, and the appropriate measurement wavelength is delected. The double-wavelength coefficient ratio results in a linear relationship after the elimination of the interference of nitrite and organic matter. Huang et al. (2009) discussed the use of the double-wavelength coefficient ratio method for determining the nitrate nitrogen content of soil and compared this method with the widely used reflectometry and flow analyzer measurement methods. The present study found that the results obtained using the double-wavelength coefficient ratio method highly correlate with those obtained using the reflectometry and flow analyzer measurement methods. However, no significant difference was observed between the results obtained by the two methods. The double-wavelength coefficient ratio method can help determine the content and application of nitrate nitrogen in soil, and provides the basis for the supply ability index for soil nitrogen. Further studies using different soil types, soil textures, and test temperatures, among other conditions, are necessary to promote the widespread use of this technology. Tu and Wen (2006) measured the nitrate nitrogen content of soil using double-wavelength spectrophotometry. The sample recovery rates ranged from 90% to 109%. Double-wavelength spectrophotometry has many advantages, including high sensitivity, low interference, wide measurement range, and effective elimination of the interference of nitrite, chloride ions, and organic compounds.

5.2.2.4 *Agro service international (ASI) method*

ASI is an efficient, rapid, and accurate method developed over many years for testing soil nutrients and determining appropriate fertilizers. This method, also called the soil nutrient status systematic approach, is suitable for neutral, acidic, and calcareous soils. Many domestic researchers have used this ASI method to measure the nitrogen content of soil and culture medium. Sun et al. (2006) measured trace elements in soil using the ASI method and showed that it is significantly correlated with the conventional method. The ASI and improved nitrate nitrogen determination methods, with better accuracy, can be used for extraction and determining the available nutrients from soil. These methods also correlate well with conventional methods. Xiong et al. (2007) compared the ammonium nitrogen content of soil obtained using the ASI method with that obtained using a conventional method. Ding et al. (2007) analyzed the effective nitrogen in an agricultural soil sample obtained from the Xinjiang Oasis using the ASI method and a conventional method. Comparison of results showed a good correlation between the two methods. At present, the ASI method is also used for nitrate nitrogen analysis, and it provides a good indication of the organic nitrogen supply conditions in soil.

5.2.3 Nitrogen content determination using the near-infrared spectrum method

Near-infrared spectrum technology, which was developed recently, is used for nondestructive testing. The underlying principle is that the flexible stretching vibration frequency of the hydrogen groups on compound molecular groups, such as O-H, C-H, and N-H, reflects the fundamental vibration frequency. Only vibration frequencies of >2000 cm^{-1} are suitable for detection and absorption in the near-infrared region. Qualitative and quantitative analyses can be performed on the basis of the near-infrared absorption spectrum peak placement and absorption strength. Nitrogen, which contains an N-H group, falls within the infrared vibration frequency. Therefore, quantitative analyses can be performed using the near-infrared absorption spectrum of the N-H group. Using spectral information obtained by aviation imaging spectrometry, Song et al. (2004) investigated the effect of soil nitrogen and variable fertilization on the growth of winter wheat. Bronson et al. (2003) determined the nitrogen content of cotton during irrigation using spectral information because it provides more accurate qualitative and quantitative analysis for nitrogen content.

5.3 Modeling and analysis of the nitrogen content of the medium

5.3.1 Materials and methods

5.3.1.1 Preparation of the medium sample

The medium used in this experiment was Murashige and Skoog (MS) containing 0.1 mg · L^{-1} 1-naphthaleneacetic acid (NAA), 2 mg · L^{-1} 6-benzylaminopurine (6-BA), and 3% sucrose. The nitrogen concentration in our medium ranged from 0.2847 g · L^{-1} to 2.8364 g · L^{-1}, and the nitrate nitrogen concentration ranged from 0.2847 g · L^{-1} to 1.5303 g · L^{-1}. The ammonium nitrogen concentration ranged from 0 g · L^{-1} to 1.3061 g · L^{-1}. These values are within the normal range of nitrogen concentration in plant tissue culture. The standard sample set contained 80 samples at different nitrogen concentrations, each with 50 ml medium. These samples were refrigerated (0°C to 4°C) to maintain sample stability.

5.3.1.2 Collection of spectral data of medium samples

To obtain the spectral data of the medium samples, a FieldSpec 3 portable spectrum radiometer (ASD) was used. The spectral range selected for scanning was 350 nm to 2500 nm in a full-wave mode to span the entire near-infrared spectrum and obtain as much information as possible from the samples. The appropriate processing method was used for the spectral data. Sampling resolution, which refers to the capacity of the spectrometer to distinguish two spectral lines with extremely close wavelengths, directly influences the fine structure of the spectrogram. When this resolution is high, the spectral structure is detailed, but the sampling time and data quantity are increased. When this resolution is low, the sampling time is shorter and the data quantity is reduced, but the spectral structure appears relatively rough. In the present study, the resolution settings were selected depending on the specific experiment. Figure 5.1 shows the principle of near-infrared reflection (NIR).

In this experiment, a sampling resolution of 10 nm was used with a sampling interval of 2 nm. Sampling frequency refers to the number of scans performed by a spectral instrument on a sample to complete spectral collection. When this frequency is low, the random effect in sampling during a scan increases, resulting in a spectral curve with high noise and low spectral precision. When this frequency is too high, a smooth spectral curve is obtained after averaging. However, because of the increased sampling time and elevated interference of the system error produced by the instrument and other random factors, it becomes difficult to guarantee the accuracy of the obtained spectrum. Therefore, considering that spectral precision and spectral curve smoothness increase with fewer samples, the number of samples used in the experiment should be reduced to save sampling time. The number of samples used in this experiment was 30.

The experiment was performed in the Agricultural Engineering Research Laboratories of Jiangsu University. The environmental temperature and relative humidity were maintained at 25°C ± 0.5°C and 60% ± 5%, respectively. In this test, the artificial light source supported by the instrument was used. To eliminate the influence of ambient light during spectral data acquisition, we avoided other light sources. The measurement method was a diffuse reflection type. The optical fiber probe was set at 25° viewing angle and fixed by a lampstand, forming a 45° angle around

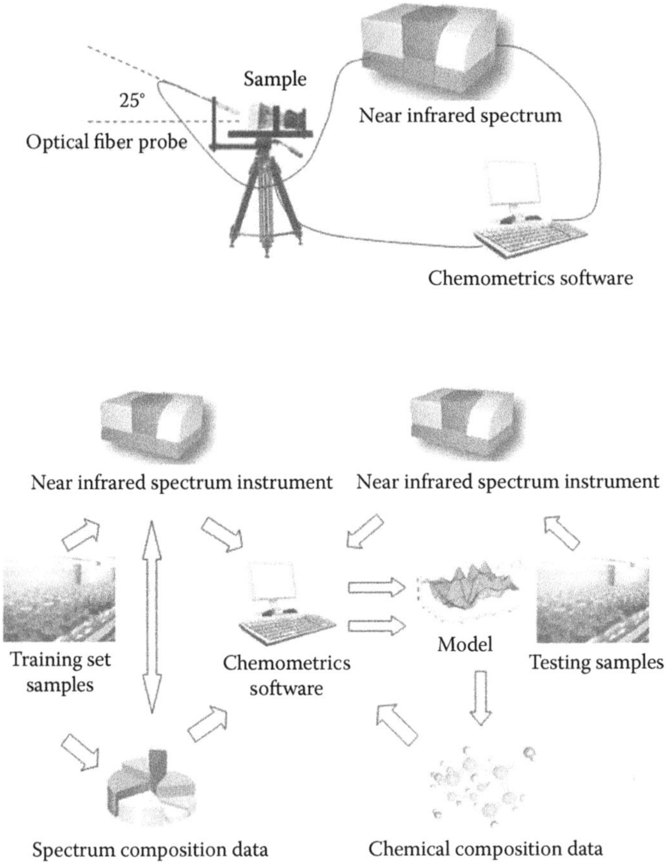

Figure 5.1 The principle of near-infrared reflection analysis.

the light source to reduce the sample mirror reflection. To ensure that the sample was within the viewing field, the probe was placed at a distance of 3 cm to 5 cm from the sample surface being measured.

We performed the system configuration, optimization, and the white board correction before sampling, collected five points in the center and around each sample, and scanned each point 10 times to collect three spectra each time. We first scanned the sample group containing only medium in glass dishes, and then scanned the medium in the glass dishes covered by a uniform glass cap and obtained their spectra. We obtained the spectra of the medium in the glass flasks by scanning through the bottom of the bottle. We collected the spectral data from three groups (each group contained 77 samples): medium samples in glass flasks, medium samples in glass dishes covered by glass caps, and bare medium samples in glass dishes without covers. Figure 5.2 shows the spectrogram of the samples.

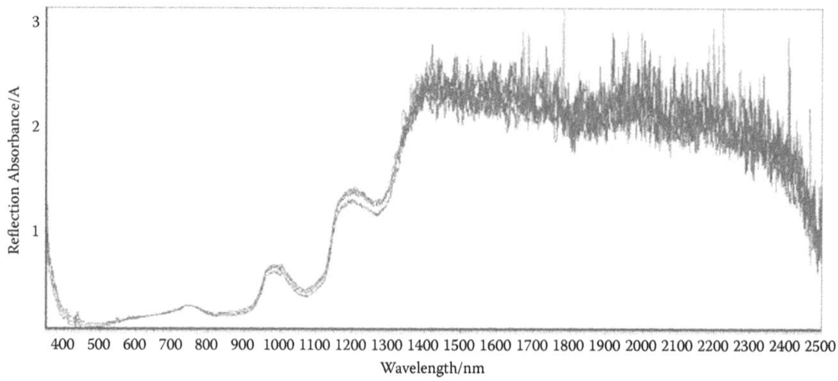

Figure 5.2 The near-infrared spectra of the samples.

5.3.2 *Effects of different pretreatments on the model*

Near-infrared spectrometry is extremely vulnerable to interference by some factors unrelated to the properties of the tested samples, such as the sample state, light scattering, stray light, sample loading conditions, and response of the instrument. The collected sample spectra contain not only sample information, but also irrelevant data, such as electrical noise, sample background, and stray light. These factors shift the near-infrared spectrum baseline and promote spectrum nonrepetition, resulting in more complicated sample spectra. Therefore, establishment of a spectral pretreatment method for the elimination of irrelevant information and noise from the collected spectral data is important. The commonly used spectral pretreatment method involves smoothing, obtaining first- and second-order derivatives, and performing standard normal variable transformation, multiple scattering corrections, and spectral normalization. This section will discuss the processing methods and their effects on test sample spectra to determine the most suitable pretreatment method for this study (Zheng et al., 2004; Zhao and Yan, 2006; Chu et al., 2008).

5.3.2.1 *Smoothing*

Spectral smoothing mainly eliminates the interference due to high-frequency random noise on the sample signal using a low-pass filter for spectral curves. Hence, the high-frequency component is removed, retaining only the useful low-frequency information. Many fitting methods are used for data smoothing. Moving average (MA) smoothing and Savitzky-Golay (SG) smoothing are the most commonly used smoothing methods in signal processing. Specific conditions, such as instrument and analytical requirements, must be considered while selecting the appropriate size of the smoothing window, and then the instrument noise can be removed.

Figure 5.3a and b shows the near-infrared spectral curves of the samples after the MA and SG smoothing, respectively.

Smoothing sacrifices a certain amount of spectral resolution to improve the signal-to-noise ratio of the spectral signal. It is related to the window size or the number of points. Different window sizes produce different smoothing effects, and the use of numerous smoothing points improves the signal-to-noise ratio but increases the spectral resolution, leading to signal distortion. Figure 5.3a shows the spectrogram after five-point MA smoothing, which is much smoother and has significantly reduced noise compared with the spectrogram shown in Figure 5.2. With respect to the complicated background and large noise spectra of the medium samples in the glass flasks, smoothing improved the signal-to-noise ratio of these spectra to a certain extent. On the basis of software design, we selected six smoothing windows with widths of 5. When the window width was 5, the model accuracy was relatively high. We then performed moving average filter (MAF) smoothing and Savitzky-Golay filter (SGF) smoothing pretreatments and used the correction method of partial least squares (PLS) modeling. Table 5.1 shows the results.

Table 5.1 shows the related coefficients were higher for the model using SGF smoothing than for the model using MAF smoothing. The corresponding root mean squared errors of prediction (RMSEP) and the corresponding average relative error percentages (AREPs) for the model were lower after SGF smoothing than after MAF smoothing. Thus, the model using SGF smoothing was more accurate than that using MAF smoothing. However, smoothing pretreatment is rarely used alone and is often combined with derivative treatment to improve the model accuracy.

5.3.2.2 Derivative

Obtaining a spectrum derivative can eliminate the effects of baseline drift and gentle background interference. In addition, a higher-resolution and clearer spectrum than the original spectrum can be achieved.

Figure 5.3c and d shows the spectrograms after first- and second-order differential treatments based on five-point SG smoothing. The first-order differentiation can effectively differentiate overlapping spectra in the near-infrared spectrum, resulting in a clearer spectral profile. After obtaining the second-order spectral derivative, the band features are also strengthened.

Table 5.2 shows the PLS model with window widths of 5 combined with MAF or SGF smoothing to perform first- and second-order differential treatments on the test sample data. The model accuracy increased to a certain extent after smoothing and first-order differential treatments. After smoothing and second-order differential treatments, the accuracy of the model improved. This occurs primarily because differential treatment

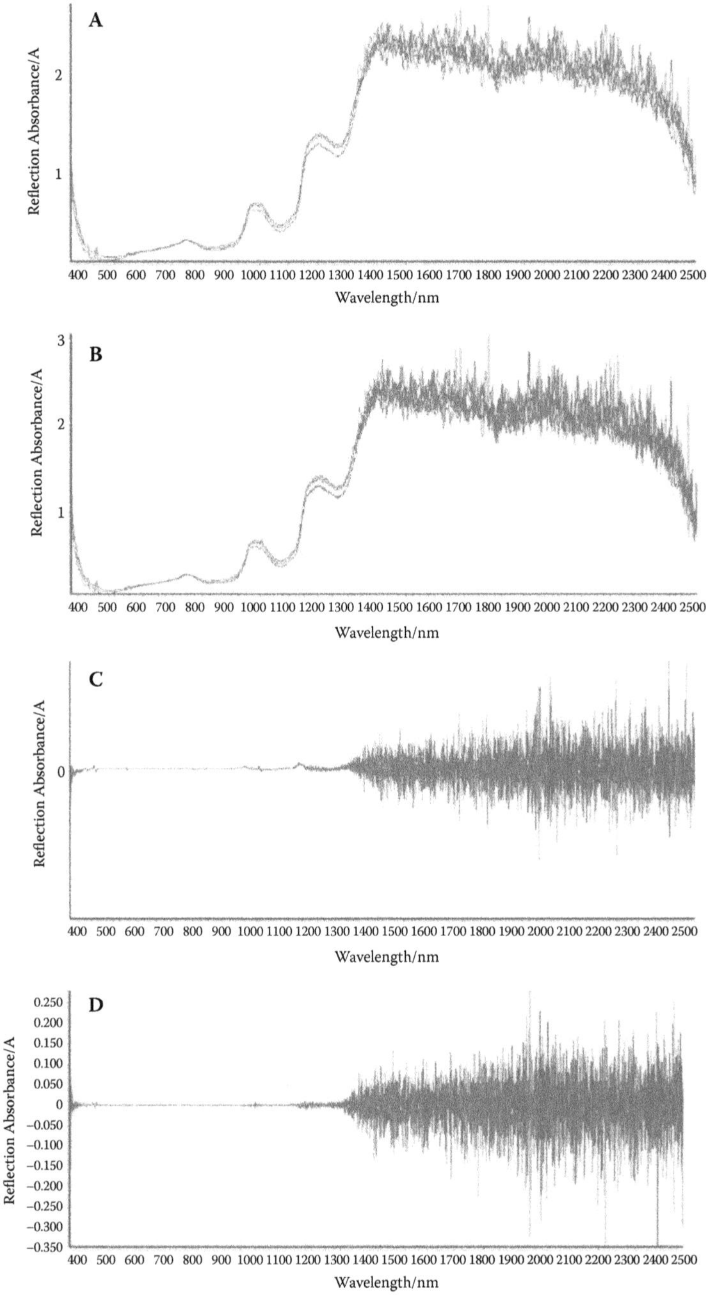

Figure 5.3 The spectra of samples after (a) MA smoothing, (b) after S-G smoothing, (c) after first-derivation smoothing, and (d) after second-derivation smoothing.

Table 5.1 Calibration Result of Nitrogen Content by PLS Using Different Smoothing Methods

Smoothing method	R	RMSEP	AREP (%)
MA	0.9589	0.1487	9.369
SG	0.9614	0.1428	9.015

Note: PLS, partial least squares; MA, moving average filter; SG, Savitzky-Golay filter; RMSEP, root mean squared errors of prediction; AREP, average relative error percentages.

Table 5.2 Calibration Results of the Models from Spectral Data under Derivation Pretreatment by PLS

Pretreatment method	R	RMSEP	AREP (%)
First-order differential + MA	0.9704	0.1226	5.483
Second-order differential + MA	0.9066	0.2389	12.057
First-order differential + SG	0.9721	0.1198	5.096
Second-order differential + SG	0.9071	0.2346	10.773

Note: PLS, partial least squares; MA, moving average filter; SG, Savitzky-Golay filter; RMSEP, root mean squared errors of prediction; AREP, average relative error percentages.

can effectively eliminate the effects of baseline drift and gentle background interference, refine overlapping peaks, and provide clear spectral profile changes. However, this method also enlarges the noise signal of the high-frequency spectrum, thereby intensifying the amplification degree of the second-order differentiation. Table 5.2 shows the spectral modeling of the samples subjected to derivative pretreatment.

Table 5.2 shows that the modeling effect of smoothing, combined with first-order differentiation, is generally better than the second-order modeling effect. The first-order differential treatment increased the predictive ability of the model. The correlation coefficient (R) increased from 0.9614 to 0.9721, and RMSEP and AREP decreased from 0.1428% and 9.015% to 0.1198% and 5.096%, respectively. The prediction precision of modeling showed a greater degree of improvement.

5.3.2.3 *Mean centering and normalization*

In near-infrared spectral analysis, a data enhancement algorithm is used to eliminate redundant information before establishing the quantitative or qualitative model and increasing the differences between samples. This algorithm can improve the robustness and predictive ability of the model, and commonly used algorithms employ mean centering and normalization.

Centralization normally involves changing the spatial coordinates and origin of a data set by obtaining the average of the data set around the origin and that of the origin. These changes immediately result in clearer spectra.

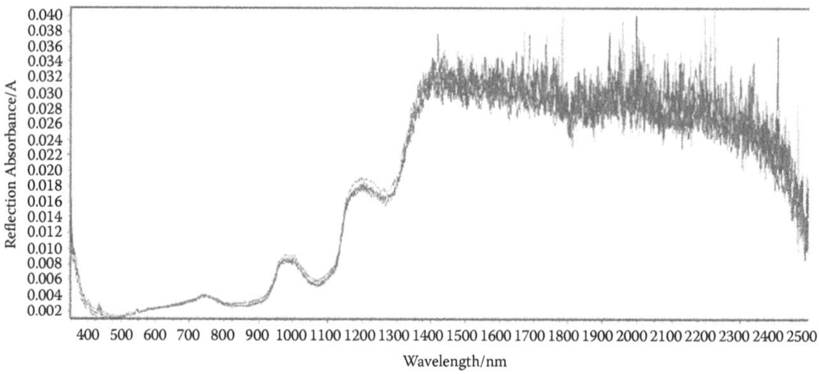

Figure 5.4 The spectra of samples after normalization.

Table 5.3 Calibration Results of the Models from the Normalized Spectral Data by PLS

Pretreatment method	R	RMSEP	AREP (%)
First-order differential + MA + normalization	0.9615	0.1425	6.785
Second-order differential + MA + normalization	0.9053	0.2347	11.834
First-order differential + SG + normalization	0.9709	0.1233	5.432
Second-order differential + SG + normalization	0.9104	0.2248	10.223

Note: PLS, partial least squares; MA, moving average filter; SG, Savitzky-Golay filter; RMSEP, root mean squared errors of prediction; AREP, average relative error percentages.

Normalization involves setting the same lengths for all vectors in a data set and providing similar squares for each element in each vector. Figure 5.4 shows the near-infrared spectral curves of the samples processed using normalization.

We also used normalization for the pretreatment of the sample spectra in Table 5.2 and performed a modeling analysis by combining the pretreatment conditions with PLS. Table 5.3 shows the results.

Table 5.3 shows the normalization does not improve the prediction precision. Compared with the pretreatment using a window width of 5, first-order differentiation, and SGF smoothing, the R of the model decreased from 0.9721 to 0.9709, and RMSEP and AREP increased from 0.1198% and 5.096% to 0.1233% and 5.432%, respectively (Table 5.2). With normalization, the precision of the predictive ability of the model decreased to some degree.

5.3.2.4 Comparison of PLS and PLS + artificial neural network (ANN) modeling

The medium samples had >95% moisture content, resulting in high light absorbance. The samples in the glass flasks produced a spectral reflection signal, indicating that the optical path of the samples and direction of light refraction changed. Therefore, the sample spectral data were nonlinear. Considering the linear and nonlinear information from the spectral data, we set the window width as 5 and used the PLS and PLS + ANN methods for model analysis.

The different pretreatment methods, such as smoothing, derivation, and normalization, had different influences on the predictive ability of the sample spectral models (Table 5.4). The related coefficient of first-order derivative processing combined with SGF smoothing pretreatment was higher for PLS modeling than for the other models. Furthermore, RMSEP and AREP were lower than those of the other models. The *R*, RMSEP, and AREP of the model were 0.9721, 0.1198, and 5.096, respectively. Compared with the other pretreatment methods, the predictive ability of the model greatly improved. These results show that first-order derivative processing combined with SGF smoothing pretreatment before applying PLS modeling resulted in higher accuracy and was more suitable for the pretreatment of nitrogen near-infrared spectra in the medium.

5.3.2.5 *Prediction effect and validation of the model*

5.3.2.5.1 Prediction model. R values between the spectral data and actual chemical values were calculated using nirsa software (computer software copyright registration number 2007sr06801). Following smoothing and first-order differentiation, pls was used to establish a mathematical model for relating the nitrogen content of the sample and spectral data. To reestablish the model, the same set of calibration and prediction data was used in each measurement. Figure 5.5a shows the prediction model.

As shown in the prediction model (Figure 5.5a), the model R ranged from 0.9289 to 0.9822. RMSEP and the average error ranged from 0.052 to 0.091and 4.301% to 7.665%, respectively. Therefore, these models meet the requirements for prediction precision. Figure 5.5a shows the prediction model. The fitting equation between the tested and chemical values was $y = 0.9802x + 0.018$ ($R^2 = 0.983$, $n = 60$) and had a slope of 0.98, which is close to 1. The spectral data from the prediction value of the standard medium samples were entered into the prediction model, and then the model was employed to predict the nitrogen content in standard medium samples. Figure 5.5b shows the determination of the nitrogen content of the standard medium sample prediction set.

The average relative error between the tested value of the nitrogen content within the standard medium samples in the prediction set and their chemical values was only 5.096% (Figure 5.5b). The fitting equation between the tested and chemical values was $y = 0.9388x + 0.079$ ($R^2 = 0.945$, $n = 20$). The results show that the determination of the nitrogen content in the tested sample medium was better when the spectral data were entered into the prediction model. The tested values were extremely close to the actual chemical values. Therefore, the visible and near-infrared spectral

Table 5.4 Results of the Models from Spectral Data after Different Modeling Methods by Different Pretreatments

Method of correction	Pretreatment method	R	RMSEP	AREP (%)
PLS	Untreated	0.9620	0.1424	8.953
	MA	0.9589	0.1487	9.369
	SG	0.9614	0.1428	9.015
	First-order differential	0.9724	0.1194	5.550
	Second-order differential	0.9173	0.2178	11.883
	First-order differential + MA	0.9704	0.1226	5.483
	Second-order differential + MA	0.9066	0.2389	12.057
	First-order differential + SG	0.9721	0.1198	5.096
	Second-order differential + SG	0.9071	0.2346	10.773
	First-order differential + MA + normalization	0.9615	0.1425	6.785
	Second-order differential + MA + normalization	0.9053	0.2347	11.834
	First-order differential + SG + normalization	0.9709	0.1233	5.432
	Second-order differential + SG + normalization	0.9104	0.2248	10.223
PLS + ANN	Untreated	0.9370	0.1948	10.584
	MA	0.9348	0.1929	11.583
	SG	0.9090	0.2362	12.126
	First-order differential	0.9730	0.1422	8.780
	Second-order differential	0.8549	0.3231	16.244
	First-order differential + MA	0.9623	0.1565	8.582
	Second-order differential + MA	0.8479	0.3196	16.108
	First-order differential + SG	0.9543	0.2078	10.998
	Second-order differential + SG	0.8575	0.3269	15.610
	First-order differential + MA + normalization	0.9628	0.1441	7.381
	Second-order differential + MA + normalization	0.8995	0.2746	14.070
	First-order differential + SG + normalization	0.9557	0.1574	8.347
	Second-order differential + SG + normalization	0.8767	0.3126	15.325

Note: PLS, partial least squares; ANN, artificial neural network; MA, moving average filter; SG, Savitzky-Golay filter; RMSEP, root mean squared errors of prediction; AREP, average relative error percentages.

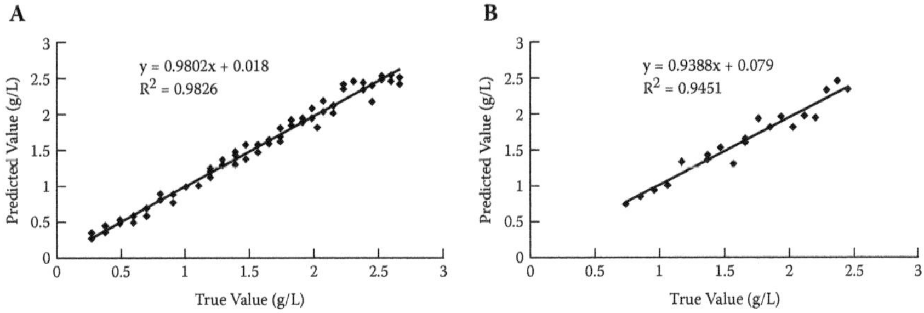

Figure 5.5 (a) Prediction model and (b) prediction results, where y denotes the predicted value and x denotes the true value.

analysis method can be used to dynamically, sterilely, and nondestructively test the nitrogen content of the culture medium.

5.3.2.5.2 Validation of model applicability. The purpose of spectral modeling was to test the nitrogen content of the culture medium in empty bottles and to nondestructively investigate the culture medium of plantlets growing in vitro at different growth stages. The culture medium used in our experiments was highly transparent. When the plantlets were transplanted too deeply into the medium, the penetrability of the visible light and short waves' near-infrared spectrum was high, which may carry a small amount of spectral information of the root and influence the predictive ability of the model. Therefore, model adaptability after plantlet transplantation was verified. This was performed to predict the spectral outcomes for the medium of the transplanted plantlets. On the day of plantlet transplantation, the tested nitrogen and initial nitrogen contents in the MS medium were 42.55 mg ± 0.05 mg (mean ± standard error, n = 9) and 42.05 mg, respectively. Both nitrogen values were relatively consistent. Thus, plantlet transplantation had no influence on the determination of the nitrogen content of the medium. Furthermore, the spectral data obtained from near-infrared scanning and the actual chemical values were adaptable. This indicates that the near-infrared analysis model can be applied to nondestructive testing of the nitrogen content of the plantlet culture medium.

5.4 Sterile and dynamic determination of the in vitro nitrogen content of the plantlet-culturing medium

5.4.1 Materials and methods

These experiments were performed in an artificial climate chamber, with the culture room temperature, CO_2 concentration, and air humidity controlled at 25°C ± 0.5°C, 360 μmol · mol^{-1}, and 45%, respectively. The tested plantlets grown in vitro were *Orychophragmus violaceus*, *Morus alba*, *Atractylodes lancea*, and *Brassica napus*. Before the experiments, these plantlets were first subcultured for three to four generations. For culturing, 150 ml culture glass bottles were used. The light period consisted of a 12 h/12 light-dark cycle. Each tissue culture bottle contained MS medium (50 ml) containing 3% nitrogen, 0.1 mg · L^{-1} NAA, and 2 mg · L^{-1} 6-BA. The light intensity was set at 45 μmol · m^{-2} · s^{-1}.

The initial nitrogen content of the medium was low (0.356 g · L^{-1}), moderate (0.632 g · L^{-1}), or high (1.159 g · L^{-1}). The nitrogen content of the control medium was 0.908 g · L^{-1}. *O. violaceus*, *M. alba*, *A. lancea*, and *B. napus* plantlets grown in vitro and subcultured for three to four generations were transplanted into culture media containing the three nitrogen concentrations. The culture period was 30 days. Table 5.5 shows the initial nitrogen concentrations in the medium with the different plantlet types.

Standard samples with 0.2847 g · L^{-1} to 2.8364 g · L^{-1} nitrogen were produced on the basis of all nitrogen concentrations used. The nitrate nitrogen and ammonium nitrogen concentrations ranged from 0.2847 g · L^{-1} to 1.5303 g · L^{-1} and 0 g · L^{-1} to 1.3061 g · L^{-1}, respectively. These samples were refrigerated (0°C to 4°C) to maintain the stability of the sample properties.

Given that several plantlets were transplanted at regular intervals, the diffuse reflection spectra of the 80 standard medium samples and tested plantlet samples were collected using a portable spectrometer. The standard samples were divided into two sets: a correction set (60 samples) and

Table 5.5 Initial Nitrogen Concentration in the Culture Medium of Various Subcultured Plantlets In Vitro

Sample number	Plant species	Nitrogen level (g/L)
1	*Orychophragmus violaceus*	Low nitrogen, 0.356
2	*Orychophragmus violaceus*	Moderate nitrogen, 0.632
3	*Orychophragmus violaceus*	Control, 0.908
4	*Orychophragmus violaceus*	High nitrogen, 1.159
5	*Morus alba*	Moderate nitrogen, 0.632
6	*Morus alba*	Control, 0.908
7	*Atractylodes lancea*	Low nitrogen, 0.356
8	*Atractylodes lancea*	Moderate nitrogen, 0.632
9	*Atractylodes lancea*	Control, 0.908
10	*Atractylodes lancea*	High nitrogen, 1.159
11	*Brassica napus*	Moderate nitrogen, 0.632
12	*Brassica napus*	Control, 0.908
13	*Brassica napus*	High nitrogen, 1.159

a prediction set (20 samples). To reestablish the model, the same correction and prediction sets were used in each measurement.

5.4.2 In vitro nitrogen content of the plantlet-culturing medium

5.4.2.1 Correction of moisture dissipation in the culturing medium

Factors such as evaporation, transpiration, and air circulation cause moisture dissipation from the culture medium. This artificially increases the measured nitrogen concentrations in the medium. Therefore, moisture dissipation was corrected.

5.4.2.2 Calculation method for the nitrogen content of the medium

During transplantation, the plantlets were placed in the center of a culture bottle at a depth of 0.5 cm. When the spectrometer collected the spectral information of the medium, the points were collected in a row (from the periphery to the center) at the bottom radius of the bottle. These points were collected at the center of the bottom of the glass bottle at circumference distances of 2.5, 1.5, and 0.5 cm. The spectral information of the test sample medium was collected at the same distances each time. The medium thickness within the bottle was 2.5 cm, and the plantlet depth in the medium was approximately 0.5 cm. The actual distances between the measuring point and the root of the plantlets were calculated using the Pythagorean theorem. These distances were 3.2, 2.5, and 2.05 cm. We obtained the medium concentration by determining the average value of many surrounding points.

The test results show a nitrogen concentration gradient within the medium. The concentration near the root was lower than the surrounding concentration. With that, the roots were constantly absorbing the nutrients present in the medium, and the nutrient concentration of the rhizosphere decreased. Furthermore, the nutrient concentration gradient differed both horizontally and vertically. Nutrient absorption by the roots only occurred when the rhizosphere nutrients were maintained at a certain level. The rhizosphere medium had an extremely different nutrient concentration than the original medium. These differences were mainly controlled by the comprehensive effects of root absorption and nutrient migration rates. The nutrient supply intensity within the rhizosphere directly influenced the plant nutritional status. Therefore, the measurement results coincided with the physiological characteristics of plant growth.

The nitrogen nutrient content of the culture medium showed horizontal and vertical gradients. Therefore, the nitrogen concentration could not be obtained by simply averaging many measurement points. The medium in the culture bottle was approximated to be a cylinder having a radius and height of 2.523 cm and 2.5 cm, respectively. The plantlet depth in the medium was approximately 0.5 cm. On the basis of these data, we established a model Y for estimating the average nitrogen concentration as a function of its distance from the roots. The nitrogen content within the entire medium in the bottle was expressed as the integral over the cylinder:

$$Y = \int_0^2 y\pi R^2 dx + \int_0^{0.5} y\pi R^2 dx$$

x is the distance from root.

We then fitted the nitrogen content using a linear model on the basis of the root distance. The fitting equation was $y = 0.1184x + 0.323$ ($R^2 = 0.997$, $n = 9$).

5.4.2.3 *Consumption of nitrogen from the medium by different plantlets*

R between the spectral data and actual chemical values was calculated using NIRSA software. Following smoothing and first-order differentiation, PLS was used to establish the mathematical model for relating the nitrogen content of the samples and spectral data. To reestablish the model, the same set of calibration and prediction data was used in each measurement. The model then predicted the nitrogen content of the plantlet medium. In addition, moisture loss was considered. The corrected nitrogen concentration during a culture period was calculated using the nitrogen concentration gradient model.

During the first 5 days of culture, the nitrogen content of the medium of plantlets 3, 4, 8, 12, and 13 increased. From day 5 to day 15, the nitrogen content for all plantlets significantly decreased. From day 15 to day 30, the nitrogen content homogeneously decreased with culture time. After 30 days, the tested nitrogen content was considerably lower than the initial nitrogen content. In general, changes in the nitrogen content of the medium of each plantlet type were relatively consistent.

The root organization of plantlets being transplanted is damaged to a certain extent. Moreover, the nitrogen released in the liquid that flows from a root in response to an injury increases the nitrogen content of the culture medium (Lin and Chang, 1998). Therefore, the nitrogen content was higher in the medium of transplanted plantlets than its initial content for a short time. Subsequently, the nitrogen content of the medium gradually decreased as the plant utilized nitrogen and the liquid released in response to the injury ceased to flow. After 30 days, the observed nitrogen content of the medium was considerably lower than the initial nitrogen content. The nitrogen use efficiency of each plantlet was different because each plantlet species had a different root system.

Figure 5.6 shows the variation in the consumption (%) and content of nitrogen (mg) in the culture medium of *O. violaceus, M. alba, A. lancea,* and *B. napus* plantlets grown in vitro with subculture days.

The trend of nitrogen consumption for the four plantlet types was basically consistent and increased with subculture days (Figure 5.6). After 30 days of culture, as reflected through the observed nitrogen consumption, 80% to 90% of the nitrogen content of the medium was consumed. Normal plantlet growth was affected when the nitrogen content of the medium was considerably low and almost completely consumed. Therefore, the measurement of nitrogen consumption from the culture medium provides information for the determination of an optimal transplant time.

Nitrogen content significantly decreased when the initial nitrogen concentrations were extremely high (1.159 $g \cdot L^{-1}$). At low initial nitrogen concentrations, the decreases in the nitrogen content of the medium were smaller (0.356 $g \cdot L^{-1}$). Later, during the culture period, the nitrogen content of the medium of plantlets 1 and 7 decreased slightly compared with the nitrogen content of the medium of other plantlets. These results show that the plantlets could not absorb or use nitrogen when the nitrogen content of the medium decreased to a certain extent. Reasonable nitrogen addition results in efficient plantlet use and prevents resource wastage.

5.5 *In vitro nitrogen use efficiency of plantlets grown in medium containing different nitrogen concentrations*

5.5.1 *Consumption of nitrogen from the medium and biomass change*

O. violaceus, M. alba, A. lancea, and *B. napus* plantlets grown in vitro were transplanted into media with low (0.356 $g \cdot L^{-1}$), moderate (0.632 $g \cdot L^{-1}$), or high (1.159 $g \cdot L^{-1}$) nitrogen content. The control medium had a nitrogen content of 0.908 $g \cdot L^{-1}$. The culture period was 30 days. Plantlet biomass was measured using the internal standard method, and the nitrogen content of the medium was measured by near-infrared spectroscopy. Figure 5.7 shows the results.

Figure 5.7 shows that the nitrogen content of the medium was different for different types of plantlets with different growth rates. We found that the plantlet biomass increased the most in the middle culture period (10 days to 20 days). In the initial culture period, plantlet transplantation likely caused a certain extent of damage. Given that the plantlets required some time for recovery, they showed low nutrient utilization. Therefore, the plantlets grew at a slower pace during the initial culture period. In

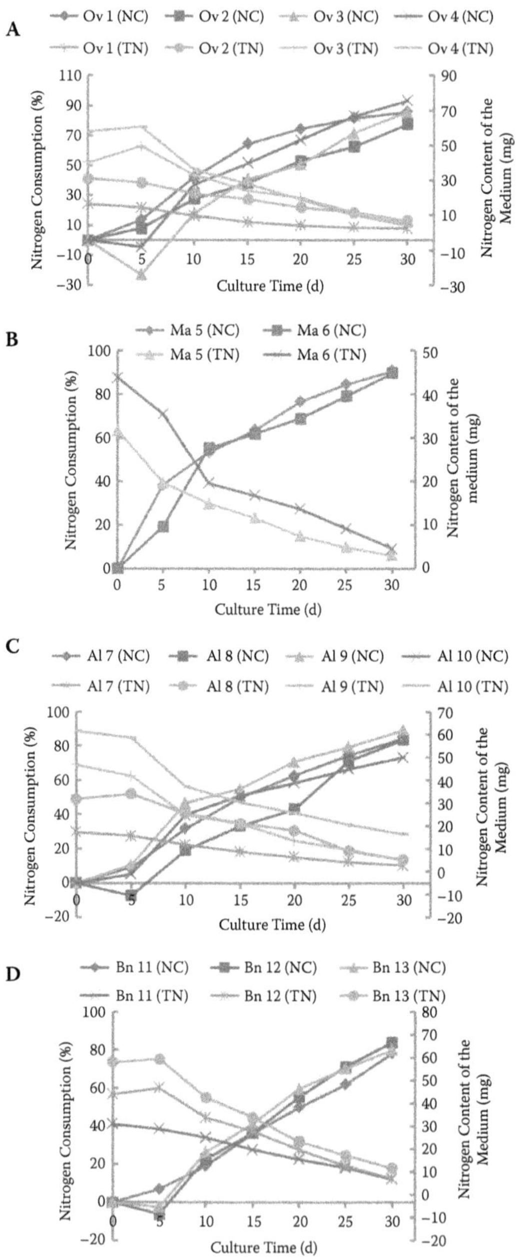

Figure 5.6 The variation in the consumption and content of nitrogen in the culture medium subcultured various plantlets in vitro with subculture days. Ov, *Orychophragmus violaceus*; Ma, *Morus alba*; Al, *Atractylodes lancea*; Bn, *Brassica napus*; NC, nitrogen consumption; TN, total nitrogen.

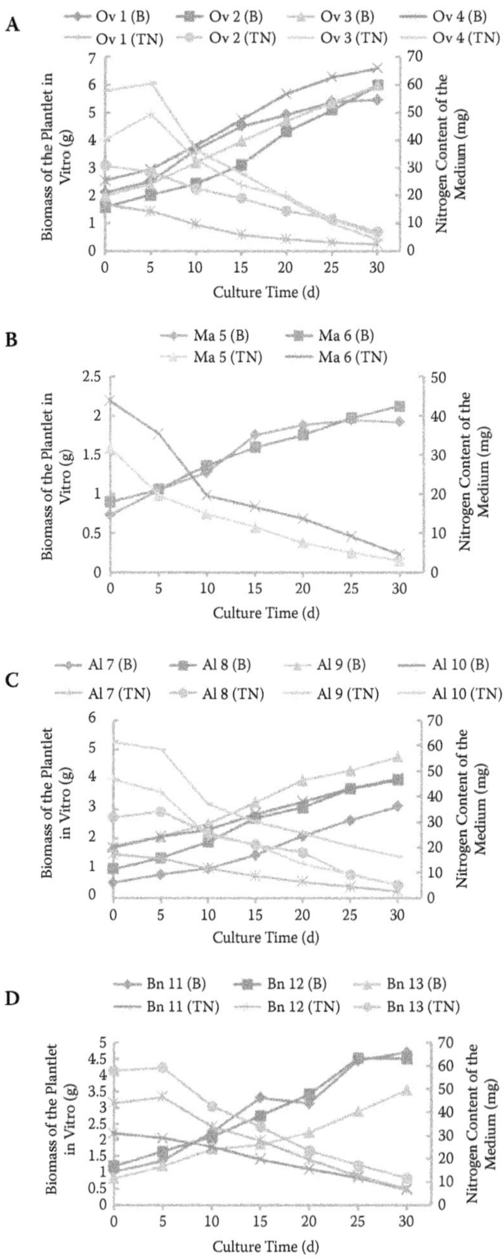

Figure 5.7 The variation of biomass in plantlets in vitro and the content of nitrogen in the culture medium subcultured with subcultured days. Ov, *Orychophragmus violaceus*; Ma, *Morus alba*; Al, *Atractylodes lancea*; Bn, *Brassica napus*; B, biomass; TN, total nitrogen.

the later culture period, plantlet growth was hindered slightly because of nutrient shortage. Compared with the middle culture period, the plantlet growth was also slow in the later culture period (Song, 2010).

On the basis of the experimental data, we determined that the increase in plantlet biomass was relatively small under low nitrogen conditions. The biomass of *O. violaceus* 1 only increased by 0.057 g during 20 days to 25 days. When the nitrogen content of the medium decreased to a certain extent, the nitrogen was not absorbed and used by the plantlet (Figure 5.7). Under low nitrogen conditions, as in the later culture period, plantlet growth was inhibited to a certain degree. Therefore, the biomass of plantlet 1 did not increase.

Plantlets derive energy from each nutritional element in the medium. However, each nutritional element in the medium was limited. The nutritional element contents in the medium continuously decreased with culture time. After 30 days, the nutritional shortage in the medium could affect normal plantlet growth. Therefore, when the culture time was 25 days to 30 days, plantlets 5 and 12 showed negative growth. The nutrient content of the medium is a good indicator of an appropriate time for plantlet transplantation. This helps to avoid the economic losses that result from delayed transplantation.

The plantlet biomass at different nitrogen concentrations also differed. The in vitro growth of *O. violaceus* plantlets follows the order *O. violaceus* 4 > *O. violaceus* 1 > *O. violaceus* 3 > *O. violaceus* 2 (Figure 5.7). At high nitrogen concentrations, the in vitro growth of these *O. violaceus* plantlets was the fastest. The *O. violaceus* plantlets adapted efficiently to different nutrition environments and could grow in poor nutrient conditions. These plantlets exhibited the best growth in a nutrient-rich environment. The in vitro growth of *M. alba* plantlets was the same at moderate and high nitrogen concentrations, indicating that the growth of *M. alba* plantlets was not sensitive to nitrogen content provided their growth requirements were met. The in vitro growth of *A. lancea* plantlets indicates that the growth of *A. lancea* 9 was the fastest. Furthermore, the growth patterns of *A. lancea* 8 and *A. lancea* 9 were the same, but they showed faster growth than *A. lancea* 7. The growth of *A. lancea* plantlets was the fastest at a moderate nitrogen concentration. The in vitro growth of *B. napus* plantlets was the fastest at moderate and high nitrogen concentrations. The growth rate at these concentrations was the fastest at the low nitrogen concentration.

Therefore, the required optimum nitrogen concentration for each plantlet type was different. The nitrogen content of the culture medium affected plantlet growth. When the nitrogen concentration in the medium is too low, taproot growth is usually accelerated and lateral root growth is inhibited, which results in less nutrient absorption by the root system. In this case, the plant growth rate is relatively slow. When the nitrogen concentration in the medium is too high, root cells lose water, which reduces the ability of the root cells to absorb nutrients and eventually leads to root cell death. Therefore, culture medium containing moderate nitrogen content ensures that nitrogen is effectively used and resource wastage is prevented.

5.5.2 *Analysis of the nitrogen use efficiency of different plantlet types*

In the tissue culture, the plantlet volume was small and photosynthesis was weak. Therefore, the plantlets mainly relied on consumption of nutrients from the medium for their growth. The nutrients entered the plant body through slow diffusion. Therefore, the quantity of consumed nitrogen per unit of plant growth may represent the nitrogen use efficiency of the plantlets.

We determined the nitrogen use efficiency of plantlets during culture on the basis of the plantlet biomass increment and the decrease in the nitrogen content of the medium. During plantlet transplantation, the amount of injury liquid flow is related to tissue injury severity. However, the determination of the degree of injury is difficult. However, we could ensure the nitrogen use efficiency of the plantlets in a specific culture period.

The nitrogen use efficiency differed for plantlets grown at each nitrogen concentration. At low nitrogen concentrations (0.356 $g \cdot L^{-1}$), the nitrogen use efficiencies of *A. lancea* and *O. violaceus* plantlets were 235.7 $g \cdot g^{-1}$ N and 176.9 $g \cdot g^{-1}$ N, respectively, from day 0 to day 30. At high nitrogen concentrations (1.159 $g \cdot L^{-1}$), the nitrogen use efficiencies were 75.4 $g \cdot g^{-1}$ N and 51.6 $g \cdot g^{-1}$ N. In the tested nitrogen concentration range, the nitrogen use efficiency of the plantlets was higher at lower nitrogen concentrations. When nitrogen availability was lower, the plantlets used nitrogen more efficiently. This finding is in agreement with the nitrogen reduction law of the medium at different nitrogen concentrations and coincided with the plant growth rule. Figure 5.8 shows the nitrogen use efficiency (g biomass g^{-1} N) of several plantlets cultured in vitro and nitrogen content (mg) in the culture medium with culture time.

The nitrogen use efficiency was higher at low nitrogen concentrations than at other nitrogen concentrations within the tested nitrogen concentration range (Figure 5.8). In particular, the lower the nitrogen content, the more efficiently the plantlet uses nitrogen. At low nitrogen concentrations, the amount of nitrogen used was the least; therefore, the nitrogen use efficiency was the highest. Similarly, the maximum amount of nitrogen was used at high nitrogen concentrations. Therefore, the most efficient nitrogen use occurred at the lowest nitrogen concentrations.

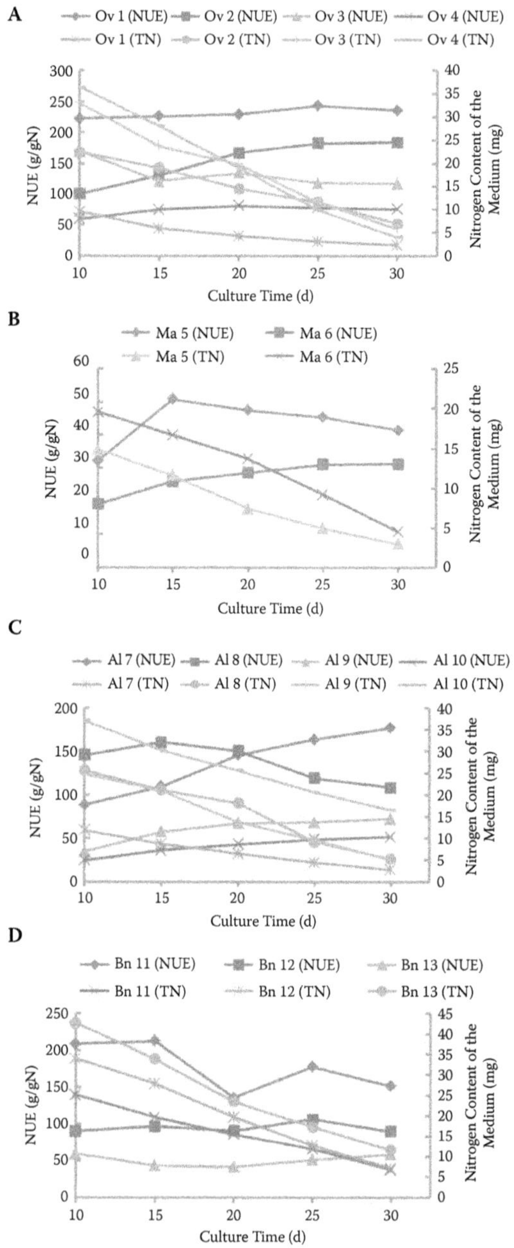

Figure 5.8 The nitrogen use efficiency of several plantlets in vitro and the content of nitrogen in the culture medium with culture time. Ov, *Orychophragmus violaceus*; Ma, *Morus alba*; Al, *Atractylodes lancea*; Bn, *Brassica napus*; NUE, nitrogen use efficiency; TN, total nitrogen.

5.6 In vitro nitrogen use efficiency of plantlets in the presence of different nitrogen sources

A growing plant uses nitrogen more than other mineral elements. Therefore, nitrogen is often the main factor that limits plant growth. During growth, plants use inorganic forms of nitrogen, primarily ammonium nitrogen and nitrate nitrogen. Plants can absorb both nitrogen forms. However, the influence of these forms on plant development is different because of their different shapes (Aerts and Chapin, 1999; Cui and Song, 2005; Liu et al., 2011). The influence of the nitrogen content of the medium on the ability of the plantlets to efficiently use nitrogen was discussed in Section 5.5. The influence of different nitrogen sources on the nitrogen use efficiency of the plantlets is discussed in this section.

5.6.1 Materials and methods

5.6.1.1 Experimental design

For these experiments, *A. lancea* plantlets grown in vitro and subcultured for four generations were used as test materials. These plantlets were cultured in MS medium containing different proportions of inorganic ammonium nitrogen and nitrate nitrogen. In all treatments, the total amount of nitrogen in the medium was kept constant. The ammonium nitrogen and nitrate nitrogen concentrations were adjusted by adding NH_4Cl and KNO_3, respectively. Treatments 1 to 4 involved changing the nitrate nitrogen content while keeping the ammonium nitrogen content constant. These treatments consisted of a control sample and samples with low, moderate, and high nitrate nitrogen contents. Treatments 5 to 8 involved changing the ammonium nitrogen content while keeping the nitrate nitrogen content constant. These treatments consisted of a control sample and samples with low, moderate, and high ammonium nitrogen contents. Table 5.6 shows the treatments.

5.6.1.2 Culture conditions

The experiment was performed in an artificial climate chamber, with the culture room temperature, CO_2 concentration, and air humidity controlled at 25°C ± 0.5°C, 360 μmol · mol^{-1}, and 45%, respectively. A 100 ml glass culture bottle was used for the experiment. The light period consisted

Table 5.6 Experimental Treatment of *Atractlodes lancea* Plantlets In Vitro

Treatment number	1	2	3	4	5	6	7	8
Nitrate nitrogen (g·L^{-1})	0.185	0.370	0.555	0.740	0.555	0.555	0.555	0.555
Ammonium nitrogen (g·L^{-1})	0.289	0.289	0.289	0.289	0.096	0.192	0.289	0.385

of a 12 h/12 h light-dark cycle. Each tissue culture bottle contained MS medium (50 ml) containing 3% nitrogen, 0.1 mg · L^{-1} NAA, and 2 mg · L^{-1} 6-BA. The light intensity was set at 45 μmol · m^{-2} · s^{-1}.

5.6.2 *Measurement indexes and methods*

5.6.2.1 *Determination of the biomass of* A. lancea *plantlets grown in vitro*

Plantlet biomass was determined using a conventional method of image processing technology. In brief, we placed the experimental *A. lancea* plantlets between six different plantlets, whose biomasses were known. We obtained a color image, converted it to a grayscale image using Photoshop, adjusted the brightness and contrast of the image, and inverted the image color. We statistically analyzed the pixel points for each plantlet using the inverted image. Internal standards were developed using the six plantlets. To determine the linear regression equation, a linear regression plot was constructed using the biomass and pixel information from the internal standard plantlets. We then used the pixel information of the *A. lancea* plantlet, whose biomass was unknown, to calculate its biomass. We calculated the *A. lancea* plantlet biomasses of M_{t0}, M_{t1}, M_{t2}, and M_{tn} at t_0, t_1, t_2, and t_n days, respectively.

5.6.2.2 *Measurement of nitrogen consumed by* A. lancea *plantlets*

The nitrogen content of the medium of the *A. lancea* plantlets was determined by near-infrared spectroscopy. Refer to the content of Sections 5.4 and 5.5 in this chapter for more details.

5.6.3 *Establishment of a nitrogen consumption model for plantlets*

5.6.3.1 *Relationship model between the plantlet biomass (M) and culture time (t)*

A four-parameter logistic equation was used to fit and establish the model between the biomass (M) of a plantlet and its culture time (t) (Wu et al., 2009):

$$M = M_0 + \frac{a}{1+(t/T_0)^b} \tag{5.1}$$

where M_0 is the plantlet biomass at the initial logarithmic growth phase, a is the largest increment in the plantlet biomass, b is the index in the fitting parameters, and T_0 is the time halfway through the logarithmic growth phase. The instantaneous biomass of a plantlet at each time was calculated using the model for the relationship between the plantlet biomass (M) and culture time (t).

5.6.3.2 Relationship model between nitrogen use (N) and culture time (t)

A four-parameter logistic equation was used to fit and establish the relationship between nitrogen use by the plantlet (N) and the plantlet culture time (t):

$$N = N_0 + \frac{c}{1+(t/S_0)^d} \tag{5.2}$$

where N_0 represents nitrogen use by the plantlet at the initial logarithmic growth phase, c is the greatest amount of nitrogen used by the plantlet, d is the index in the fitting parameters, and S_0 is the time halfway through the logarithmic growth phase. The instantaneous nitrogen use of the plantlet at each time was calculated using the model for the relationship between nitrogen use by the plantlet (N) and the plantlet culture time (t).

5.6.3.3 Relationship model between nitrogen use by plantlets (N) and plantlet biomass (M)

The relationship between nitrogen use by a plantlet (N) and plantlet biomass (M) was fitted using the following nonlinear hyperbolic equation:

$$M = \frac{x \cdot N}{y+N} \tag{5.3}$$

where M is the independent variable representing the plantlet biomass, and x and y are the fitting parameters. The first-order derivative was obtained by constructing the nonlinear hyperbolic equation with M to N for the derivative. The water use efficiency equation was expressed as follows:

$$M' = \frac{x \cdot y}{(y+N)^2} \tag{5.4}$$

5.6.4 Calculation method for the nitrogen use efficiency of plantlets

The formula

$$N = N_0 + \frac{c}{1+(t/S_0)^d}$$

was entered into the equation

$$M' = \frac{x \cdot y}{(y+N)^2}$$

resulting in

$$M' = \frac{x \cdot y}{\left[y + N_0 + \frac{c}{1+(t/S_0)^d}\right]^2}$$

The nitrogen use efficiency at each time was calculated using this equation.

5.6.5 *Nitrogen use model using experimental data from* A. lancea *plantlets*

Plantlet biomass (*M*) and nitrogen use by the plantlets (*N*) under different treatment conditions at different culture times were calculated (Table 5.7).

5.6.6 *Measurement of the nitrogen use efficiency of plantlets*

5.6.6.1 *Establishment of a nitrogen use efficiency model for* A. lancea *plantlets*

Nitrogen use efficiency models for the eight treatment conditions using *A. lancea* plantlets were established. Four-parameter logistic equations fit the models for the plantlet biomass (*M*) and culture time (*t*) ($R^2 > 0.99$), as confirmed previously. These models accurately predicted the plant growth conditions (Wu et al., 2009). Moreover, the models for nitrogen use (*N*) by a plantlet and plantlet culture time (*t*) were fitted by four-parameter logistic equations (treatments 1, 2, 3, 5, 6, and 7). Considering that conditions under treatments 4 and 8 were in the logarithmic growth phase, linear equations were used for both treatments to fit the relationship between biomass (*M*)/nitrogen use (*N*) and culture time (*t*). The relationship between biomass (*M*) and nitrogen use (*N*) was a better fit by a nonlinear hyperbolic equation. The nitrogen use efficiency at each time t_i was calculated using the first-order derivative of that equation (Table 5.8).

5.6.6.2 *Calculated values of the nitrogen use efficiency of* A. lancea *plantlets*

The nitrogen use efficiency generally decreased with prolonged culture time (Table 5.9). The different ratios of ammonium nitrogen and nitrate

Table 5.7 Experimental Data for the Model of Nitrogen Use Efficiency of *Atractlodes lancea* Plantlets In Vitro

Treatment number		Culture time *t* (days)					
		5	10	15	20	25	30
1	Biomass (*M*)/g	0.241	0.295	0.404	0.559	0.655	0.704
	Nitrogen use (*N*)/mg	0.056	1.116	5.264	8.760	10.382	12.028
2	Biomass (*M*)/g	0.280	0.355	0.606	0.662	0.704	0.751
	Nitrogen use (*N*)/mg	5.512	6.158	9.993	13.638	20.740	24.316
3	Biomass (*M*)/g	0.286	0.321	0.442	0.519	0.601	0.684
	Nitrogen use (*N*)/mg	10.933	20.255	28.444	31.020	34.963	38.234
4	Biomass (*M*)/g	0.303	0.368	0.439	0.491	0.641	0.777
	Nitrogen use (*N*)/mg	9.172	24.916	29.492	35.626	39.501	44.271
5	Biomass (*M*)/g	0.299	0.362	0.440	0.597	0.760	0.821
	Nitrogen use (*N*)/mg	1.627	1.860	3.018	4.016	6.319	7.982
6	Biomass (*M*)/g	0.341	0.429	0.691	0.817	0.989	1.244
	Nitrogen use (*N*)/mg	5.444	6.012	8.480	14.408	20.322	24.279
7	Biomass (*M*)/g	0.312	0.374	0.486	0.572	0.663	0.735
	Nitrogen use (*N*)/mg	16.400	18.383	24.623	32.166	36.236	38.736
8	Biomass (*M*)/g	0.332	0.432	0.491	0.535	0.620	0.655
	Nitrogen use (*N*)/mg	25.129	30.748	36.619	38.082	47.353	51.650

nitrogen significantly influenced *A. lancea* plantlets. These plantlets used ammonium nitrogen at a higher rate than nitrate nitrogen. Nitrogen use was higher under the treatments 1 (2) than under the treatments 3 (4). When the nitrogen content of the medium was low, the nitrogen use rate of the plantlet was high, allowing the plant to adapt to the nutritional environment. The nitrogen use efficiency decreased with increasing ammonium nitrogen concentrations. This finding indicates that low ammonium nitrogen concentrations significantly affected nitrogen use by the plantlets. Given that the samples were subcultured for four generations, the propagating ability of the plantlets was also higher. Therefore, at high nitrogen concentrations, the plantlets displayed a constant nitrogen use efficiency and showed the same nitrogen use ability because the plantlets were in the logarithmic growth phase.

Table 5.8 Various Models for Biomass, Nitrogen Use of *Atractlodes lancea* Plantlets In Vitro, and Culture Time ($n = 9, p < 0.001$)

Treatment number	Biomass M and culture time t	Nitrogen use N and culture time t	Biomass M and nitrogen use N
1	$M = 0.2417 + \frac{0.5268}{1+(t/18.1322)^{-3.9608}}$ ($R^2 = 0.9997$)	$N = -0.1807 + \frac{13.0573}{1+(t/16.5510)^{-3.9938}}$ ($R^2 = 0.9987$)	$M = 0.2347 + 0.0385\,N$ ($R^2 = 0.9945$)
2	$M = 0.2727 + \frac{0.4551}{1+(t/12.8956)^{-5.4453}}$ ($R^2 = 0.9946$)	$N = 33.0053 - \frac{27.5652}{1+(t/24.1830)^{3.7725}}$ ($R^2 = 0.9972$)	$M = \frac{1.1788 \cdot N}{12.7121 + N}$ ($R^2 = 0.9504$)
3	$M = 0.2656 + \frac{0.6535}{1+(t/23.9509)^{-2.4691}}$ ($R^2 = 0.9974$)	$N = 61.4377 - \frac{69.7302}{1+(t/14.3891)^{0.9186}}$ ($R^2 = 0.9975$)	$M = 0.0717 + 0.0148\,N$ ($R^2 = 0.9571$)
4	$M = 0.0185t + 0.1792$ ($R^2 = 0.9790$)	$N = 1.2879t + 7.9584$ ($R^2 = 0.9621$)	$M = 0.1130 + 0.0128\,N$ ($R^2 = 0.9059$)

5	$M = 0.3098 + \frac{0.6353}{1 + (t/20.6738)^{-4.0357}}$ ($R^2 = 0.9978$)	$N = 1.5888 + \frac{13.1788}{1 + (t/30.4235)^{-3.2412}}$ ($R^2 = 0.9969$)	$M = \frac{1.2666 \cdot N}{4.8828 + N}$ ($R^2 = 0.9869$)
6	$M = 758.2999 - \frac{758.0664}{1 + (t/4915.3742)^{1.3010}}$ ($R^2 = 0.9949$)	$N = 5.4515 + \frac{23.0305}{1 + (t/22.0256)^{-4.3803}}$ ($R^2 = 0.9999$)	$M = \frac{2.7217 \cdot N}{31.5889 + N}$ ($R^2 = 0.9621$)
7	$M = 0.2866 + \frac{0.7271}{1 + (t/24.1646)^{-2.1675}}$ ($R^2 = 0.9995$)	$N = 40.8673 - \frac{24.5855}{1 + (t/17.4870)^{4.2589}}$ ($R^2 = 0.9999$)	$M = \frac{6.6491 \cdot N}{322.4575 + N}$ ($R^2 = 0.9882$)
8	$M = 0.0127t + 0.2880$ ($R^2 = 0.9913$)	$N = 1.0507t + 20.5779$ ($R^2 = 0.9988$)	$M = \frac{4.1351 \cdot N}{269.5010 + N}$ ($R^2 = 0.9920$)

Table 5.9 Calculated Values of Nitrogen Use Efficiency of *Atractlodes lancea* Plantlet in vitro

	Treatment number	Culture Time (days) 5	10	15	20	25	30
Nitrogen use efficiency/(g/g) (biomass)	1	38.50	38.50	38.50	38.50	38.50	38.50
	2	45.12	41.06	30.80	20.26	13.93	10.80
	3	14.80	14.80	14.80	14.80	14.80	14.80
	4	12.80	12.80	12.80	12.80	12.80	12.80
	5	146.00	133.00	104.84	73.66	50.82	37.10
	6	52.00	50.69	44.31	33.90	26.47	22.94
	7	18.67	18.46	17.79	17.07	16.64	16.44
	8	12.78	12.33	11.91	11.52	11.14	10.77

A. lancea plantlet biomass was significantly higher with ammonium nitrogen than with nitrate nitrogen (Table 5.7). Therefore, *A. lancea* plantlet growth was affected by the selective use of ammonium nitrogen and nitrate nitrogen. The results for the nitrogen use efficiency show that the *A. lancea* plantlets utilized ammonium nitrogen well. These findings provide an important theoretical basis for configuring the nutrients in culture medium for the *A. lancea* plantlets and have important implications for the industrial production of the *A. lancea* plantlets.

5.7 *Conclusions and prospects*

Nitrogen plays an important role in in vitro plant tissue culture. The concentration and form of nitrogen available in vitro to the plantlets can significantly influence their growth and growth rate. Therefore, the measurement of the nitrogen content of the culture medium provides an important theoretical basis for understanding tissue culture. Near-infrared spectrum technology, a rapid and sterile method, can determine the nitrogen content of a culture medium without disrupting the closed and sterile environment of the plantlets.

The amount of nitrogen available for the plant can also be determined by determining the nitrogen content of the culture medium. At different nitrogen concentrations, the plantlets showed different nitrogen use efficiencies. The nitrogen consumption rate was obviously lower for the medium with low nitrogen concentrations than for the medium with high nitrogen concentrations. The nitrogen use rate of the plantlets was correlated with culture time, which was determined using the plantlet biomass. At different nitrogen source concentrations, *A. lancea* plantlets showed a higher ability to use ammonium nitrogen than nitrate nitrogen. The plantlet biomass was higher with more available ammonium nitrogen than with more available nitrate nitrogen. In addition, the plantlets use nitrogen efficiently when it is available at low concentrations. These results provide an important theoretical basis for rational nitrogen utilization in large-scale tissue culture activities.

Nitrogen consumption by plantlets at different nitrogen concentrations and forms was nondestructively measured by near-infrared spectral technology. Plantlet biomass was determined using image processing technology. On the basis of the results of these two methods, the nitrogen use efficiency of the plantlets cultured in media containing different nitrogen concentrations was determined. This study provides information for investigating the plantlet growth rate and improving the culture medium composition, which can be easily applied to industrial propagation practices. Different plant species selectively use different nitrogen resources. Some plant species prefer to use nitrate nitrogen, whereas other plant species prefer to use ammonium nitrogen. Most plants grow well in an environment containing mixed nitrogen resources. Determination of nitrogen consumption and the nitrogen use efficiency in a sterile and dynamic manner helped us to quickly identify the nitrogen source preferred by the plant species. These outcomes serve as a theoretical basis for repairing ecological systems (e.g., karst ecosystems) and for selecting species for coastal, beach, and wetland ecological systems.

The use of sterile, dynamic methods for the determination of nitrogen consumption and the nitrogen use efficiency also contributes to studies on nitrogen metabolism, which is related to nitrogen absorption, nitrite reduction, and ammonium nitrogen assimilation. These methods may help in improving our knowledge on nitrogen metabolism. Nitrogen and carbon metabolisms are closely interrelated. Dynamic determination of nitrogen consumption and the nitrogen use efficiency, as well as investigation on sucrose metabolism, may further expand our understanding of the metabolic processes in plantlets.

Acknowledgment

The authors gratefully acknowledge the financial support of the Priority Academic Program Development of Jiangsu Higher Education Institutions (PAPD).

References

Aerts, R., and Chapin III, F.S., The mineral nutrition of wild plant revisited: a reevaluation of processes and patterns, *Adv. Ecol. Res.*, 30, 1–67, 1999.

Bobelyn, E., Serban, A.S., Nicu, M., Lammertyna, J., Nicolai, B.M., and Saeys, W., Postharvest quality of apple predicted by NIR-spectroscopy: study of the effect of biological variability on spectra and model performance, *Postharvest Biol. Technol.*, 55, 133–143, 2010.

Bronson, K.F., Chua, T.T., Booker, J.D., Keeling, J.W., and Lascano R.J., In-season nitrogen status in irrigated cotton. II. Leaf nitrogen and biomass, *Soil Sci. Soc. Am.*, 67, 1439–1488, 2003.

Chen, H.C., Yang, Z.G., Li, H.Y., and Chen, X.D., Feasibility study for rapid determination of cholesterol concentration in human serum by using Fourier transform near infrared spectroscopy, *Acta Laser Biol. Sin.*, 13, 429–432, 2004.

Chio, J.M., Shin, C.Y., Chung, H.J., Choi, J.M., Shim, C.Y., and Chung, H.J., Effect of liming fertilization on changes of nutrient concentration in rice-hull based media, crop growth and uptake of chrysanthemum, *J. Korean Soc. Sci.*, 42, 553–556, 2001.

Chu, X.L., Xu, Y.P., and Lu, W.Z., Research and application progress of chemometrics methods in near infrared spectroscopic analysis, *Chin. J. Anal. Chem.*, 36(5), 702–709, 2008.

Cui, X.Y., and Song, J.F., Soil, NH_4^+/NO_3^- nitrogen characteristics in primary forests and the adaptability of some coniferous species, *Acta Ecol. Sin.*, 25(11), 3082–3092, 2005.

Ding, Y., Liu, D.J., and Zhang, Y., Application of ASI and general analytical method to soil testing in Xinjiang, *Xinjiang Agric. Sci.*, 44(6), 820–823, 2007.

Fan, X., Hang, L.J., and Huang, C.J., Determination of nutrient contents in beef manure with near infrared reflectance spectroscopy, *Trans. Chin. Soc. Agric. Mach.*, 37(3), 76–79, 2006.

Gao, J., Xu, Y.Y., and Yao, C., Determination of aromatic hydrocarbons in gasoline by near-infrared spectroscopy, *J. Nanjing Univ. Technol.*, 27(3), 51–53, 2005.

Han, D.H., Liu, X.X., Zhao, L.L., and Tu, R.L., Color, tissue and near-infrared spectrum characteristics of bruised apples, *Trans. Chin. Soc. Agric. Mach.*, 34(6), 112–115, 2003.

Huang, C.B., Fang, W.M., Yang, Y., Chen, F., and Xu, Y., Effects of different nitrogen levels and forms on the appearance quality and photosynthesis characters of potted chrysanthemum with small inflorescences, *Acta Agric. Zhejiangensis*, 22(1), 45–50, 2010.

Huang, Y.F., Ye, Y.L., and Yang, S.Q., Feasibility of NO_3-N determination by dual wavelength spectrophotometric method, *Chin. Agric. Sci. Bull.*, 25(02), 43–45, 2009.

Huang, Y.H., Li, Y.T., and Hou, C., Introduction of Kjeldahl nitrogen determination macro method, *Chin. J. Hosp. Pharm.*, 18(4), 185, 1998.

Jiang, J.F., Li, L., and Zhao, M.Y., Rapid determination of main chemical components in tobacco leaf by NIR diffuse reflectance spectroscopy, *Acta Tab. Sin.*, (4), 8–12, 2006.

Kong, Y., Han, L.J., Jia, G.R., and Zhu, S.P., Rapid near infrared prediction of broiler manure nutrient contents, *Trans. Chin. Soc. Agric. Eng.*, 20(6), 251–254, 2004.

Li, C., Liu, S.Y., and Li, Y.Y., The development of chemiluminescence and their application, *Anal. Test. Technol. Instrum.*, 12(2), 75–81, 2006b.

Li, Y.X., Zhu, Y., Tian Y.C., Yao, X., Qin, X.D., and Cao, W.X., Quantitative relationship between leaf nitrogen accumulation and canopy reflectance spectra in wheat, *Acta Agron. Sin.*, 3(2), 203–209, 2006a.

Lin, C.S., and Chang, W.C., Micropropagation of *Bambusa edulis* through nodal explants of field-grown culms and flowering of regenerated plantlets, *Plant Cell Rep.*, 17, 617–620, 1998.

Lin, X., and Niu, Z.Y., Fast discrimination of tea species based on near infrared spectroscopy (NIRS), *J. Huazhong Agric. Univ.*, 27(2), 326–330, 2008.

Liu, C.N., Cui, X.Y., Guo, Y.F., and Zheng, H., Effects of different ratios of NH_4^+-N/NO_3^--N on growth of *Larix gmelinii* seedlings, *J. Northeast Forest. Univ.*, 39(1), 28–30, 2011.
Liu, S., Zhu, H., Chu, X.L., Yuan, H.F., and Lu, W.Z., Study on the thermodynamic interaction between host compounds and guest substances by gas chromatographic method, *J. Instrum. Anal.*, 21(1), 40–43, 2002.
Lu, Y.T., Zhang, X.M., Ye, S.H., Qi, Y.B., Yan, W.C., and Jin, Q.S., Measuring of seven by near amino acid contented in japonica rice infrared spectroscopy technique, *Acta Agric. Nucl. Sin.*, 14(5), 478–482, 2007.
Luypaert, J., Massart, D.L., and Heyden, Y.V., Near-infrared spectroscopy applications in pharmaceutical analysis, *Talanta*, 72(3), 865–883, 2007.
Nicolai, B.M., Beullens, K., Bobelyn, E., Peirs, A., Saeys, W., Theron, K.I., and Lammertyna, J., Nondestructive measurement of fruit and vegetable quality by means of NIR spectrscopy: a review, *Postharvest Biol. Technol.*, 46(2), 99–118, 2007.
Pan, Y.C., Estimation of soil nitrogen accumulation based on remotely-sensed monitoring of winter-wheat growth status, *Trans. Chin. Soc. Agric. Eng.*, 23(9), 58–64, 2007.
Peng, Y., Wei, H., Zhu, W., and Gao, X.M., Estimating methods analysis on crop nitrogen using hyperspectral remote sensing, *J. Yuannan Nat. Univ. (Nat. Sci. Ed.)*, 15(1), 45–49, 2006.
Shiroma, C., and Rodriguez-Saona, L., Application of NIR and MIR spectroscopy in quality control of potato chips, *J. Food Compos. Anal.*, 22(6), 596–605, 2009.
Song, X.Y., Wang, J.H., Xue, C.Z., Niu, L.Y., Chen, L.P., and Zhao, C.J., Assessment of the influence of soil nitrogen supplies and variable fertilization on winter wheat growth condition using airborne hyperspectral image, *Trans. Chin. Soc. Agric. Eng.*, 20(4), 45–49, 2004.
Song, Y.J., Sterile measurement the sucrose content in the medium and application based on near infrared spectroscopy analysis, Master's thesis, Jiangsu University, Zhenjiang, China, 2010.
Sun, S.Y., Zhang, G.Y., and Wang, L., Study on the relations of nutrients contents determined by different methods for typical soil samples of Hebei province, *Acta Agric. Boreali-Sin.*, 21, 51–53, 2006.
Tang, Y.L., Wang, R.C., and Zhang, J.H., Study on determining nitrogenous levels of barley by hyperspectral and chlorophyll meter, *J. Triticeae Crops*, 23(1), 63–66, 2003.
Tu, C.Q., and Wen, X.R., Determination of nitric nitrogen in soil extracts by dual wavelength spectrophotometric method, *Soil Fert. Sci. China*, 1, 50–52, 2006.
Uddin, M., Ishizaki, S., Okazaki, E., and Tanaka, M., Near-infrared reflectance spectroscopy for determining end-point temperature of heated fish and shellfish meats, *J. Sci. Food Agric.*, 82, 286–292, 2002.
Uddin M., and Okazaki E., Classification of fresh and frozen-thawed fish by near-infrared spectroscopy, *J. Food Sci.*, 69(8), 665–668, 2004.
Wang, H.X., Zhao, D.Y., Ma, H.Y., Niu, G.C., Lv, D.G., and Qin, S.J., Preliminary study on nitrogen metabolism in sweet cherry during flower bud differentiation period, *Acta Agric. Boreali-Sin.*, 24, 201–204, 2009.
Wu, Y.Y., Li, P.P., Liang, Z., Yang, X.Y., Mao, H.P., and Huang, D.K., Acquiring the dynamic growth and development information on *Atractylodes lancea*, in *BMEI'09*, 2009, pp. 1676–1679.

Xiong, G.Y., Niu, D.B., and Chen, F., Study on the correlation of ASI and China routine method for soil available P, K and N determination, *China Soils Fert.*, 3, 73–76, 2007.

Xue, L.H., Cao, W.X., Luo, W.H., Jiang, D., Meng, Y.L., and Zhu, Y., Diagnosis of nitrogen status in rice leaves with the canopy spectral reflectance, *Sci. Agric. Sin.*, 36(7), 807–812, 2003.

Yue, X.J., Cao, X.J., Pu, S.S., and Lv, T.R., Preliminary research of identifying the property of nerve fiber by near-infrared spectroscopy and cluster analysis, *Acta Univ. Med. Nanjing (Nat. Sci.)*, 26(5), 313–317, 2006.

Zhang, J.H., and Wang, K., New vegetation index for estimating nitrogen concentration using fresh leaf spectral reflectance, *Trans. Chin. Soc. Agric. Eng.*, 24(3), 158–161, 2008.

Zhao, H.H., and Yan, Y.L., The effects of noise on NIR analysis and related mathematic pretreatments and models, *Spectrosc. Spec. Anal.*, 26(5), 842–845, 2006.

Zheng, C.S., Oba, S., Matsui, S., and Hara, T., Effects of calcium and magnesium treatments on growth, nutrient contents, ethylene production, and gibberellin content in *Chrysanthemum* plants, *J. Jpn. Soc. Hortic. Sci.*, 74(2), 144–149, 2005.

Zheng, Y.M., Zhang, T.Q., Zhang, J., Chen, X.D., and Shen, X.G., Influence of smooth, 1st derivative and baseline correction on the near-infrared spectrum analysis with PLS, *Spectrosc. Spec. Anal.*, 24(12), 1546–1548, 2004.

chapter six

Noninvasive measurement of in vitro growth of plantlets by image analysis

Yanyou Wu and Kaiyan Zhang

Contents

6.1 Introduction

Plant tissue culture is a basic technology in current biotechnology. Plant tissue culture techniques are widely applied in the areas of gardening, agriculture, and forestry. Plant tissue culture rapidly propagates plants under controlled conditions, such as temperature, light, moisture, nutrition, and hormones. Compared with pot culture and field cultivation, plant tissue culture has numerous advantages, such as eliminating the need for weeding, watering, spreading manure, and performing pest control. This technique can conserve numerous resources, including human energy expenditure and vast amounts of land and materials required for large-scale planting. Tissue culture methods are applied in commercial micropropagation, virus-free plantlets production and preservation of germplasm resources, genetic breeding, organelle transplantation, genetic engineering, and protein engineering. Plant tissue culture is the development direction of agricultural industrialized seedling, which will continue to have an increasingly important function in combating the environmental problems and fuel and food shortages of the twenty-first century, which will worsen with increasing population.

During plant tissue culture, we need to maintain the appropriate temperature, light, moisture, nutrition, and hormone conditions for plant growth and development. Plantlet growth is highly dependent on artificial culture conditions and nutrient supply. The culture and nutrition conditions can directly affect the efficiency of plantlet biomass production.

Sterile monitoring of the plantlet growth is a key factor during large-scale production of in vitro plantlets. To control plantlet growth and development in vitro, dynamic monitoring was employed during plantlet growth, including rooting, photoautotrophic ability, and nutrition consumption. Dynamic monitoring can lead to a greater understanding of the optimal nutrition and environmental conditions under different growth phases. Controlling plantlet development in vitro will greatly reduce the costs, improve production planning, as well as intensify and automate industrial propagation. All of these factors suggest that controlling plantlets' in vitro growth will greatly improve the efficiency of industrial propagation practices.

To obtain plantlet biomass, traditional methods were employed, including contact measurement or drainage. These methods are simple and precise, but have many inherent shortcomings. First, contact measurement requires plantlet handling, which disrupts the sterile environment and can lead to contamination. Second, the growth environment for each individual plantlet is disturbed; therefore, information about the growth of a group of plantlets can be obtained only once, as opposed to the acquisition of dynamic growth information about individual plantlets. Finally, the results acquired by these traditional methods cannot be validated or permanently preserved and cannot be applied online. Thus, the development of a method that includes noncontact, sterile measurements of plantlets in vitro is essential. Digital image processing techniques are computer-based methods used to reduce the noise in an image while improving image quality and recovering, dividing, and extracting the features of an image. With increasing reliance on computer technologies and calculation abilities in different fields, image processing techniques have also rapidly advanced. Image processing technology is widely applied in biological research and agricultural production (Wang et al., 2003; Li et al., 2003; Brosnan and Sun, 2004; Chen et al., 2005; Rodrigues and Fernandes, 2007). In plant tissue culture, image processing techniques are used to evaluate plantlets' growth and to monitor the effects of stressors on plantlets' growth. Using image processing technology, in vitro growth and development of the plantlets can be monitored aseptically. Smith and his colleagues used image processing technology for sterile monitoring of in vitro growth and development of plantlets, with salt stress (Smith et al., 1989). In a study by Olofsdotter (1993), image processing was used to measure the growth of carrot callus in a sterile manner. In other applications, Ibaraki and Kenji (2001) used image processing technology to evaluate cells cultured in suspension, whereas Supalkova and his colleagues (2007) used image processing to monitor the development of a spruce somatic embryoid growing in a heavy metal-polluted environment. Image processing technology has also been employed to monitor the genetic expression of an in vitro culture cultured in a Petri dish (Buenrostro-Nava et al., 2005). Zhao and his colleagues (2005) used image processing to measure the biomass of plantlets in vitro and to monitor the growth conditions of plantlets. However, this method has low precision, and these techniques are limited by the plantlet itself. At present, reports on the sterile dynamic measurement of in vitro plantlet growth are meager. In this chapter, we discuss different image analysis techniques for noninvasive measurements of plantlets' growth in vitro. We analyze the dynamic growth of different types of plantlets cultured in a medium containing different combinations of growth hormones to provide a knowledge base for the methods applicable to noninvasive measurement of plantlets in vitro. The information provided in this chapter has great significance for the large-scale production of plantlets.

6.2 Sterile measurement of plantlets in vitro by labeling

6.2.1 Plant materials and culture conditions

Orychophragmus violaceus and *Atractylodes lancea* plantlets in vitro were provided by the Institute of Agricultural Engineering of Jiangsu University. *O. violaceus* plantlets were cultured in a Murashige and Skoog (MS) medium (Murashige and Skoog, 1962) with 6.0 g · L^{-1} of agar and a sucrose concentration of 30 g · L^{-1}. We used a triangular 120 ml culture bottle with more than 85% light transmittance for tissue culture. The CO_2 concentration inside the growth chamber was adjusted to 360 μmol · mol^{-1}, and the relative humidity was maintained at 45%. The cultures were maintained at a 14 h photoperiod using a common fluorescent lamp with photosynthetic photon flux density of 147 μmol · m^{-2} · s^{-1} at 25 ± 0.5°C. The pH was adjusted to 5.8 before autoclaving.

6.2.2 Sterile image analysis of plantlets

6.2.2.1 The reference set

We chose appropriately sized *O. violaceus* plantlets as the experimental subjects. A uniform specification label was pasted in the middle of each

Figure 6.1 Label as the standard reference object.

culture bottle. The label was used as a standardized reference. It was difficult to maintain the same focal distance when each image was acquired, resulting in possible disturbance of pixel points. Using the same label size, we acquired the standard pixel points and eliminated the image distortion because of distance (Figure 6.1).

6.2.2.2 *Image acquisition*

In total, 60 bottles containing plantlets were selected, and identical labels were pasted on each bottle according to the plantlet size, from the largest to smallest. The plantlet was placed in a fixed position and a picture was taken using a digital camera. To analyze in vitro plantlet growth, we took three pictures per bottle at an angle of 60° between the bottle and the camera, keeping the same object distance and focal distance.

6.2.2.3 *Correlation analysis between relative pixel number and biomass/volume of* O. violaceus *plantlets in vitro*

A digital image is composed of pixels. Thus, pixel deviation caused by the object distance and the focal distance, while taking a picture, was

Table 6.1 Correlation between Relative Pixels and Biomass/Volume of *O. violaceus* Plantlets In Vitro

		Pixels	Weight (g)	Volume (ml)
Pixels	Pearson correlation	1.000	0.695(**)	0.709(**)
	Significance (two-tailed)	—	0.000	0.000
	N	111	111	111
Weight (g)	Pearson correlation	0.695(**)	1.000	0.992(**)
	Significance (two-tailed)	0.000	—	0.000
	N	111	111	111
Volume (ml)	Pearson correlation	0.709(**)	0.992(**)	1.000
	Significance (two-tailed)	0.000	0.000	—
	N	111	111	111

** Significant at the level of 0.01 (two-tailed).

considered for using the pixels of the label as a fixed reference. Using SPSS statistical software, the correlation between the number of relative pixels and the biomass/volume of *O. violaceus* plantlets in vitro was determined (Table 6.1).

The relationship between pixel number and biomass of *O. violaceus* plantlets in vitro is shown in Figures 6.2 and 6.3. This relationship was fitted with a linear equation, quadratic equation, composite quadratic

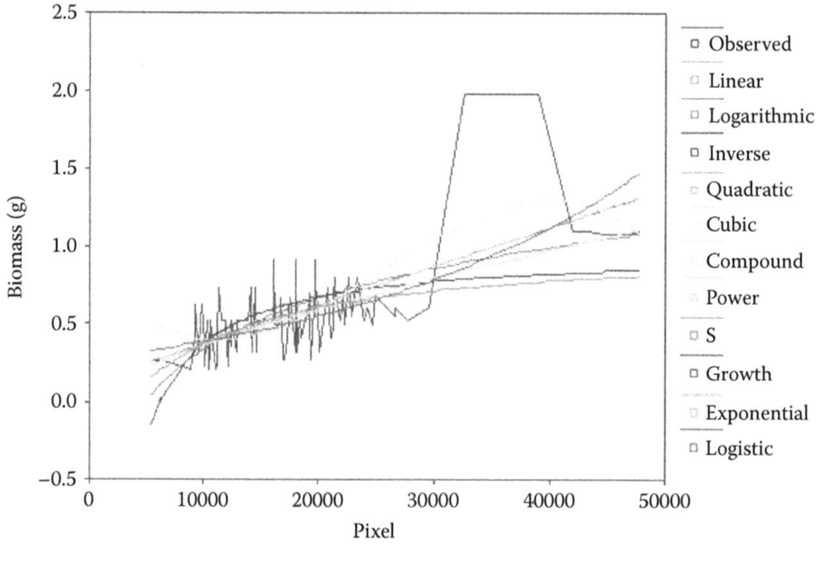

Figure 6.2 **(See color insert.)** Simulation on the relationship between pixel and biomass of *O. violaceus* plantlets in vitro.

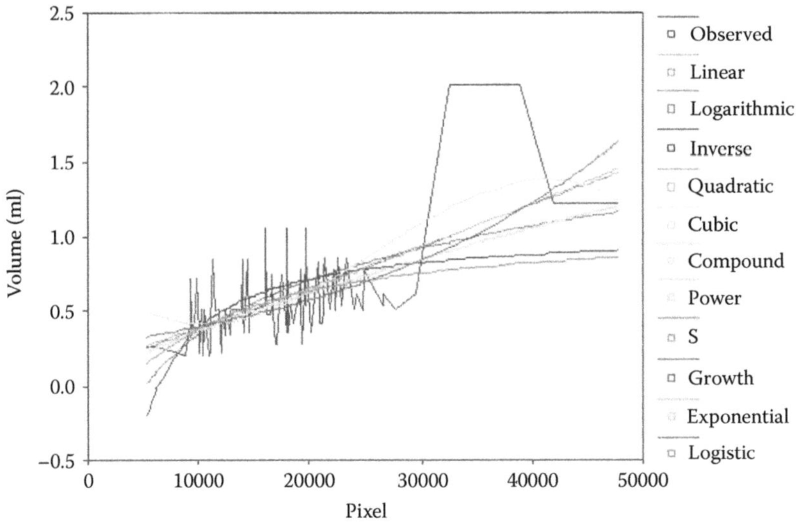

Figure 6.3 **(See color insert.)** Simulation on the relationship between pixel and volume of *O. violaceus* plantlets in vitro.

equation, logarithmic equation, cubic equation, S-type equation, index equation, reciprocal equation, power equation, and logistic equation.

The figures show that when plantlet biomass was measured between 0.4 and 1.0 g, no difference was detected between any of the equations given that the points were very dense. When the biomass was <0.4 g, the plantlet volume was small and enabled the acquisition of a limited number of pixels. The difference between the plantlet and the reference label led to large errors while fitting a biomass equation. When plantlet biomass was >1 g, there were large differences between the prediction models. *O. violaceus* belongs to the family Cruciferae; this plant has a relatively small volume. The presence of a sample with a biomass >1 g is rare, accounting for the larger differences between models as biomass increases above 1 g. The cubic equation fits well with the observed measurements of *A. lancea* plantlets in vitro; thus, it was acceptable to use that curve to fit the data. The corresponding polynomial equation is as follows:

$$Y = 0.8708 - 10^{-4} X + 5.5 \times 10^{-9} X^2 - 7 \times 10^{-14} X^3 \quad (6.1)$$

where Y denotes biomass and X denotes pixels.

6.2.2.4 Correlation analysis between relative pixel number and biomass/volume of A. lancea *plantlets in vitro*

We analyzed the relationship between the relative pixel number and biomass/volume of *A. lancea* plantlets in vitro. We fitted the biomass and volume data to linear, quadratic, composite quadratic, logarithmic, cubic, S-type, index, reciprocal, power, and logistic equations, following which we chose the most appropriate model on the basis of the standard error and R^2 values.

The results from fitting the pixel data obtained from images of *A. lancea* plantlets in vitro to the measured biomass/volume of plantlets are shown in Table 6.2. As shown in Table 6.2, the linear and cubic equations are the best prediction models. Each prediction model fit different ranges of plantlet biomass and volume. When plantlets ranged between 0 and 2.0 g, only the linear equation provided a good fit; when the biomass was <1 g, no predicting model fit the plantlet data well. The linear equation provided a good fit to plantlet biomass, and the cubic equation provided a good fit to plantlet volume when the biomass was between 0.4 and 2.0 g. When the plantlet biomass was between 0.5 and 1.5 g, the linear and cubic equations fit well with plantlet volume and biomass. However, the cubic equation was more beneficial for fitting volume than for fitting biomass. When plantlet biomass was between 0.6 and 1.5 g, the linear equation was the best model for both biomass and volume data. These equations are as follows:

$$M = 0.52 + 0.45P \ (R^2 = 0.78, n = 55, p < 0.01)$$

and

$$V = 0.58 + 0.96P \ (R^2 = 0.89, n = 55, p < 0.01)$$

We can thus determine the number of pixels associated with plantlets in vitro in a sterile environment. The relationship between relative pixel number and biomass/volume enables the creation of appropriate models to fit the associated biomass or volume data. Different types of plantlets cultured in vitro fit different predicting models. The cubic equation fit well with the values observed from *O. violaceus* plantlets in vitro, whereas the linear and cubic equations fit well with the values observed from *A. lancea* plantlets in vitro. The linear equation was the best model to fit the biomass and volume data obtained from *A. lancea* plantlets that had a biomass between 0.6 and 1.5 g, with associated R^2 values of 0.92 and 0.89, respectively. Overall, we found that image processing technology is applicable to monitoring of plantlets in vitro during rapid growth periods. When the labeling method was used, measurements of small plantlets in vitro under sterile conditions provided imprecise results. Therefore, new methods must be developed to meet this shortcoming.

Table 6.2 Developed Models between Relative Pixel and Volume/Biomass of *A. lancea* Plantlets In Vitro

Mass range	Model	S.E.	*F* value	DF	R^2
0–2.0 g	$M = 0.19 + 0.75P$ S.E.(0.05)*** (0.05)***	0.29***	195.61	118	0.62
	$M = 0.55 - 1.27P + 2.93P^2 - 1.11P^3$ S.E.(0.19) **(0.71) (0.80)*** (0.27)***	0.27***	85.18	116	0.69
	$V = 0.09 + 1.39P$ S.E.(0.08) (0.08)***	0.45***	277.28	118	0.70
	$V = 0.90 - 2.62P + 5.34P^2 - 1.93P^3$ S.E.(0.29)* (1.10) (1.24)*** (0.42)***	0.41***	117.69	116	0.75
Less than 0.5 g	$M = 0.27 + 0.25P$ S.E.(0.04)*** (0.72)*	0.08*	11.62	37	0.24
	$M = -0.42 + 5.37P - 11.05P^2 + 7.29P^3$ S.E.(0.19) (1.40)*(3.19)* (2.26)*	0.07**	10.35	35	0.47
	No significant linear model between pixel and volume				
	$V = -1.00 + 10.32P - 21.47P^2 + 14.10P^3$ S.E.(0.29)* (2.17)***(4.96)** (3.50)**	0.10***	11.98	35	0.51
Less than 1.0 g	No significant linear or cubic model between pixel and biomass				
0.4–2.0 g	$M = 0.27 + 0.70P$ S.E.(0.07)*** (0.06)***	0.30***	125.49	110	0.56
	$M = 1.17 - 3.11P + 4.60P^2 - 1.58P^3$ S.E.(0.32) ***(1.09) (1.12)*** (0.35)***	0.27***	59.67	98	0.65
	$V = 0.16 + 1.34P$ S.E.(0.10) (0.10)***	0.48***	186.57	110	0.65
	$V = 1.96 - 5.83P + 8.29P^2 - 2.76P^3$ S.E.(0.49)*** (1.68)*** (1.73)*** (0.55)***	0.42***	87.80	98	0.73
0.5–1.5 g	$M = 0.43 + 0.49P$ S.E.(0.04)*** (0.04)***	0.17***	152.20	67	0.69
	$M = 1.31 - 2.46P + 2.85P^2 - 0.83P^3$ S.E.(0.21) ***(0.71)*** (0.72)*** (0.22)***	0.15***	68.42	65	0.76
	$V = 0.39 + 1.04P$ S.E.(0.08)*** (0.07)***	0.29***	225.19	67	0.77
	$V = 2.19 - 4.85P + 5.58P^2 - 1.60P^3$ S.E.(0.36)*** (1.18)*** (1.20)*** (0.37)***	0.25***	110.65	65	0.84

Continued

Table 6.2 (Continued) Developed Models between Relative Pixel and Volume/Biomass of *A. lancea* Plantlets In Vitro

Mass range	Model	S.E.	*F* value	DF	R^2
0.6–1.5 g	$M = 0.52 + 0.45P$ S.E.(0.03)*** (0.03)***	0.13***	191.08	55	0.78
	$M = 0.93 - 1.17P + 1.84P^2 - 0.61P^3$ S.E.(0.18) **(0.61) (0.61)* (0.19)*	0.12***	78.21	53	0.82
	$V = 0.58 + 0.96P$ S.E.(0.05)*** (0.05)***	0.19***	435.34	55	0.89
	$V = 1.43 - 2.27P + 3.57P^2 - 1.15P^3$ S.E.(0.24)*** (0.80) (0.80)*** (0.25)***	0.16***	206.94	53	0.92

Note: *P*, the relative pixel of plantlets; *M*, plantlet biomass (g); *V*, plantlet volume (ml); S.E., standard error; DF, degree of freedom.

* Significant at 0.5%.
** Significant at 0.1%.
*** Significant at 0.01%.

6.3 *Sterile measurement of plantlets in vitro by single-reference and double-reference objects*

6.3.1 *Materials and methods*

Appropriate *O. violaceus* plantlets in vitro without roots were selected as the reference objects for the present image analysis study. Images were compared using a single-reference object and double-reference objects, and the results are shown in Figures 6.4 and 6.5.

A moderately sized in vitro plantlet was selected for single-reference images, while one large and one small in vitro plantlet were selected for double-reference images.

The method employed for using a single-reference object (Figure 6.4) involved the adjacent placement of two plantlets, followed by image acquisition; the biomass of one plantlet was known and used as a standard for comparison with the unknown biomass. For double-reference objects, a plantlet with unknown biomass was placed adjacent to two different plantlets with known biomasses used as double-reference objects, following which the image was acquired. The number of pixels associated with each plantlet grown in vitro was obtained with SigmaScan Pro, an image analysis software. Plantlet biomass was calculated by the relationship between the relative pixel number of plantlets of unknown biomass and the pixel number of the reference plantlets and the biomass of the reference plantlets. The equation for this relationship is as follows: biomass of plantlet by image analysis = biomass of the reference plantlet × number of pixels of the tested plantlet/number of pixels of the reference plantlet.

Figure 6.4 Single-reference object.

Figure 6.5 Double-reference object.

6.3.2 Correlation between relative pixel number and biomass of a plantlet grown in vitro

The number of pixels and the biomass of a plantlet grown in vitro were obtained by analysis of the image of the plantlet in comparison with the single- and double-reference objects using SigmaScan Pro. The correlation between the pixel number and actual/relative biomass of plantlets in vitro was determined using the statistical software SPSS. The results are shown in Table 6.3.

The correlation coefficient between the number of image pixels associated with a plantlet and its actual biomass reached 0.971 and was significant (Table 6.3). The correlation coefficient between the number of plantlet pixels and the plantlet relative biomass reached 0.958, which was also significant.

We performed univariate regression analysis between the plantlet mass obtained through image analysis and the actual mass of the plantlet grown in vitro using the single-reference object. The linear regression equation was as follows (Figure 6.6a):

$$y = 0.9488x - 0.2327 \ (R^2 = 0.7213, n = 20, p < 0.001)$$

A moderately sized in vitro plantlet was selected for single-reference images, and one large and one small in vitro plantlet were selected for double-reference images. Univariate regression analysis was then performed between the plantlet mass calculated by image analysis and the actual mass of the plantlet grown in vitro when double-reference objects were used. The linear regression equation was as follows (Figure 6.6b):

$$y = 0.9775x + 0.0111 \ (R^2 = 0.9108, n = 29, p < 0.001)$$

Table 6.3 Correlation between Pixel and Actual/Relative Biomass of Plantlets In Vitro

		Pixel	Actual biomass/g	Relative biomass/g
Pixel	Correlation	1.000	0.971**	0.958**
	Significance (two-tailed test)	0.000	0.000	0.000
	Samples	29	29	29
Actual biomass (g)	Correlation	0.971**	1.000	0.993**
	Significance (two-tailed test)	0.000	0.000	0.000
	Samples	29	29	29
Relative biomass (g)	Correlation	0.958**	0.993**	1.000
	Significance (two-tailed test)	0.000	0.000	0.000
	Samples	29	29	29

** Significant at 0.01 (two-tailed test).

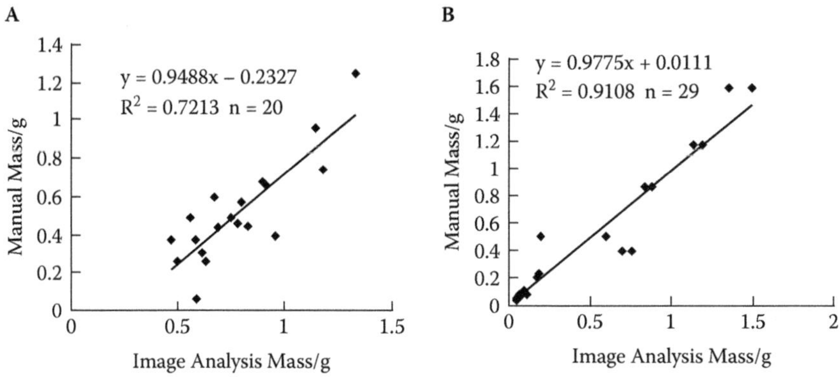

Figure 6.6 The correlation of the biomass derived from image analysis (y) to the manual biomass (x) of plantlet in vitro at (a) single-reference object and (b) double-reference object.

Both the methods resulted in linear correlations (Figure 6.6a and b). The correlation coefficient associated with the double-reference object method was higher than that associated with the single-reference object method. When the double-reference object method was used, the predicted values were closer to the actual values. Thus, the predictions of the double-reference object method were more precise than those of the single-reference object method. The use of a single-reference object cannot account for different growth periods of a plantlet, and this leads to large errors during image analysis because the image sharpness of a plantlet grown in vitro differs during different growth periods. However, these shortcomings will be overcome and the precision of image processing techniques will be improved using the double-reference objects. Thus, the use of double-reference objects enables appropriate sterile measurements of *O. violaceus* plantlets in vitro.

6.3.3 Dynamic monitoring of plantlets in vitro when different combinations of hormones are used in culture medium

6.3.3.1 Dynamic monitoring using double-reference objects

Dynamic and continuous monitoring of plantlets in vitro was performed when the plantlets were exposed to different hormone conditions. Using the double-reference object method, four groups that received different hormone combinations were subjected to image analysis. Based on the images, plantlet biomass was calculated. SPSS was then used to construct a curve relating the calculated biomass of the plantlet to the growth time. The hormone combinations are shown in Table 6.4.

Table 6.4 Hormone Combinations in the Culture Medium Used for *O. violaceus* Plantlets

Hormone combination	NAA concentration (mg/L)	6-BA concentration (mg/L)
R_1	0.02	0.4
R_2	0.1	0.4
R_3	0.1	2
R_4	0.5	10

Note: NAA, naphthalene acetic acid; 6-BA, 6-benzylaminopurine.

A logistic equation that fit these curves well was used to fit curves for the growth and development of plantlets in vitro. An indicial equation relating image analysis mass (Y) to time (X) was obtained: $Y = 1/[1/a + d(b^X)]$, where a represents the maximum mass, b represents the index growth coefficient, and d represents the initial mass when the plantlet entered the logarithmic growth phase. The growth coefficient reflects the growth rate of the plantlet during the logarithmic phase. The indicial equation corresponds to the curve shown in Figure 6.3. Second-order differentiation was performed on the curve $Y = 1/[1/a + d(b^X)]$ to obtain the second derivative:

$$Y'' = \frac{50(\ln b)^2 b^{2X} - 5(\ln b)^2 b^X (a + 5b^X)}{(a + 5b^X)^3} \tag{6.2}$$

where $Y'' = 0$ and

$$X = \log_b \frac{a}{d} \tag{6.3}$$

The calculated X value is substituted in the index equation $Y = 1/[1/a + d(b^X)]$. The Y value is as follows:

$$Y = \frac{a}{a^2 + 1} \tag{6.4}$$

The knee point is

$$\left(\log_b \frac{a}{d}, \frac{a}{a^2 + 1}\right)$$

The model relating plantlet mass to time during the logarithmic growth phase was represented by this curve. The duration of the plantlet logarithmic growth phase was obtained from the X value.

6.3.3.2 *Comparison of simulation models of* O. violaceus *plantlets in vitro using different hormone combinations*

We obtained four groups of logistic equations using SPSS to simulate plantlet growth with different hormone combinations. The simulation equations are as follows:

$$R_1: Y = 1/[1/0.22 + 1.648\ (0.854^X)]\ (R^2 = 0.909, n = 7, p < 0.001)$$

$$R_2: Y = 1/[1/1.28 + 4.684\ (0.913^X)]\ (R^2 = 0.983, n = 7, p < 0.001)$$

$$R_3: Y = 1/[1/0.91 + 12.148\ (0.815^X)]\ (R^2 = 0.982, n = 7, p < 0.001)$$

$$R_4: Y = 1/[1/0.487 + 5.064\ (0.726^X)]\ (R^2 = 0.948, n = 7, p < 0.001)$$

a and b values can be obtained from the above-mentioned equations:

$$R_1: a = 0.220, b = 0.854$$

$$R_2: a = 1.280, b = 0.913$$

$$R_3: a = 0.910, b = 0.815$$

$$R_4: a = 0.487, b = 0.726$$

As shown by the above-mentioned equations, each simulation model exhibited a high R^2 value, indicating that the logistic equations provided a good fit. However, differences were observed in the simulation models for different hormone combinations. The simulation models for hormone combinations R_2 and R_3 were the best; however, the simulation models for hormone combinations R_1 and R_4 were not as good, although all the models reached significance. The growth processes of these in vitro plantlets appeared to follow S shapes, indicating that the plantlets followed a slow-rapid-slow growth pattern (Figure 6.7a–d). Logistic equations were appropriate for simulating this plantlet growth process.

6.3.4 *The effects of different hormone combinations on* O. violaceus *plantlets in vitro*

6.3.4.1 *Hormone combinations of the medium for* O. violaceus *plantlets in vitro*

The medium of *O. violaceus* plantlets was supplemented with different concentrations of the hormones naphthalene acetic acid (NAA) and 6-benzyladenine (6-BA) in the hormone combinations used in the

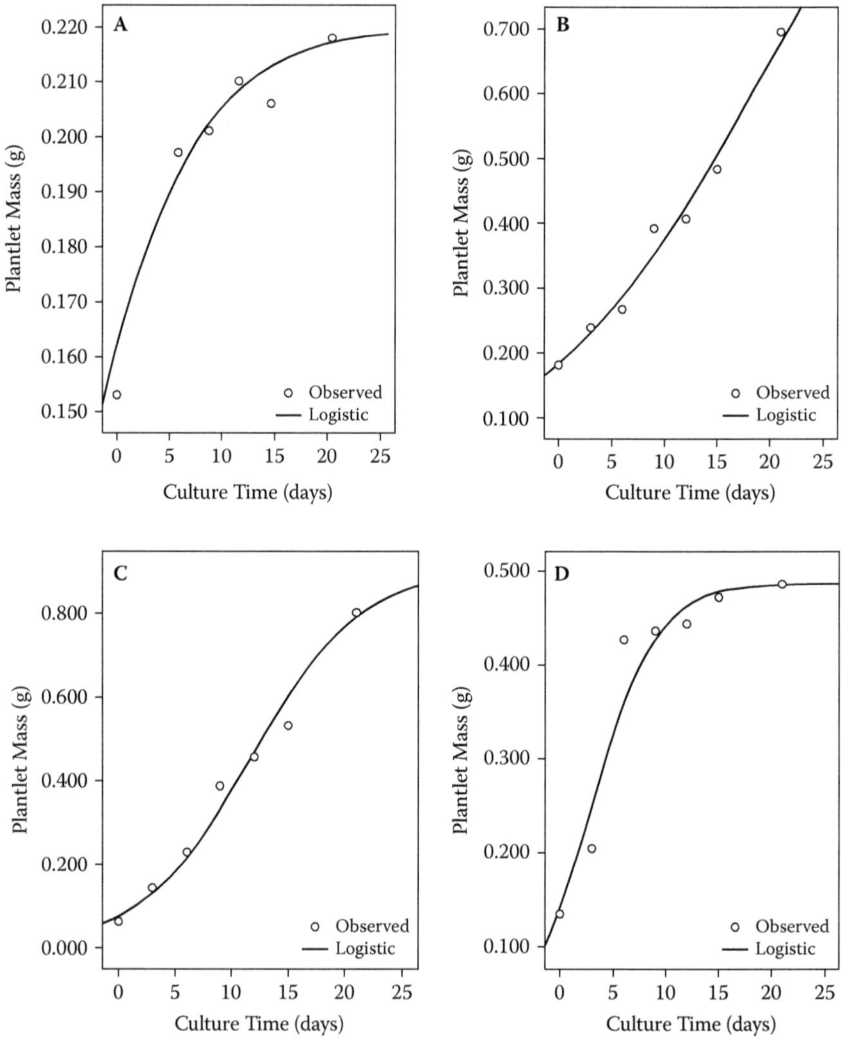

Figure 6.7 Logistic simulation equation of *O. violaceus* biomass under (a) R_1, (b) R_2, (c) R_3, and (d) R_4 hormone combinations.

Table 6.5 Combinations of NAA and 6-BA for the Media a–i

Hormone	a	b	c	d	e	f	g	h	i
NAA (mg/L)	0.10	0.10	0.10	0.02	0.02	0.02	0.50	0.50	0.50
6-BA (mg/L)	0.40	2.00	10.0	0.40	2.00	10.0	0.40	2.00	10.0

present experiment. The effects of different hormone concentrations on *O. violaceus* plantlets were examined. An experiment using two factors and three levels was designed to obtain a group of nine different hormone-level combinations. The hormone combinations are shown in Table 6.5.

6.3.4.2 *Dynamic models of* O. violaceus *plantlet growth under different hormone concentrations*

Dynamic and continuous monitoring of *O. violaceus* plantlets under different hormone concentrations was performed using image analysis technology. We obtained curves representing the relationship between plantlet mass calculated by image analysis and time. Logistic equations with three parameters were used to simulate plantlet growth in terms of mass. The results are shown in Table 6.6, while the model curves are shown in Figure 6.8 (a–i).

The R^2 values for the regression of image analysis mass on the growth time of *O. violaceus* reached 0.911–0.991, with all equations having $p < 0.0001$ (Table 6.6). Thus, all the correlations between these factors were significant. The logistic equations provided satisfactory curve fitting.

Table 6.6 Simulated Logistic Equation of Plantlet Biomass

Sample number	*p* value	Degree of simulation	Model	*n* value
1	<0.0001	$R^2 = 0.982$	$Y = 1/(1/0.91 + 12.148(0.815^X))$	7
2	<0.0001	$R^2 = 0.983$	$Y = 1/(1/1.28 + 4.684(0.913^X))$	7
3	<0.0001	$R^2 = 0.911$	$Y = 1/(1/0.31 + 6.568(0.792^X))$	7
4	<0.0001	$R^2 = 0.964$	$Y = 1/(1/1.10 + 11.150(0.800^X))$	7
5	<0.0001	$R^2 = 0.971$	$Y = 1/(1/1.56 + 8.335(0.824^X))$	7
6	<0.0001	$R^2 = 0.909$	$Y = 1/(1/0.22 + 1.648(0.854^X))$	7
7	<0.0001	$R^2 = 0.984$	$Y = 1/(1/6.70 + 6.458(0.905^X))$	7
8	<0.0001	$R^2 = 0.991$	$Y = 1/(1/0.98 + 12.735(0.823^X))$	7
9	<0.0001	$R^2 = 0.948$	$Y = 1/(1/0.487 + 5.064(0.726^X))$	7

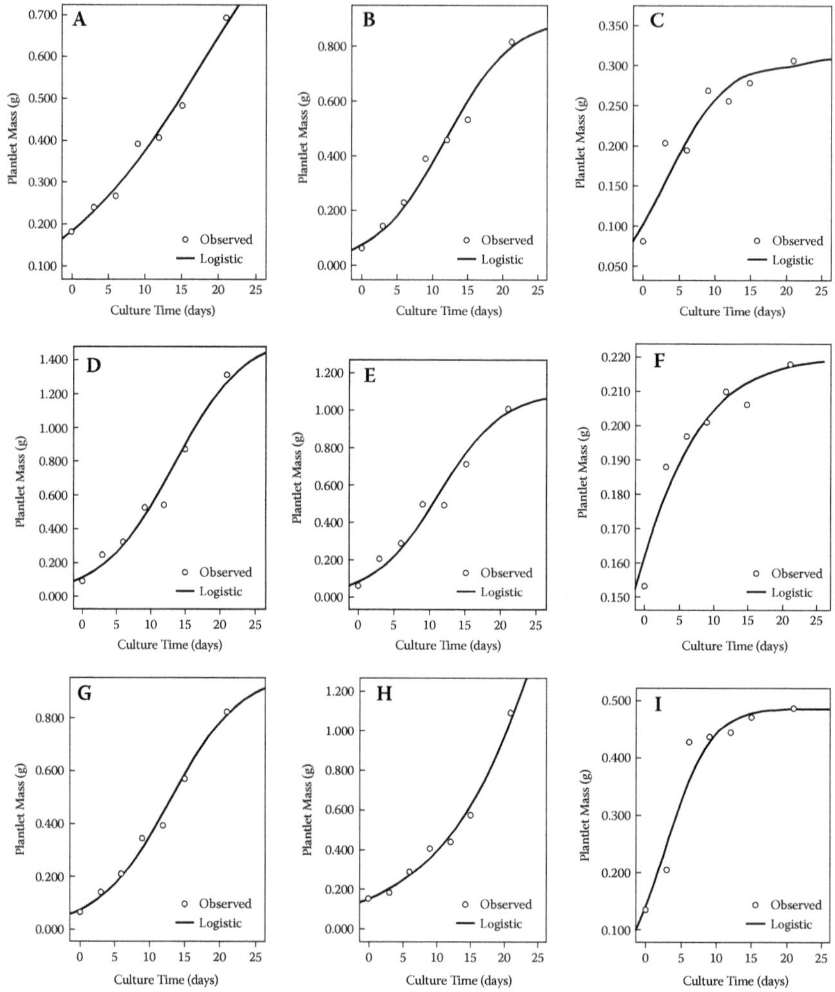

Figure 6.8 Curve estimation on growth of plantlets in vitro. a–b–c–d–e–f–g–h–i correspond to the hormone combinations from Table 6.5.

6.3.4.3 *The effect of different hormone combinations on the biomass increment of* O. violaceus *plantlets during the logarithmic growth phase*

The growth rate and maximum biomass of *O. violaceus* plantlets was estimated in vitro during the logarithmic growth phase using predictive models. The models provided the exponential growth coefficients, maximum increments, and durations of the logarithmic phase of *O. violaceus* plantlet growth at the nine different hormone combinations used. The results are shown in Table 6.7.

Table 6.7 Influence of Different Hormone Combinations on the Logarithmic Growth of *O. violaceus* Plantlets In Vitro

Sample number	Maximum increment *a* (g)	Logarithmic phase GR_{50} (day)	Exponential growth coefficient
1	0.91	11.74	0.815
2	1.28	19.66	0.913
3	0.31	3.050	0.792
4	1.10	11.24	0.800
5	1.56	13.25	0.824
6	0.22	–5.028[a]	0.854
7	6.70	37.74	0.905
8	0.98	12.96	0.823
9	0.487	2.820	0.726

[a] The plantlet had been in the logarithmic growth period before inoculation; the day was set as 0 days.

Different combinations of NAA and 6-BA had different effects on the growth parameters of *O. violaceus* plantlets during the logarithmic growth phase (Table 6.7). When the NAA concentration was kept constant, different 6-BA concentrations had different effects on the growth of *O. violaceus* plantlets. The maximum growth rate of plantlets in vitro was 41.8% lower and the duration of the logarithmic phase was 2.01 days shorter when the 6-BA level was low than the values when the 6-BA level was moderate and the NAA level was low (0.02 mg · L^{-1}). When the 6-BA level was high, the maximum rate of plantlet growth was only 14.1% of the value when the 6-BA level was moderate. Thus, increasing the 6-BA concentration is beneficial to plantlet growth; however, excess 6-BA concentration has the opposite effect when NAA concentrations are maintained at a low level (0.02 mg · L^{-1}). When NAA concentrations were maintained at a moderate level (0.1 mg · L^{-1}), the application of extremely high or extremely low 6-BA concentrations did not promote plantlet growth. The maximum rate of plantlet growth was 42.2% less and the duration of the logarithmic growth phase was 7.92 days shorter when 6-BA concentration was low (0.4 mg · L^{-1}) than the values when 6-BA concentration was moderate (2 mg · L^{-1}). The maximum rate of plantlet growth was very small when 6-BA concentration was high (10 mg · L^{-1}), and it accounted for only 24.2% of the maximum increment when 6-BA concentration was moderate, and the duration of the logarithmic growth phase was 7.92 days shorter. Thus, increasing 6-BA concentration promotes plantlet growth to a certain degree; however, when NAA concentrations are maintained at a moderate level, excess 6-BA concentration has the opposite effect. The plantlet mass was the highest, reaching 6.7 g, and the duration of the logarithmic phase was the longest, reaching 37.74 days, when the 6-BA concentration was at its lowest and the NAA concentration was at its highest (0.5 mg · L^{-1}). With this hormone combination, the leaves of *O. violaceus* plantlets curled and the stem appeared thick and transparent. Thus, a high NAA concentration causes vitrification of the plantlet and abnormal plantlet growth. When the 6-BA concentration was moderate (2 mg · L^{-1}) or high (10 mg · L^{-1}), the plantlet mass was relatively small; however, no hyperhydricity was observed. Thus, low 6-BA levels can cause hyperhydricity and an increase in 6-BA concentration can potentially inhibit hyperhydricity. Furthermore, the application of moderate 6-BA levels provided better outcomes than the application of high 6-BA levels. When the concentration of 6-BA was kept constant, different NAA concentrations had similar effects on plantlets. The maximum rate of plantlet growth decreased when the NAA concentration increased; however, the duration of the logarithmic phase increased when the NAA concentration increased and 6-BA was maintained at low (0.4 mg · L^{-1}) and moderate (2 mg · L^{-1}) levels. The maximum rate of plantlet growth and the duration of the logarithmic growth phase were both higher under a moderate concentration of 6-BA than those under a low concentration. Thus, a moderate level of 6-BA promoted plantlet growth; however, high levels of both 6-BA and NAA did not have a similar effect.

O. violaceus plantlets showed continuous growth at a hormone combination of 0.1 mg · L^{-1} NAA and 2 mg · L^{-1} 6-BA and showed rapid growth at a hormone combination of 0.02 mg · L^{-1} NAA and 2 mg · L^{-1} 6-BA.

6.4 *Sterile measurement on biomass of* A. lancea *plantlets in vitro using the internal standard method of multiple references*

6.4.1 *The prediction of biomass for* A. lancea *plantlets in vitro*

The Institute of Agricultural Engineering, Jiangsu University, provided three types of *A. lancea* plantlets in vitro. Sterile measurement of plantlets in vitro using the double-reference object method has several shortcomings. When two plantlets are chosen as reference objects, their biomasses must be moderate, without large differences. The biomass and volume of a moderately sized plantlet can be precisely predicted using double-reference objects. However, the accuracy of the prediction suffers when predicting the biomass and volume of a relatively small or large plantlet. To overcome this drawback, we accordingly designed a reference system consisting of six to eight plantlets, ranging in size from small to large.

This internal standard method employed more reference objects in order to place the plantlet with unknown biomass near the middle of the

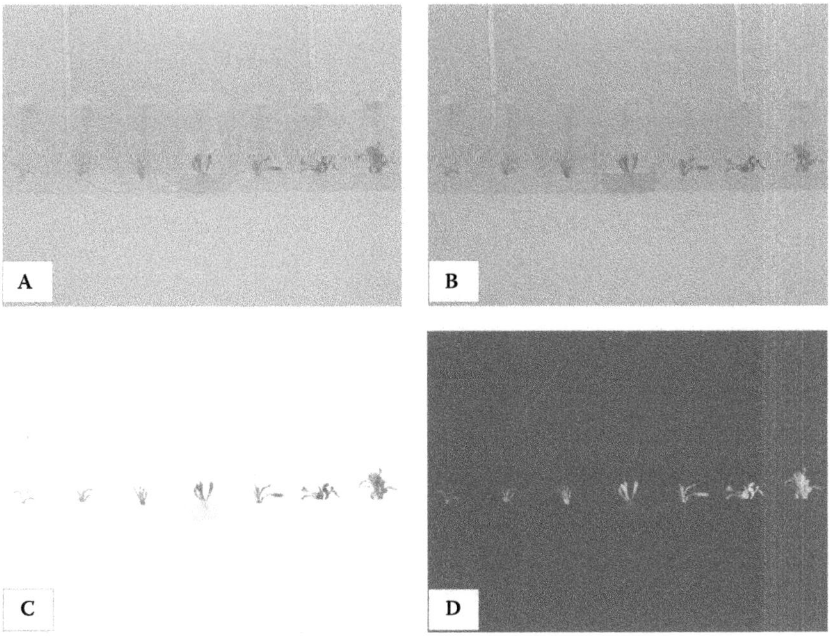

Figure 6.9 **(See color insert.)** Steps in the image analysis procedure: (a) original picture, (b) grayscale image, (c) image after brightness and contrast of (b) adjusted, and (d) image after the color (c) reversed.

size range of the standard, which had known biomass. Image processing involved the acquisition of a color image (Figure 6.9a), converting the image to a grayscale image (Figure 6.9b), adjusting the brightness and contrast of the grayscale image (Figure 6.9c), and inverting the image colors (Figure 6.9d). Pixel point statistics for each plantlet were then performed according to the inverted image.

Linear regression analysis relating the biomass and pixel data of the internal standards with known biomass was then performed, and we acquired a linear regression equation (Figure 6.10). Following this, the pixel information of the plantlet with unknown biomass was used to calculate the biomass of the unknown sample.

The linear relationship between pixel number and biomass of the internal standard plantlets was strong ($R^2 = 0.983$, $p < 0.001$) (Figure 6.10). Statistical analysis is shown in Table 6.8. Among all coefficients of variation of biomass, <5% accounted for 37.5%. The average coefficient of variation was only 8.2%. The accuracy of prediction using this model can meet the requirements for measurements on *A. lancea* plantlets in vitro under sterile conditions. In addition, this method is applicable to the monitoring of small plantlets in vitro.

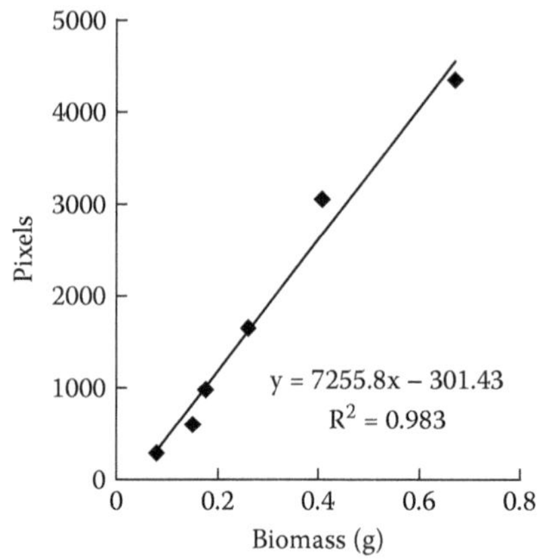

Figure 6.10 The relationship between biomass (*x*) and pixels (*y*) of plantlets in vitro.

Table 6.8 Biomass of *A. lancea* Plantlets from Eight Flasks Estimated via Image Processing (g)

	1	2	3	4	5	6	7	8
Average	0.215	0.124	0.752	0.168	0.343	0.211	0.718	0.317
Standard error	0.002	0.012	0.121	0.005	0.018	0.030	0.028	0.043
Coefficient of variation (%)	0.7	9.3	16.1	3.1	5.3	14.1	3.9	13.4

6.4.2 *Capturing dynamic growth information of* A. lancea *plantlets in vitro*

Using the internal standard method, we took pictures of *A. lancea* plantlets in vitro at 7-day intervals during their in vitro culture process. First, we determined the best-fit linear equation created using the internal standard plantlets. Following this, we calculated the biomass of each plantlet using image analysis at each time point. The biomass determined at each time point was considered to represent the dynamic growth of *A. lancea* plantlets in vitro. Logistic equations with four parameters were used to fit the calculated biomass at different culture times (Figure 6.11).

The logistic equation containing four parameters is as follows:

$$Y = Y_0 + \frac{a}{1+(X/X_0)^b} \tag{6.5}$$

where Y_0 is the initial biomass during logarithmic phase. Y_0 did not represent the starting biomass of the plantlet because of its subculture effect.

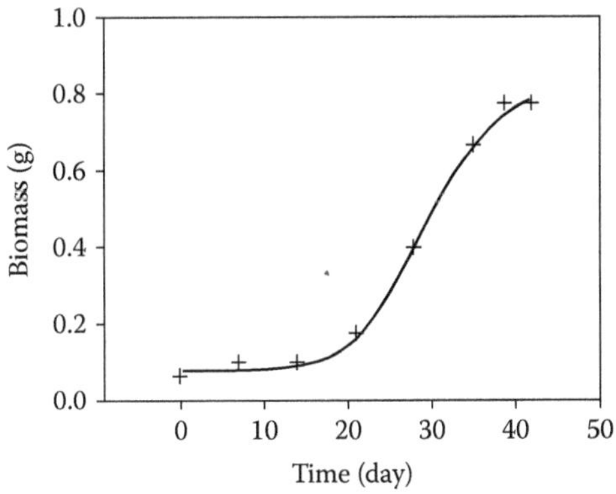

Figure 6.11 The changes in plantlets' biomass in vitro with the culture times.

α is the maximum biomass of the plantlet during one growth period, and X_0 is the required time (d) for the biomass to reach half of its maximum growth when the plantlet enters the logarithmic growth phase. b is the index in the fitting parameters. The growth rate at half of the logarithmic growth phase was marked as GR_{50} ($GR_{50} = -ab/4X_0$). The end of the logarithmic growth phase was represented by $DT_{\log}$ ($DT_{\log} = -4X_0/b$). The durative time of the growth starting stage is defined as $DT_s = X_0 + 2X_0/b$.

6.4.3 *Dynamic comparison of the growth of different types of* A. lancea *plantlets in vitro*

We selected three types of *A. lancea* plantlets, namely, incision type, long-needle type, and short-needle type, as subjects of the present experiment. These different types of plantlets were cultured using a medium supplemented with three different hormone combinations. The type of *A. lancea* plantlets used and their associated hormone combinations are listed in Table 6.9.

We used image analysis methods to obtain the biomass of *A. lancea* plantlets cultured in different hormone combinations. We used logistic equations with four parameters to fit the calculated biomass and obtained nine model parameters. We found that each model fitted the biomass data well, indicating that the biomass of tested plantlets was predicted using the logistic equations.

The logistic curve equations simulating the biomass of three types of *A. lancea* plantlets cultured in the S_1 hormone combination are as follows ($n = 8$):

The incision type:

Table 6.9 Types of *A. lancea* and Culture Media Used for Their Growth

Sample number	Type of *A. lancea* plantlets	Medium
1	Incision type	0.02 mg/L NAA + 0.4 mg/L 6-BA (S_1)
2	Short-needle type	0.02 mg/L NAA + 0.4 mg/L 6-BA (S_1)
3	Long-needle type	0.02 mg/LNAA + 0.4 mg/L 6-BA (S_1)
4	Incision type	0.10 mg/L NAA + 2.0 mg/L 6-BA (S_2)
5	Short-needle type	0.10 mg/L NAA + 2.0 mg/L 6-BA (S_2)
6	Long-needle type	0.10 mg/L NAA + 2.0 mg/L 6-BA (S_2)
7	Incision type	0.50 mg/LNAA + 10.0 mg/L 6-BA (S_3)
8	Short-needle type	0.50 mg/L NAA + 10.0 mg/L 6-BA (S_3)
9	Long-needle type	0.50 mg/LNAA + 10.0 mg/L 6-BA (S_3)

$$Y = 0.091 + \frac{0.627}{1 + (X/25.758)^{-9.042}},\ R^2 = 0.987\ (p < 0.0001)$$

The short-needle type:

$$Y = 0.0108 + \frac{0.572}{1 + (X/25.250)^{-9.440}},\ R^2 = 0.990\ (p = 0.0001)$$

The long-needle type:

$$Y = 0.055 + \frac{0.410}{1 + (X/25.957)^{-11.381}},\ R^2 = 0.984\ (p = 0.0001)$$

The corresponding growth parameter values were obtained from the above-mentioned equations:

The incision type: $a = 0.627$, $X_0 = 25.758$, $GR_{50} = 0.0550$, $DT_{log} = 11.395$, $DT_S = 20.061$

The short-needle type: $a = 0.572$, $X_0 = 25.250$, $GR_{50} = 0.0535$, $DT_{log} = 10.669$, $DT_S = 19.900$

The long-needle type: $a = 0.410$, $X_0 = 25.957$, $GR_{50} = 0.0449$, $DT_{log} = 9.123$, $DT_S = 21.395$

The logistic curve equations simulating the biomass of the three types of *A. lancea* plantlets cultured in the S_2 hormone combination are as follows ($n = 8$):

The incision type:

$$Y = 0.0357 + \frac{1.297}{1+(X/39.679)^{-3.153}},\ R^2 = 0.986\ (p = 0.0004)$$

The short-needle type:

$$Y = 0.0370 + \frac{0.679}{1+(X/34.703)^{-3.192}},\ R^2 = 0.985\ (p < 0.0001)$$

The long-needle type:

$$Y = 0.155 + \frac{0.710}{1+(X/14.444)^{-4.275}},\ R^2 = 0.991\ (p < 0.0001)$$

The corresponding growth parameter values were determined from the above-mentioned equations:

The incision type: $a = 1.297$, $X_0 = 39.679$, $GR_{50} = 0.0258$, $DT_{log} = 50.337$, $DT_S = 14.510$
The short-needle type: $a = 0.679$, $X_0 = 34.703$, $GR_{50} = 0.0156$, $DT_{log} = 43.491$, $DT_S = 12.958$
The long-needle type: $a = 0.710$, $X_0 = 14.444$, $GR_{50} = 0.0525$, $DT_{log} = 13.515$, $DT_S = 7.686$

The logistic curve equations simulating the biomass of three types of *A. lancea* plantlets cultured in the S_3 hormone combination are as follows ($n = 8$):

The incision type: $R^2 = 0.992$ ($p < 0.0001$)
The short-needle type: $R^2 = 0.998$ ($p < 0.0001$)
The long-needle type: $R^2 = 0.999$ ($p < 0.0001$)

The corresponding growth parameter values were determined from the above-mentioned equations:

The incision type: $a = 0.613$, $X_0 = 32.578$, $GR_{50} = 0.0163$, $DT_{log} = 37.572$, $DT_S = 13.792$

The short-needle type: $a = 0.381$, $X_0 = 23.722$, $GR_{50} = 0.0185$, $DT_{log} = 20.594$, $DT_S = 13.425$

The long-needle type: $a = 0.433$, $X_0 = 26.243$, $GR_{50} = 0.0271$, $DT_{log} = 16.016$, $DT_S = 18.235$

As the equations and parameters were obtained for different types of *A. lancea* plantlets cultured under different hormone combinations, *a*, GR_{50}, and DT_{log} values of the incision-type *A. lancea* plantlets were higher than the values of the other two types when plantlets were cultured in S_1 medium. Thus, when grown in a medium with low hormone concentrations, incision-type *A. lancea* plantlets show rapid and constant growth. GR_{50} of long-needle-type plantlets was higher than that of incision-type and short-needle-type plantlets when the plantlets were grown in S_2 and S_3 media. However, *a* follows the pattern *a* (incised) > *a* (long needle) > *a* (short needle), whereas DT_{log} follows the pattern DT_{log} (incised) > DT_{log} (short needle) > DT_{log} (long needle). The incised *A. lancea* plantlets show rapid and constant growth in S_2 and S_3 media. Thus, *a* and DT_{log} values of incised *A. lancea* plantlets were found to be higher than those of the other types of *A. lancea* plantlets when all types were grown in S_1, S_2, and S_3 media. *a* represents the maximum biomass during one growth period and reflects the growth capacity of the plantlet. DT_{log} is the duration of the logarithmic growth period and reflects the growth period of the plantlet. *A. lancea* plantlets had the greatest growth capacity and the longest duration of fast growth. Thus, the growth rate of incised *A. lancea* plantlets was the fastest among the three types of *A. lancea* plantlets studied.

The growth parameters of the three types of *A. lancea* plantlets cultured in different medium conditions show that $a\ (S_2) > a\ (S_1) > a\ (S_3)$, and $DT_{log}\ (S_2) > DT_{log}\ (S_3) > DT_{log}\ (S_1)$. These findings indicate that S_2 medium ($0.1\ mg \cdot L^{-1}$ NAA + $2\ mg \cdot L^{-1}$ 6-BA) was superior to S_1 and S_3 media. Thus, *A. lancea* plantlet growth during reproduction is best promoted by a hormone combination consisting of $0.1\ mg \cdot L^{-1}$ NAA + $2\ mg \cdot L^{-1}$ 6-BA.

Thus, we found that conducting sterile measurements of *A. lancea* plantlets using the internal standard method, which included multiple references, was extremely effective. The use of a combination of the internal standard method and a four-parameter logistic equation enabled the accurate prediction of the growth conditions of *A. lancea* plantlets in vitro.

6.5 *Sterile measurement on biomass of* O. violaceus *plantlets in vitro by the internal standard method of multiple references*

6.5.1 *Predicting the biomass of* O. violaceus *plantlets*

For the present experiment, a reference system consisting of six to eight internal standards for measurements of the biomass of *O. violaceus*

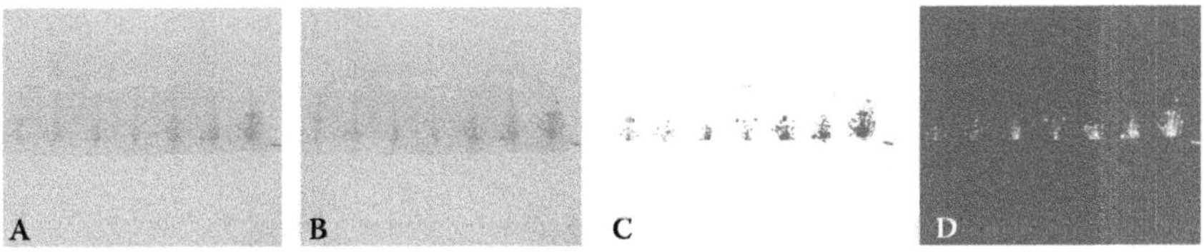

Figure 6.12 **(See color insert.)** Steps in the image treatment procedure: (a) original picture, (b) grayscale image, (c) image after brightness and contrast of (b) adjusted, and (d) image after color of (c) reversed.

plantlets under sterile conditions was designed. First, the effects of different hormone combinations on *O. violaceus* plantlet growth were investigated. The hormone combinations used in the present experiment are shown in Table 6.5.

A plantlet with unknown biomass was included with six different internal standard plantlets of known biomass. We acquired a color image (Figure 6.12a), converted the image to grayscale (Figure 6.12b), adjusted the brightness and contrast of the grayscale image (Figure 6.12c), and inverted the image color (Figure 6.12d). Following this, we performed pixel point statistics on the inverted image of each plantlet.

We constructed a linear regression of the biomass on the pixel data of the six internal standard plantlets with known biomass (Figure 6.13a). The biomass of the experimental plantlet was determined using its pixel information. Finally, the biomass of *O. violaceus* plantlet samples was calculated and a linear regression of the calculated biomass on the actual biomass measured by artificial damage was constructed (Figure 6.13b). The linear regression equation is as follows:

$$y = 0.9497x + 0.0018 \ (R^2 = 0.9861, n = 29) \qquad (6.6)$$

Equation (6.6) shows that the slope of the line was close to 1, with a value of 0.9497. The intercept value was 0.0018 g and was negligible. This plot shows that calculation of the biomass using image analysis methods was a reliable estimation strategy.

6.5.2 *Capturing the dynamic growth information of* O. violaceus *plantlets in vitro*

Using the internal standard method, we acquired images of *O. violaceus* plantlets in vitro at 3-day intervals during their culture process. We calculated the plantlet biomass using image analysis at each time period by constructing a best-fit linear equation according to the internal standard plantlets. The biomass calculated at each time was considered to be the dynamic growth information of *O. violaceus* plantlets in vitro.

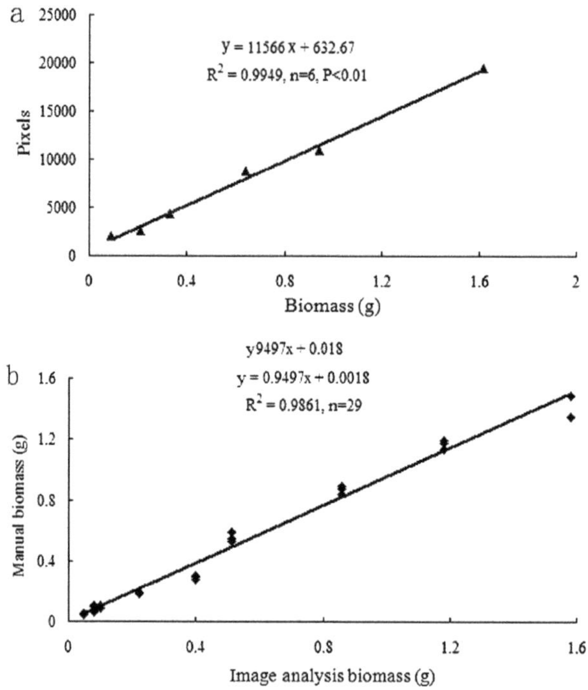

Figure 6.13 (a) The linear relationship between the biomass (x) and pixels (y) of *O. violaceus* plantlet in vitro. (b) The correlation between the image analysis biomass (x) and the manual biomass (y) of *O. violaceus* plantlet in vitro.

The linear equation (6.7) and the four-parameter logistic equation (6.8) represent the relationship between plantlet biomass and subculture time:

$$Y = Y_0 + kX \tag{6.7}$$

$$Y = Y_0 + \frac{a}{1+(X/X_0)^b} \tag{6.8}$$

In Equation (6.7), k represents the growth rate and Y_0 represents the initial subculture biomass. The parameters in Equation (6.8) are the same as those in Equation (6.5).

6.5.3 *The effect of different hormone combinations on the growth rate of* O. violaceus *plantlets cultured in vitro*

Depending on the curve relating biomass to culture time, we used Equations (6.7) and (6.8), respectively, to predict the dynamic growth relationship of *O. violaceus* plantlets cultured in nine different hormone combinations and to calculate the growth rate of plantlets cultured using these different hormone combinations. The results are shown in Table 6.10.

Although the growth rate in the logarithmic growth phase reached 93.1 mg · day^{-1} in the medium containing MS + 0.5 mg · L^{-1} NAA + 10 mg · L^{-1} 6-BA (combination 9), the duration of the logarithmic phase was only 3.56 days (Table 6.10). Thus, the average rate of growth increase was <10 mg · day^{-1}. The growth rate of plantlets was the fastest in the medium containing MS + 0.02 mg · L^{-1} NAA + 2 mg · L^{-1} 6-BA. Thus, the medium containing MS + 0.02 mg · L^{-1} NAA + 2 mg · L^{-1} 6-BA was the most suitable for bud multiplication. These results are in accordance with the conclusion presented in Section 6.3. The effect on the growth of *O. violaceus* plantlets cultured in the medium containing MS + 0.02 (or 0.5) mg · L^{-1} NAA and 0.4 mg · L^{-1} 6-BA was more advantageous than the other six combinations. The medium containing a high 6-BA concentration (10 mg · L^{-1}) severely inhibited plantlet growth, as observed for combinations 3, 6, and 9. Thus, both 6-BA and NAA have apparent hormonal effects. Low concentrations

Table 6.10 Growth Rates of *O. violaceus* Plantlets Cultured in Different Hormone Combinations

Combination	k/GR_{50} (mg/day)	DT_{log} (day)	Equation and R^2
1	34.9		$Y = 0.0444 + 0.0349X$ $R^2 = 0.990$ ($n = 7, p < 0.0001$)
2	23.7		$Y = 0.1566 + 0.0237X$ $R^2 = 0.968$ ($n = 7, p < 0.0001$)
3	9.3		$Y = 0.1393 + 0.0093X$ $R^2 = 0.777$ ($n = 7, p = 0.0087$)
4	43.9		$Y = 0.0508 + 0.0439X$ $R^2 = 0.980$ ($n = 7, p < 0.0001$)
5	56.2		$Y = 0.0291 + 0.0562X$ $R^2 = 0.957$ ($n = 7, p = 0.0001$)
6	2.5		$Y = 0.1722 + 0.0025X$ $R^2 = 0.746$ ($n = 7, p = 0.0122$)
7	42.0		$Y = 0.0526 + 0.0420X$ $R^2 = 0.899$ ($n = 7, p = 0.0011$)
8	36.1		$Y = 0.0219 + 0.0361X$ $R^2 = 0.979$ ($n = 7, p < 0.0001$)
9	93.1	3.56	$Y = 0.1329 + \frac{0.3313}{1 + (X/3.9651)^{-4.4576}}$ $R^2 = 0.989$ ($n = 7, p = 0.0020$)

of these hormones promote plant growth, whereas high concentrations inhibit plant growth (Woodward and Bartel, 2005).

6.6 Conclusions and prospects

This chapter describes a sterile monitoring technique based on image analysis for plantlets in vitro under different culture conditions. The dynamic growth of different types of plantlets cultured in medium containing different hormone combinations was also analyzed in the present study. Sterile monitoring technology for plantlets was established using the labeling system, the single-reference object method, the double-reference object method, and the internal standard method with multiple references. Using the label method on different plantlets in vitro resulted in different fitted equations. The linear equation fit the plantlet biomass most appropriately and was the best model for fitting calculations of plantlet volume and biomass. The fitted equations used for the single-reference and double-reference methods showed linear correlations; however, the double-reference method was more accurate than the single-reference method. The internal standard method using multiple references was applied to determine the growth of various types of *A. lancea* plantlets within a wide range of biomasses. This study contributes to fundamental research on sterile measurement of plantlet growth and has implications for the standardization of plantlets in vitro during industrial production.

Sterile dynamic measurement of plantlet growth enables convenient alteration and optimization of a medium. During tissue culture, the medium is the foundation for plantlet growth and development. Plantlets' growth strongly depends on the composition of the culture media. The sterile monitoring technique can be employed to dynamically adjust the composition of the medium.

Sterile dynamic measurements of plantlets in vitro will enable convenient physiological and biochemical research on plantlets. During tissue culture, the in vitro environment of plantlets and the physiological factors affecting plantlet growth besides the composition of the medium must be considered. To conduct intensive research in plant physiology, molecular biology, and biochemistry, it is essential to refer to changes in biomass of plantlets, and these changes could be determined by image analysis. Thus, noninvasive measurement promotes physiological and biochemical research on plantlets grown in vitro.

Acknowledgment

The authors gratefully acknowledge the financial support of the Priority Academic Program Development of Jiangsu Higher Education Institutions (PAPD).

References

Brosnan, T., and Sun, D.W., Improving quality inspection of food products by computer vision—a review, *J. Food Eng.*, 61, 3–16, 2004.

Buenrostro-Nava, M., Ling, P., and Finer, J., Development of an automated image acquisition system for monitoring gene expression and tissue growth, *Trans. ASAE*, 48, 841–847, 2005.

Chen, Q.C., Zhao, J.W., Zhang, H.D., and Fang, M., Identification of tea color by using computer vision, *J. Jiangsu Univ. (Nat. Sci.)*, 26, 461–464, 2005.

Ibaraki, Y., and Kenji, K., Application of image analysis to plant cell suspension cultures, *Comput. Electron. Agric.*, 30, 193–203, 2001.

Li, C.Y., Teng, G.H., Zhao, C.J., Qiao, X.J., and Wu, C., Development of non-contact measurement on plant growth in greenhouse using computer vision, *Trans. Chin. Soc. Agric. Eng.*, 19, 140–143, 2003.

Murashige, T., and Skoog, F., A revised medium for rapid growth and bio assays with tobacco tissue cultures, *Physiol. Plant*, 15, 473–497, 1962.

Olofsdotter, M., Image processing: a non-destructive method for measuring growth in cell and tissue culture, *Plant Cell Rep.*, 12, 216–219, 1993.

Rodrigues, S., and Fernandes, F.A.N., Image analysis of osmotically dehydrated fruits: melons dehydration in a ternary system, *Eur. Food. Res. Technol.*, 225, 685–691, 2007.

Smith, M., Spomer, L.A., Meyer, M., and McClelland, M., Non-invasive image analysis evaluation of growth during plant micropropagation, *Plant Cell Tiss. Org. Cult.*, 19, 91–102, 1989.

Supalkova, V., Petrek, J., Baloun, J., Adam, V., et al., Multi-instrumental investigation of affecting of early somatic embryos of spruce by cadmium (II) and lead (II) ions, *Sensors*, 7, 743–759, 2007.

Wang, G.Q., Yang, Z.B., Zheng, L.M., Zhu, H., Liao, S.H., Shan, C.G., and Wu, F.N., Application of computer vision technology in detection of agricultural products, *Rev. China Agri. Sci. Technol.*, 5, 52–56, 2003.

Woodward, A.W., and Bartel, B., Auxin: regulation, action, and interaction, *Ann. Bot.*, 95, 707–735, 2005.

Zhao, Y.G., Wu, Y.Y., Wu, X.M., Qin, S., Li, J.Z., Xu, J.J., and Gong, Y., The investigation on the aseptic measurement technique of plant tissues culture tube seedlings, *J. Agric. Mech. Res.*, 7, 87–89, 2005.

chapter seven

Digital imaging of seed germination

Didier Demilly, Sylvie Ducournau, Marie-Hélène Wagner, and Carolyne Dürr

Contents

7.1 Introduction

As reproductive units, seeds are essential for the survival of plant species and for their role in stand establishment. Seeds are also a crucial input for agriculture, as they are the vector of genetic progress and of different plant protection treatments. Germination is the first step in plant development and is a complex process (Bewley and Black, 1994; Nonogaki et al., 2010; Rajjou et al., 2012). It is a critical phase in the plant life cycle during which quiescent seeds are reactivated to produce seedlings. Germination sensus stricto begins with uptake of water by the seed and terminates with the initiation of growth of the embryo axis and radicle protrusion (Côme and Thévenot, 1982; Nonogaki et al., 2007). During this initial stage, seed water content increases in direct relation with the moisture content of the environment, allowing seed reserves to be hydrolyzed and metabolism to restart. Seed water content has to reach a threshold value to allow germination and to maintain this value until the radicle emerges (Bewley and Black, 1994; Kigel and Galili, 1995). Subsequent seedling growth depends

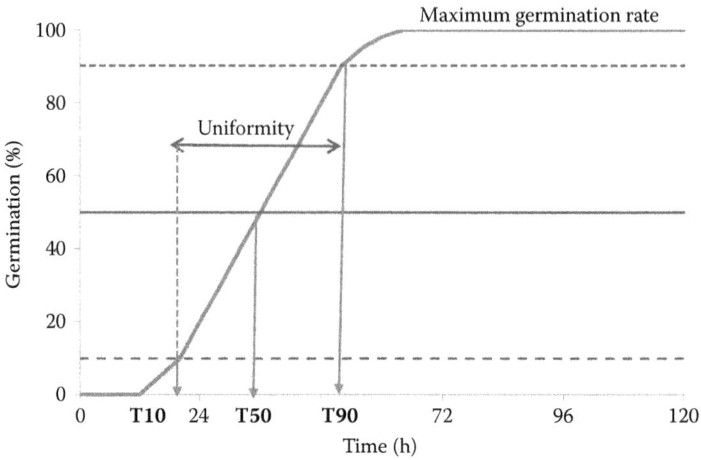

Figure 7.1 Time course curve of germination and the extracted main indicators.

on a high rate of water uptake by the radicle, on the depletion of seed reserves, and on mineral absorption (Dürr and Mary, 1998). Thus, germination and seedling growth rates depend to a great extent on environmental conditions, mainly water, temperature, and light levels. They also depend on the species (Baskin and Baskin, 2001), genotype, and seed technology (Jisha et al., 2013). Both the stages, i.e., germination and seedling heterotrophic growth, have to be considered because it is the whole process that leads to seedling emergence in field conditions. The official germination testing procedures include seedling growth as part of germination (ISTA, 2013). Seed germination experiments are widely performed because seeds are very easy to handle and provide a useful support to study fundamental physiological processes. The period of germination and seedling early growth is very short, and its characterization provides essential information on plant establishment and future growth. Germination is usually represented by a cumulative curve (Figure 7.1) where the completion of radicle protrusion is pooled as a function of time in a population of individual seeds following imbibition (Black et al., 2006). These germination curves provide several parameters that have been used for decades (Ranal and De Santana, 2006). The most important and widely used parameters are listed in Table 7.1.

When samples are fully germinated (i.e., the final germination rate is 100%), time to reach 50% germination (T50) and mean germination time (MGT) appeared to be very close. Uniformity is usually calculated using time to reach 90% germination (T90) minus time to reach 10% germination (T10), as used in commercial seed lots to give an idea of the homogeneity of germination when the rate is high (Figure 7.1).

Table 7.1 Main Measurements of Germination

Measurement	Expression	Unit
Germination rate	Cumulative germination at a defined period of time after sowing seeds	%
Mean germination time (MGT)	MGT = $\Sigma(nt)/\Sigma n$, where n = number of seeds newly germinated at time t and t = time since sowing (start of germination test)	Days or hours
T50	Time necessary to reach 50% germination; the most commonly used percentile	Days or hours
Uniformity	Time necessary for a sample to complete germination (extreme seeds, i.e., very rapid or very slow to germinate, are excluded)	Days or hours

Even though the time course curves of germination give detailed and pertinent information, there are still certain problems. Periodic counts are always time-consuming and have to be done at very precise moments, which can cause logistical problems when germination is rapid. An automated system using computer vision is an efficient alternative (Dell'Aquila, 2009). Moreover, computer quantification of radicle protrusion is more objective and repeatable than a human count. More generally, digital imaging is a noninvasive way to study the imbibition phase, germination, and seedling growth separately using the same seed material, and to analyze correlations between different seed morphological characteristics.

7.2 Existing systems for monitoring seed germination using digital imaging

In the last 15 years, a number of operational tools have been developed to monitor seed germination using digital analysis. A summary of the characteristics of existing systems is presented in Table 7.2 to allow easier comparison with the automated systems described below.

Flatbed scanners are used as a quite simple system in American horticulture departments to evaluate seed germination and seedling growth (Geneve and Kester, 2001; McDonald et al., 2001, 2008; Hoffmaster et al., 2005). The different systems using scanners are based approximately on the same procedures. Seeds are germinated and seedlings grown either on blotters or on a thin transparent cellulose layer in a germinator set at the recommended temperature for the species being tested. Digital images of the blotters or the cellulose film with seeds or seedlings are acquired using a flatbed scanner with uniform white lighting. Images of 15 to 50

Table 7.2 Summary of the Main Characteristics of Four Operational Tools for Monitoring Seed Germination Using Digital Imaging

Device Used to Monitor Seed Germination	Germination Conditions		Number of Seeds Tested at a Time	Image Acquisition Process		Image Analysis Process	References
	Substrate	Environment		Equipment	Frequency		
Seed scanner	Blotters or cellulose in transparent trays	Germination room separate from image acquisition system (20°C and 20–30°C) 272 seeds	15–50 seeds	Flatbed scanner	Manual placement of seeds for image acquisition Manual or automated image acquisition (one per day to every 30 min) Automated image acquisition	Seed germination assessed by human vision on video or images Seedling length and growth rate assessed using commercial software	Geneve and Kester, 2001; McDonald et al., 2001; Texeira et al., 2007
	Agarose in petri dishes	Thermostatic chamber includes image acquisition system (25°C, with light)	Max. 12 seeds	CCD camera	Automated image acquisition at hourly intervals	Radicle protusion and length analyzed automatically Software based on area and roundness factor visible in successive images (Image Pro Plus or ImageJ)	Dell'Aquila et al., 2000

Germinator	Blue blotters in transparent trays	Germination room separate from image acquisition system (20°C, with light)	300–1200 seeds	Digital camera	Manual placement of seeds for image acquisition 6 images for a germination curve of up to one image per hour	Radicle protusion analyzed automatically Software based on seed position and 2 color thresholds (Photoshop and ImageJ)	Joosen et al., 2010
Icarus	Blue pleated paper in transparent trays	Incubator also contains image acquisition system (15°C or 20°C)	300 seeds	Camera	Automatic seed placement by robot Image acquisition at 3 h intervals	Length of radicle > 2 mm analyzed automatically No mention of the software used	Harper and Long, 2012

seeds or seedlings can be captured simultaneously in less than 2 min. The resulting images can be saved for human visual evaluation of the seeds and seedlings. In addition to classification as normal or abnormal seedlings, seedling length, area, and growth rate can be evaluated using computer-aided analysis of digital images (Geneve and Kester, 2001), and several important characteristics for seed lot evaluation can be calculated (Oakley et al., 2004).

This system was subsequently improved to automatically acquire images at set time intervals with the possibility of monitoring individual seeds throughout germination and early seedling growth (Geneve and Dutt, 2008). Movies made from all the combined images are used to study water entry into the seed coat with and without scarification.

Flatbed scanners are also used in a quite sophisticated system developed by Teixeira et al. (2007). This system allows images to be aquired automatically at chosen time intervals and seed germination to be studied over a broad range of temperatures and using a large number of seeds (272) at a time. Quantitative analysis of each image is performed by visual observation on the computer monitor. This system has been used to study germination and growth characteristics in different flower species such as impatiens and petunia (Dutt and Geneve, 2007) or physically dormant species such as *Gleditsia* (Geneve and Dutt, 2008).

In the same period, other tools were developed by two European teams. The image analysis system developed at the Institute of Plant Genetics in Bari, Italy, has an environmentally controlled incubator in which seeds are placed for germination, and a computer imaging system (Dell'Aquila et al., 2000). Seeds are germinated in Petri dishes containing polymerized agarose. The machine vision prototype consists of a thermostatic chamber designed to include a Petri dish holder in which a maximum of 12 seeds are placed for germination, a timer-dependent white lighting system, and a charged-coupled device (CCD) camera (3.3 MP) for image acquisition. Images are processed using a high-power personal computer and the commercial software package Image Pro-Plus or the free software ImageJ (Dell'Aquila, 2004, 2007, 2009). Images are acquired automatically at hourly intervals and stored, starting from seed imbibition up to complete seed germination. The video signal from the camera is digitized into a binary image, and the software automatically records and analyzes the successive images of individual seeds. The area and roundness factors, corresponding respectively to size and shape, are measured on each seed throughout the time course, and radicle length can also be measured after radicle protrusion. The increase in area is correlated with the three-phase curve of water uptake described for the seed imbibition and germination processes. The rapid increase in the seed area is related to the radicle protrusion, which allows the real onset of radicle emergence to be estimated. Image analysis has proved to be successful for monitoring the swelling and germination process of different species, including cabbage, lentil, pepper, and tomato (Dell'Aquila, 2004).

Another automated device for image acquisition associated with a software package for image analysis called Germinator was developed by Joosen et al. (2010) to monitor seed germination. Compared to the systems described above, this tool uses a different image analysis approach for seed germination. Seeds are germinated on blue blotters placed in transparent germination trays. Six samples of 50 to 200 seeds can be scored at a time. Users can choose and record their experimental design (number of samples and replicates) using a module of the Germinator package. Germination trays are stacked in a temperature-controlled incubator with side lighting. The seeds in each tray are manually photographed at different time intervals. It has been demonstrated that at least six data points with two points in the exponential part of the germination curve are necessary to accurately predict germination parameters of the germination kinetics curve. Images are automatically identified and preprocessed using Adobe Photoshop to separate the six different samples into six individual pictures. In the second module of the Germinator software package, image analysis is then performed twice on each image using ImageJ free software, first with a color threshold that only selects the color of the seed, and then with another color threshold that selects everything in the image except the background. In both analyses, the XY position and size of each individual seed are extracted. At each time interval, seeds are considered not germinated when both the difference in the XY position and the size of the two color thresholds are still within a user-defined limit. Germination data and time intervals are recorded and a curve is fitted in the third module of the Germinator package. This tool has been used to score germination in *Arabidopsis*, for which visual detection of germination is both difficult and time-consuming, due to the small size (<1 mm) of the seed. It has also been used for the germination study of 24 genotypes of *Brassica* seeds with different colored seed coats.

Another operational machine vision system called Icarus was developed to automatically monitor seed germination (Harper and Long, 2012). This system can perform high-throughput phenotyping of sugar beet seed germination. Seeds are germinated in three blue pleated papers, each containing 100 seeds, and are placed in a transparent plastic tray in a temperature-controlled room. At 3 h intervals, the trays are automatically placed beneath a camera by a robot. Images are acquired of all 300 seeds at a time, directly within the pleats, which are sufficiently open to make radicle emergence visible. The software detects emerging roots, and seeds are scored as germinated when the radicle is more than 2 mm in length. This system is used routinely to assess the germination speed of primed sugar beet seeds and, in the future, could be applied to other species germinated in pleated paper.

Several operational systems have thus been used to monitor seed germination by image analysis. Depending on the systems described above, choices were made to automatize image acquisition or image analysis. Currently, no system, including automated seed deposition, is available. The main differences are in the germination supports and the control of environmental conditions (light and temperature), the number of seeds tested, and the image acquisition and image analysis software.

7.3 *A high-throughput seed germination phenotyping system*

The imaging system presented here includes some of the capacities described above, with several objectives. The main objectives are to achieve high throughput of a large set of species, genotypes, and seed production conditions, in varied environmental conditions. Based on the system described by Ducournau et al. (2004, 2005) with a camera installed over a Jacobsen table, a system was developed to study the imbibition, germination, and early radicle elongation in a large number of species and with a large number of seeds (Wagner et al., 2011). The process comprises three steps: image acquisition, image analysis, and data analysis. The illustrations provided here are from different species of economic interest and also include *Medicago truncatula* and *Arabidopsis thaliana*, the model species used for genomic research.

7.3.1 *Image acquisition*

Two Jacobsen tables are equipped with four CMOS color cameras (Pixelink CAM_PLA662 1.3 MP) with a Rainbow Zoom 11.5–69 mm lens (Figure 7.2). The system is piloted automatically by the software. The sequence is as follows: the light (6 × 36 W cool white lights) is switched on, the cover opens, a picture of the seeds is taken with each camera, the data are saved, the light is switched off, and the cover is closed. The closed system enables good control of water conditions. This sequence is repeated at a specified time interval, generally 2 h, but with a minimum possible interval of about 10 min. The software identifies and stores images in separate directories. The system can be used for different species by using different magnification. Seeds germinated on blue blotters are continuously watered by the Jacobsen table. The temperature can be regulated between 10 and 30°C. It is possible to analyze up to 200 seeds per camera for large seeds like *Helianthus annuus* or *Zea mays*, and 400 seeds per camera for small seeds like *Brassica napus* or *M. truncatula* (Figure 7.3). The two tables enable 200 to 300 seeds to be analyzed simultaneously. Throughput has progressively increased and there are now two additional tables with high-definition cameras that can analyze *A. thaliana*, seeds in boxes with

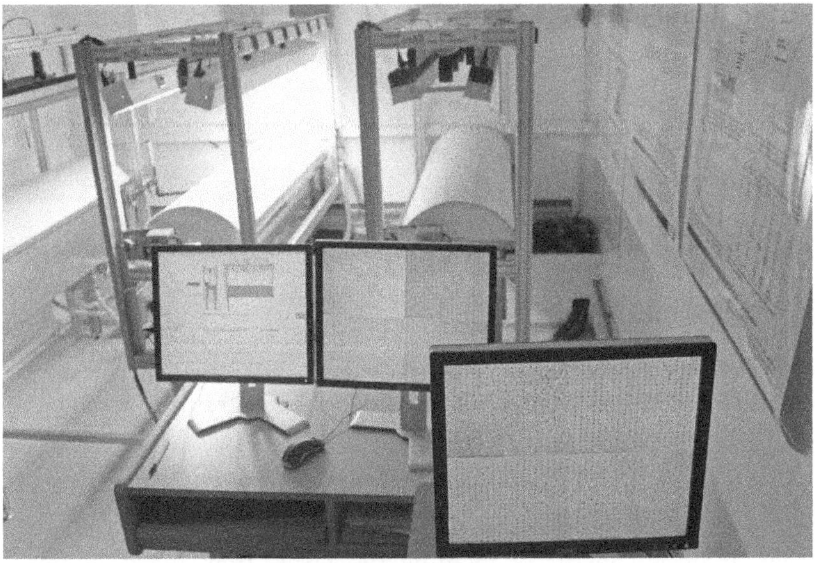

(a)

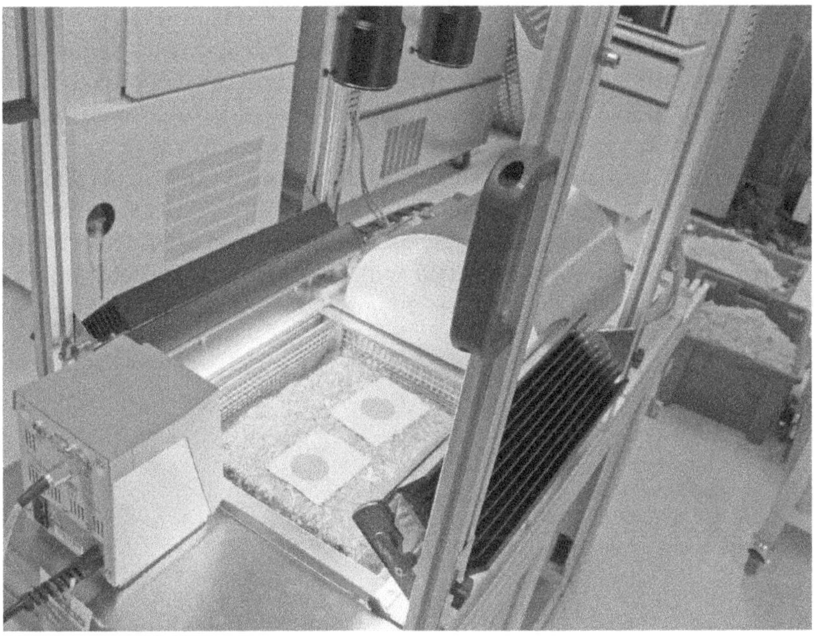

(b)

Figure 7.2 Two Jacobsen tables, each equipped with four cameras. The smaller Jacobsen table on the right is mostly used for *A. thaliana*.

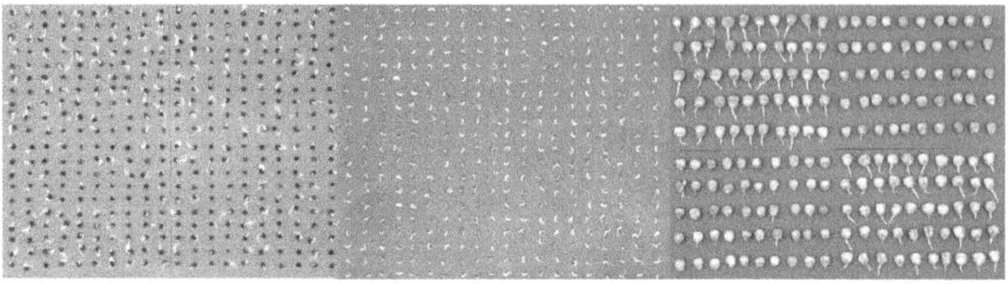

Figure 7.3 (**See color insert.**) Example of images obtained at 20°C. From left to right, oilseed rape 48 h after sowing, *M. truncatula* 24 h after sowing, and maize 72 h after sowing.

controlled water potential, or study pathogen contamination. These tables have direct or indirect light-emitting diode (LED) lighting that can be varied from visible to green (to simulate darkness).

7.3.2 *Image analysis*

Image analysis consists of merging the pixels from the color images into two classes (seeds or background) and then measuring each seed (Figure 7.4). Image analysis is performed with the Fiji image processing package based on ImageJ (http://fiji.sc/). For each species, a sequence of pretreatment, thresholding, and posttreatment was developed to obtain a binary image from the color image. A batch plug-in is used to analyze all the images from a run in one operation. The measurements of each image and each seed are collected in one matrix for subsequent data analysis. The image analysis sequence is as follows: channels are split from the RGB image, the background on the gray-level image is extracted from the red channel, segmentation is performed with the entropy threshold function, the image is inverted, and each seed is measured with the "measure ROI

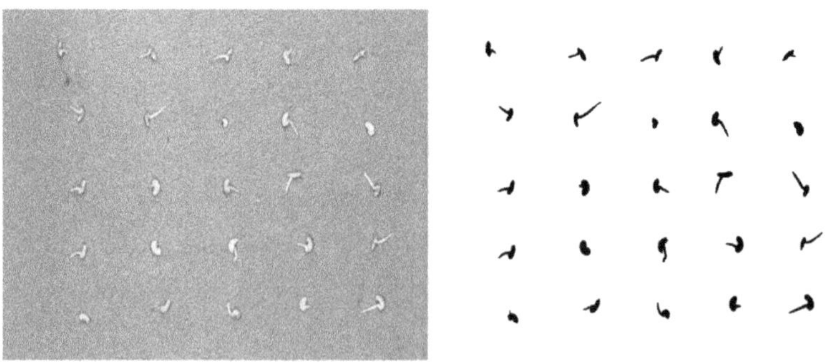

Figure 7.4 (**See color insert.**) Images obtained with a video camera and after segmentation in the red channel. (*M. truncatula* 30 h after sowing at 20°C.)

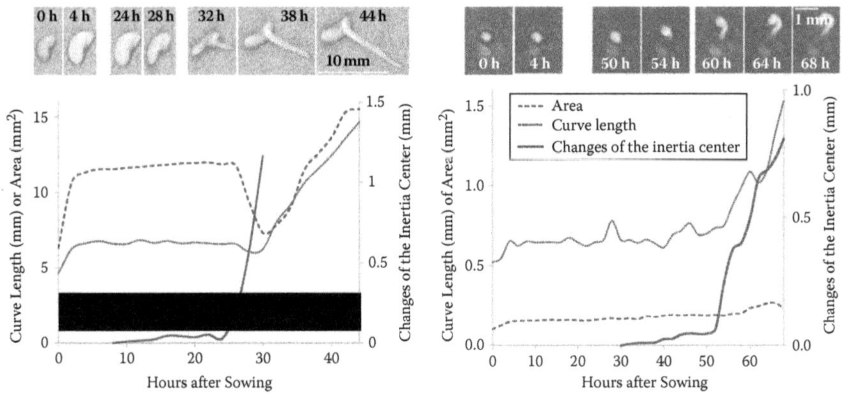

(a) *Medicago truncatula* (b) *Arabidopsis thaliana*

Figure 7.5 **(See color insert.)** Sequence of images and main image analysis parameters extracted from sowing to radicle elongation stages of germination for (a) *M. truncatula* and (b) *A. thaliana*.

curve" plug-in. The selected information and measurements that apply to each seed are the name of the image, the area, the positions of X and Y with respect to the center of inertia, the length and the width of the ellipsoid, and the length of the curve (Figure 7.5). Subsequent developments were made to improve measurements and data analysis. For example, when seeds are dark like those of oilseed rape (*B. napus*), a second segmentation was performed that allows the radicle to be distinguished from the seed, and the length of the radicle can then be precisely measured.

7.3.3 Data analysis

The Microsoft Excel software package is used for data analysis. Using a Visual Basic module, a script was developed to label the measurements according to the position of the seed on the images. With such individual labeling of seeds, it is possible to apply experimental designs to obtain germination time courses from several genotypes or seed lots on each camera. The X and Y positions are also used to calculate changes in the center of inertia with time after sowing and to detect germination (Figure 7.5). Figure 7.6 illustrates the cumulative germination curve obtained from the analysis of images of individual seeds, compared with visual counting on images or in situ.

In most species, the size of the seed increases during imbibition. Increase in seed area or volume during imbibition (expressed as percentages of the initial values at sowing) was compared to seed moisture content (Figure 7.7). A highly significant correlation was found between the area and the weight of each seed at any given time after sowing. Seed

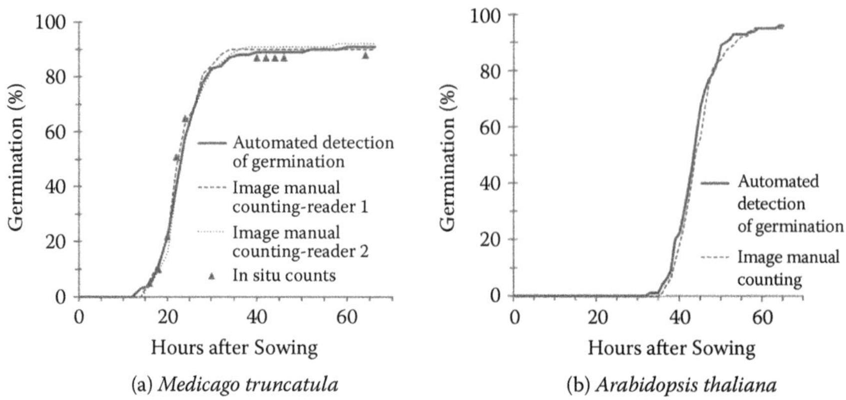

Figure 7.6 Automated detection of germination compared with in situ or manual counting on the image in (a) *M. truncatula* and (b) *A. thaliana*.

volume is estimated from the measurements on images using the formula $V = \pi \times \text{length} \times \text{width}^2/6$ (Casco and Dias, 2008).

After germination, the radicle of the young seedling elongates. The elongation of the embryonic axis of *M. truncatula* was determined for each seed from 4 to 14 h after germination by measuring the length of the curve, and the results are comparable to manual measurements made on the same images (Figure 7.8). Subsequently, the elongation rates are calculated for each seed.

From individual characterizations, means, standard deviations, and time courses are calculated for replicates, genotypes, or seed samples.

In the early 2000s the system was able to analyze only two species (sunflower and oilseed rape) with a low throughput of 25 seeds/run (Ducournau et al., 2004). The technological development of cameras,

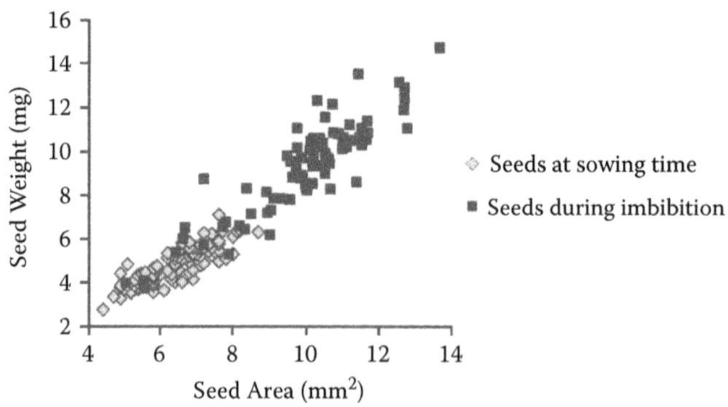

Figure 7.7 Correlation between seed area measured on images and seed mass measured on dry seeds or during imbibition of *M. truncatula*.

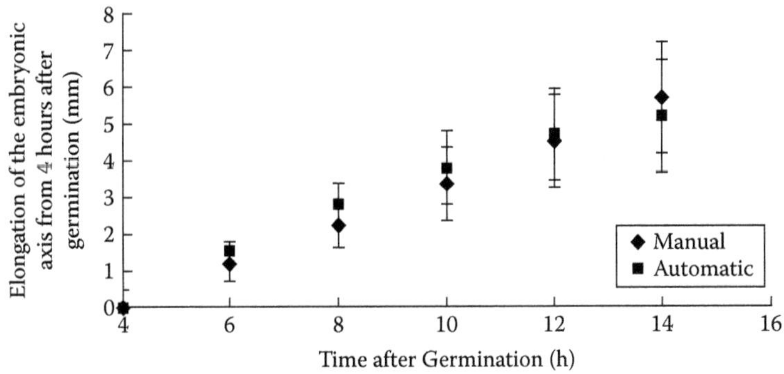

Figure 7.8 Automated measurement of embryonic axis elongation compared to manual measurements on *M. truncatula* from 4 to 14 h after germination.

combined with a high interest from seed companies and researchers for digital imaging, has led to the development of a system for high throughput on a large range of species. Figure 7.9 shows that about 50 species can now be analyzed.

With four Jacobsen tables and a total of 14 cameras, high-throughput phenotyping is now possible for a wide range of species and environmental conditions. For instance, in 2012, 150,000 images were taken,

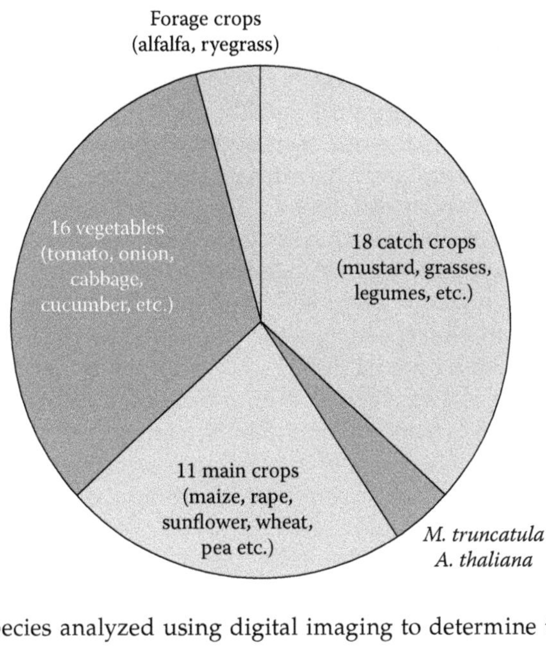

Figure 7.9 Species analyzed using digital imaging to determine their germination parameters.

stored, and analyzed to obtain germination, imbibition, and elongation characteristics of 100,000 seeds from more than 35 species.

7.4 Applications

The time course of germination can be now determined automatically, and so the results can be used to provide more detailed information about seed lots. Automation allows information to be obtained at different specific times, such as first germination time, times to x% germination, and mean germination time (Figure 7.1). Curves can also be fitted, giving equation parameter values for modeling. Base temperature values (T_b) (Dahal and Bradford, 1994) can be calculated from the results of curve fitting obtained at different temperatures, taken from the linear part of the correlation between germination rates and temperatures. This information can then be used in crop modeling and simulations of plant emergence (Dürr et al., 2001; Gardarin et al., 2012), or to compare species and genotypes.

All these indicators provide information on germination speed as a function of environmental conditions and plant characteristics. Indicators of germination speed, even those obtained in the lab in optimum conditions, have been shown to correlate well with field emergence (Wagner et al., 2007). There are two important reasons for this: First, high germination speed at optimal temperature is often correlated with the germination speed and rates at lower temperatures. When one seed lot already germinates more slowly than another one at optimal temperature, stress conditions only exacerbate differences. Second, a slower germination speed also increases the risk in the field of the seedlings being blocked under a surface crust. This is because the seedling takes longer to reach the soil surface, which increases the risk of crusting after rainfall and before seedling emergence. Consequently, germination speed can be a good predictor of emergence rates under different environmental conditions (Matthews et al., 2010, 2012). Another advantage of the automated systems is that different seed characteristics can be measured precisely and separately, and then correlated and compared, e.g., seed physical characteristics, seed imbibition, germination speed, and the elongation rate of the radicle. This contributes to better understanding of variations in seed lots' quality. Automated analysis of seed germination would be of great help in the evaluation of the initial quality and the quality over time of the stored seeds. For example, large collections of species and genotypes are maintained for the preservation of genetic resources in different centers of germplasm resources (e.g., U.S. Fort Collins Science Center or the Millennium Seed Bank in the UK). The germination ability of the large number of stored seed lots must be checked regularly during storage. Another interest for the genetic resource centers is that automated systems can facilitate the characterization of large sets of species and genotypes, which can help to predict their germination potential under different conditions, such as high temperature or water stress. The use of automated systems to characterize the germination of weeds, wild plants, forest species, and crops could help to create large databases. This information could then be used for modeling purposes, for example, when evaluating new agricultural systems or the effects of climate change. Finally, automated image acquisition and analysis can be used to characterize populations of inbred lines and reveal quantitative trait loci (QTL) correlated to variations in germination traits such as imbibition rate, germination speed, or early elongation rate of the radicle. An example is given in Figure 7.10. Different chromosomal areas associated with germination and seedling trait values are useful for breeding and improving stand establishment under varied environmental conditions.

A future use of automated systems will be to quantify the development of pathogens during germination and seedling growth. There is an increasing need for an improved evaluation of the sanitary condition of seed lots, and of the efficiency of seed treatments. However, to achieve this, further development in equipment is required.

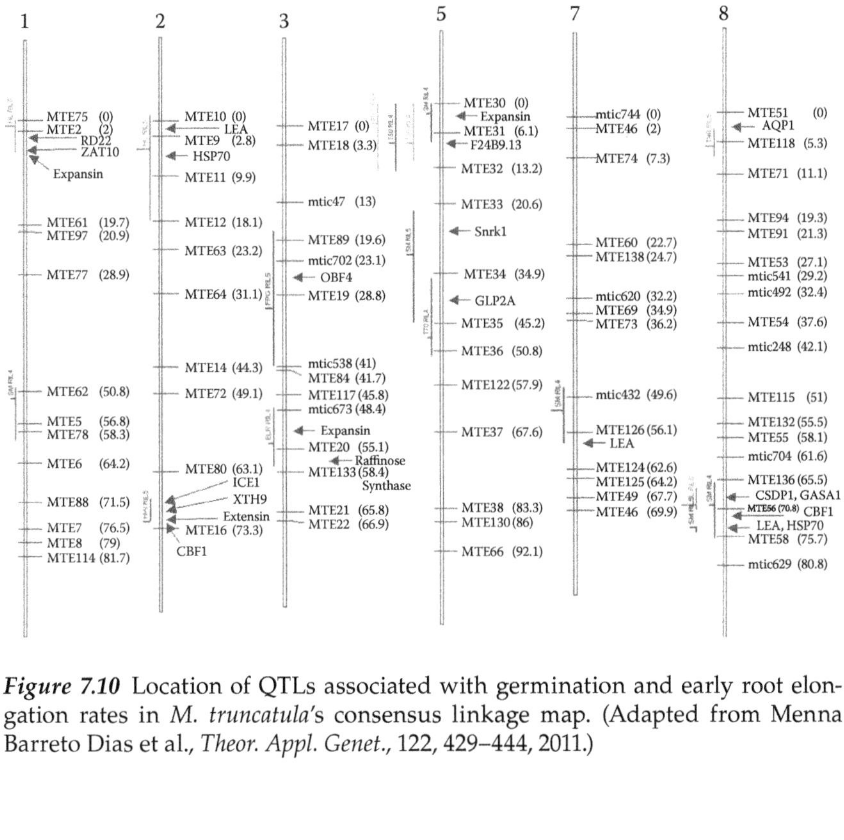

Figure 7.10 Location of QTLs associated with germination and early root elongation rates in *M. truncatula*'s consensus linkage map. (Adapted from Menna Barreto Dias et al., *Theor. Appl. Genet.*, 122, 429–444, 2011.)

7.5 Conclusion

High-throughput automated systems of image analysis on seeds and on the early stages of seedling development are relatively new technology that has been developed over the last 10 years. They have opened up many new possibilities in terms of the number and quality of the measurements of seeds and seedlings that can be taken, and with regards to the number of seed lots and genotypes that can be characterized. The depth and breadth of information available can be used to identify the reasons for variations in the quality of seed lots. An innovative future use would be to make these data available on a large set of genotypes for use in breeding programs so as to improve stand establishment in different environmental conditions. The increasing availability of thermographic and hyperspectral cameras at a more reasonable cost has led to a more widespread use of them in plant phenotyping platforms. So, it is probable that new high-throughput applications, especially for the early detection of pathogens and seed composition analyses, will be available in the near future.

References

Baskin, J.M., and Baskin, C.C., *Seeds: ecology, biogeography, and evolution of dormancy and germination*, Academic Press, New York, 2001.

Bewley, J.D., and Black, M., *Seeds: physiology of development and germination*, Plenum Press, New York, 1994.

Black, M., Bewley, J.D., and Halmer, P., *The encyclopedia of seeds: science, technology and uses*, CAB International, Wallingford, 2006.

Casco, H., and Dias, L.S., Estimating seed mass and volume from linear dimensions of seeds, *Seed Sci. Technol.*, 36, 230–236, 2008.

Côme, D., and Thévenot, C., Environmental control of embryo dormancy and germination, in *The physiology and biochemistry of seed development, dormancy and germination*, ed. A.A. Khan, 271–298, Elsevier, Amsterdam, 1982.

Dahal, P., and Bradford, K.J., Hydrothermal time analysis of tomato seed germination at suboptimal temperature and reduced water potential, *Seed Sci. Res.*, 4, 71–80, 1994.

Dell'Aquila, A., Cabbage, lentil, pepper and tomato seed germination monitored by an image analysis system, *Seed Sci. Technol.*, 32, 225–229, 2004.

Dell'Aquila, A., Towards new computer imaging techniques applied to seed quality testing and sorting. *Seed Sci. Technol.*, 35, 519–538, 2007.

Dell'Aquila, A., Digital imaging information technology applied to seed germination testing: a review, *Agron. Sustain. Dev.*, 29, 213–221, 2009.

Dell'Aquila, A., van Eck, J.W., and van der Heidjden, G.W., The application of image analysis in monitoring the imbibition process of white cabbage (*Brassica oleracea* L.) seeds, *Seed Sci. Res.*, 10, 163–169, 2000.

Ducournau, S., Feutry, A., Plainchault, P., Revollon, P., Vigouroux, B., and Wagner, M.H., An image acquisition system for automated monitoring of the germination rate of sunflower seeds, *Comput. Electron. Agric.*, 44, 189–202, 2004.

Ducournau, S., Feutry, A., Plainchault, P., Revollon, P., Vigouroux, B., and Wagner, M.H., Using artificial vision to monitor germination time course of sunflower (*Helianthus annuus* L.) seeds, *Seed Sci. Technol.*, 33, 329–340, 2005.

Dürr, C., Aubertot, J.N., Richard, G., Dubrulle, P., Duval, Y., and Boiffin, J., SIMPLE: a model for simulation of plant emergence predicting the effects of soil tillage and sowing operations, *Soil Sci. Soc.*, 65, 414–442, 2001.

Dürr, C., and Mary, B., Effects of nutrient supply on pre-emergence growth and nutrient absorption in wheat (*Triticum aestivum* L.) and sugarbeet (*Beta vulgaris* L.), *Ann. Bot.*, 81, 665–672, 1998.

Dutt, M., and Geneve, R.L., Time to radical protrusion does not correlate with early seedling growth in individual seeds of impatiens and petunia, *J. Am. S. Hort.*, 132, 423–428, 2007.

Gardarin, A., Dürr, C., and Colbach, N., Modeling the dynamics and emergence of a multispecies weed seed bank with species traits, *Ecol. Model.*, 240, 123–138, 2012.

Geneve, R.L., and Dutt, M., Using sequential digital images to study seed germination and dormancy, *Propag. Ornam. Plants*, 8(1), 13–16, 2008.

Geneve, R.L., and Kester, S.T., Evaluation of seedling size following germination using computer-aided analysis of digital images from a flat-bed scanner, *HortScience*, 36, 1117–1120, 2001.

Harper, S., and Long J., Automated analysis of primed seed, presented at ISTA Seminar "New Developments and Technologies in Seed Testing," Venlo, The Netherlands, June 11, 2012.

Hoffmaster, A.F., Xu, L.J., Fujimura, K., McDonald, M.B., Bennett, M.A., and Evans, A.F., The Ohio State University seed vigor imaging system (SVIS) for soybean and corn seedlings, *Seed Technol.*, 27, 7–24, 2005.

International Seed Testing Association (ISTA), International rules for seed testing, ISTA, Basseldorf, 2013.

Jisha, K.C., Vijayakumari, K., and Puthur, J.T., Seed priming for abotic stress tolerance: an overview, *Acta Physiol. Plant.*, 35, 1381–1396, 2013.

Joosen, R.V.L., Kodde, J., Willems, L.A.J., Ligterink, W., van der Plas, L.H.W., and Hilhorst H.W.M., Germinator: a software package for high-throughput scoring and curve fitting of *Arabidopsis* seed germination, *Plant J.*, 62, 148–159, 2010.

Kigel, J., and Galili, G., *Seed development and germination*, Marcel Dekker, New York, 237–474, 1995.

McDonald, M.B., Evans, A.F., and Bennett, M.A. Using scanners to improve seed and seedling evaluations, *Seed Sci. Technol.*, 29, 683–689, 2001.

McDonald, M.B., Evans, A.F., Bennett, M.A., Fujimura, K., Sako, Y., Xu, L., and Hoffmaster, A., Using the seed vigor imaging system for improving stand establishment, *Acta Hort.*, 782, 83–91, 2008.

Matthews, S., El-Khadem, R., Casarini, E., Khajeh-Hosseini, M., Nasehzadeh, M., and Wagner, M.H., Rate of physiological germination compared with the cold test and accelerated ageing as a repeatable vigour test for maize, *Seed Sci. Technol.*, 38, 379–389, 2010.

Matthews, S., Wagner, M.-H., Kerr, L., McLaren, G., and Powell, A.A., Automated determination of germination time courses by image capture and early counts of radicle emergence lead to a new vigour test for winter oilseed rape (*Brassica napus*), *Seed Sci. Technol.*, 40, 413–424, 2012.

Menna Barreto Dias, P., Brunel, S., Dürr, C., Demilly, D., Wagner, M.-H., and Teulah-Mérah, B., QTL analysis of seed germination and pre-emergence growth at extreme temperatures in *Medicago truncatula*, *Theor. Appl. Genet.*, 122, 429–444, 2011.

Nonogaki, H., Bassel, G.W., and Bewley, J.D., Germination—still a mystery, *Plant Sci.*, 179, 574–581, 2010.

Nonogaki, H., Chen, F., and Bradford, K.J., Mechanisms and genes involved in germination sensu strict, in *Seed development, dormancy and germination*, ed. K.J. Bradford and H. Nonogaki, 264–304, Blackwell Publishing, Oxford, 2007.

Oakley, K., Kester, S.T., and Geneve, R.L., Computer-aided digital image analysis of seedling size and growth rate for assessing seed vigor in impatiens, *Seed Sci. Technol.*, 32, 837–845, 2004.

Rajjou, L., Duval, M., Gallardo, K., Catusse, J., Bally, J., Job, C., and Job, D., Seed germination and vigor, *Ann. Rev. Plant Biol.*, 63, 507–533, 2012.

Ranal, M.A., and De Santana, D.G., How and why to measure the germination process, *Rev. Bras. Bot.*, 29, 1–11, 2006.

Teixeira, P.C.N., Coelho Neto, J.A., Rocha, H., and Oliveira, J.M.D., An instrumental set up for seed germination studies with temperature control and automatic image recording, *Braz. J. Plant Physiol.*, 19(2), 99–108, 2007.

Wagner, M.H., Demilly, D., Ducournau, S., Dürr, C., and Léchappé, J., Computer vision for monitoring seed germination from dry state to young seedlings, *Seed Testing Int.*, 142, 49–51, 2011.

Wagner, M.H., Matthews, S., Feutry, A., Demilly, D., and Ducournau, S., Germination rate monitoring, a useful way for seed vigour testing on maize (*Zea mays* L.), presented at Proceedings of the 28th ISTA Seed Symposium, Iguassu Falls, Brazil, May, 5–11, 2007.

chapter eight

Thermal imaging for evaluation of seedling growth

Étienne Belin, David Rousseau, Landry Benoit, Didier Demilly, Sylvie Ducournau, François Chapeau-Blondeau, and Carolyne Dürr

Contents

8.1 Introduction

Thermography is a noninvasive imaging method that gives access to temperature by use of the blackbody law. In the domain of plants, the measurement of the temperature is an important physical parameter. Additionally, apparent temperature provided by thermography is also indirectly related to other functional or structural parameters, like leaf orientation, heat capacity, surface properties, infrared (IR) absorption, and transpiration rate (Kana and Vass, 2008; Fiorani et al., 2012). Thermography has therefore been widely tested and shown useful on plants at various observation scales from canopy down to single leaf and in various biological contexts, including, for instance, evaluation of stomatal aperture (Leinonen et al., 2006), plant water content (Wang et al., 2010a, 2010b), plant freezing (Wisniewski et al., 2008), and the development of pathogens (Chaerle and der Staeten, 2001; Chaerle et al., 2007; Belin et al., 2013).

In this chapter, we work on the application of thermal imaging on seeds and seedlings (Dell'Aquila, 2009). The early stages of the development of the plant have only recently been analyzed with thermal imaging (Baranowski et al., 2003; Kranner et al., 2010; Belin et al., 2011). Thermal imaging was first used to evaluate the germination capacity of legume seeds (Baranowski et al., 2003). The seed storage duration and the variations in germination speed were shown to induce significant thermal differences during the initial stage of the imbibition process. Biophysical and biochemical changes during imbibition and germination were also shown to be detectable with thermal imaging and useful to predict whether a quiescent seed will germinate or die upon water uptake (Kranner et al., 2010). Thermal imaging was shown useful to discriminate hypocotyl from radicle in *Medicago truncatula* during the elongation stage of the seedling (Belin et al., 2011). Because thermal imaging can be influenced by various parameters, it is necessary when working at a given observation scale to investigate the possible impact of the measuring environment on the information to be extracted from the scene. This was done, for instance, at the leaf scale in Kümmerlen et al. (1999). We propose to transpose such an approach when working with thermal imaging at the seed scale, and this chapter is organized as follows. Section 8.2 first describes the experimental setup used in this report for monitoring seed germination and elongation with thermography. In Sections 8.3 and 8.4, results are presented regarding the seed imbibition and seedling elongation stages with seeds of various species.

8.2 Experimental setup

Thermal imaging cameras are constituted, like in conventional imaging, with a lens that focuses infrared onto a detector (Breitenstein et al., 2003; Vollmer and Mollmann, 2010). The radiation that impinges on a thermal camera comes from three different sources. The camera receives radiation W_{obj} from the target object, plus radiation W_{amb} from its ambiance that has been reflected onto the object surface. Both of these radiation components become attenuated when they pass through the atmosphere. Since the atmosphere absorbs part of the radiation, it will also radiate some itself, W_{atm}. The total radiation power received by the camera can therefore be expressed as

$$W_{tot} = \varepsilon\tau W_{obj} + (1-\varepsilon)\tau W_{amb} + (1-\tau)W_{atm}$$

where ε is the object emissivity and τ is the transmission through the atmosphere. In our experiment we have taken care to maintain at thermal equilibrium the environment surrounding the targeted seeds. The distance of observation was kept constant during the observation, at some tens of centimeters, to limit the absorption by the atmosphere and maximize the spatial resolution. As a result, the observed thermal contrasts are due to effective thermal contrasts in the seeds only. The emissivity for seeds is not yet available in literature. The estimation of an emissivity is often done by placing a thermometer on the targeted object or by placing an adhesive rubber of known emissivity on part of the object. Given the size of the object observed here, the estimation of the emissivity has not been investigated. Emissivity was assumed constant and equal to unity in our experiment. We will discuss this assumption in the elongation phase later in the chapter. The majority of thermal imaging cameras have a microbolometer type detector, mainly because of cost considerations. Microbolometers generally do not require cooling, which allows compact camera designs that are relatively low in cost. Microbolometers operate in the wavelength window of 8 to 12 μm. The typical thermal resolution is 0.1°C. For more demanding applications in terms of thermal resolution (for instance, 0.01°C), quantum detectors are used. These quantum detectors require cooling systems and operate in the wavelength window of 2 to 5 μm. In the applications of thermal imaging presented here, observed thermal contrasts were between 0.01 and 0.1°C. An IR camera with a quantum detector was therefore used.

The resulting imaging setups used in this chapter are described in Figure 8.1. The setups differ for imbibition and elongation monitoring. Figure 8.1a describes the setup as used in Belin et al. (2011) for seedling elongation monitoring. The seedlings are placed in a system with regulated temperature and water content. The opening and closure of the system can be controlled and synchronized with the image acquisition of the thermal camera. Because of mechanical limitations the minimum time interval between two acquisitions is 2 min. This is not a limitation for

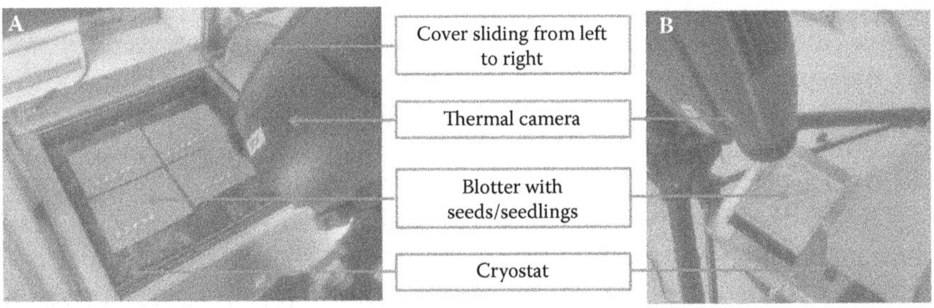

Figure 8.1 Thermal imaging computer vision system working in no-light conditions. Thermal images are acquired with a FLIR SC 5000 cooled infrared camera with a spectral range between 2.5 and 5 μm, a spatial resolution of 320 × 240 pixels, and a 12-bit dynamic. The thermal imager is positioned vertically above the seeds/seedlings. (a) Experimental setup for thermal imaging of seedlings during elongation phase. Seedlings are installed on blotter placed in a cryostat regulated at 20°C. (b) Experimental setup for thermal imaging of seeds' imbibition. Cryostat is regulated at 24°C. The entire system is placed in a growth chamber regulated at 24°C.

elongation monitoring, which is a biological process lasting several days. This is more limiting for the monitoring of imbibition, which is a faster biological process with important changes with time at hourly intervals. Figure 8.1b describes the setup used for seed imbibition and germination monitoring and reproducing the one introduced in Kranner et al. (2010).

8.3 *Thermal imaging applied to seed imbibition*

Thermal imaging was used to study seed imbibition and germination of different seeds species, such as the pea, oilseed rape, and wheat (Kranner et al., 2010). Seed imbibition is the very early stage essential for germination, and seed germination rates are dependent on imbibition rate. Seed imbibition was found to be associated with seed changes in temperature, detected in all the studied species, but with more or less large differences compared to dry seeds. This method would thus allow recording imbibition noninvasively. During the experiments carried out at 24°C, after a first and brief stage (<1 h) showing a temperature increase, a sharp temperature decrease was observed during 4–5 h, followed by a progressive reincrease in temperature. This temperature profile was more pronounced for the pea, with a drop in temperature of more than 1°C. It was also detectable, but limited to less than 0.1°C in wheat, and this drop was even lower in rape seeds. These changes in temperature were used to detect and predict nonviable seeds in the first 3 h of water uptake, a span of time that allows us to redry and store seeds after analysis without killing them. These typical temperature profiles measured for seeds were also compared for peas to a model of seed thermogenesis obtained using temperature calibration profiles measured for different soluble sugars and starch diluted at different concentrations. This approach indicated that the measured seed temperature profiles were coherent with biochemical changes and seed reserve transformation during germination. This opens roads for the use of infrared thermography to detect changes in seed and seedling metabolism noninvasively.

In the present report, as visible in Figure 8.2, we have recorded nonmonotonic temporal thermal evolution, as in Kranner et al. (2010), during the first hours following the beginning of the imbibition. This was obtained for pea seeds for which the thermal contrast was high, but we did not observe similar signatures with other species such as *M. truncatula* or sugar beet. We tried various techniques for seed imbibition, including Petri dishes on agar, sand, or blotter, like in Kranner et al. (2010). Such techniques are used for the monitoring with conventional RGB computer vision tools but did not allow recording of the short transitory period following the beginning of the imbibition. We therefore developed a setup similar to the one of Kranner et al. (2010) with polycarbonate plates, including multiple wells, each with an aperture at the bottom. When placed in a

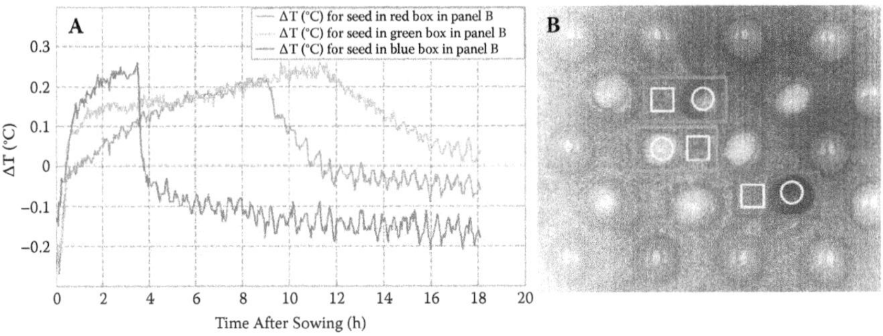

Figure 8.2 **(See color insert.)** Evolution of pea seeds' temperature during imbibition. (a) Evolution of the relative seed temperature ΔT after sowing for three different seeds. (b) Pea seeds sowed on a blotter. Studied seeds are identified with red, green, and blue boxes. Growth chamber, blotter, and water temperatures are fixed at 24°C. The relative seed temperature ΔT ($T_{seed} - T_{env.}$) is defined as the difference between mean temperature T_{seed} of individual seed (white circles) and mean temperature of close environment $T_{env.}$ of seed (white squares).

water bath (temperature kept constant at 24°C, checked with a calibrated thermometer), the plates floated and the wells were constantly filled with water. The nonmonotonic thermal signature of Figure 8.2a is more pronounced, i.e., displays a larger variation of temperature, for higher temperature of the water and larger surface of contact between the water and the seed. This is made possible by maximizing the surface contact between the seeds and the water supply.

8.4 Thermal imaging applied to seed elongation

Seedling elongation is an early stage of the development of plants. During this stage, the seed is in the soil. Following a geotropism, the seedling shoot grows to reach the light and activates photosynthesis, while the lower part of the seedling, the radicle, goes deeper, anchors in the soil, and gives larger access to water and nutrients. The rates of elongation depend on environmental conditions, and also on species and genotype. Phenotyping the genetic diversity of these plant parameters could help breeding for improved plant tolerance to environmental stresses. In addition, differences in early radicle growth rates may be predictive of differences in adult plant root growth, which are much more difficult to measure. These plant parameters have also been used in plant growth models that predict emergence and crop growth. Finally, more deterministic molecular studies on plant growth use hypocotyls as model plant parts for which different mutants are available and which need to be phenotyped (Fankhauser and Staiger, 2002; Nagy and Schafer, 2002). Computer vision tools therefore stand as interesting solutions to monitor seedling elongation. There exist a variety of approaches using conventional imaging with light in the visible spectrum (Kimura and Yamasaki, 2003; Basu et al., 2007; Armengaud et al., 2009; French et al., 2009; Wang et al., 2009; Yazdanbakhsh and Fisahn, 2010; Naeem et al., 2011; Cole et al., 2011). One limit of these computer vision tools is the use of light for image acquisition. The seedling grows in the soil in the absence of light. If exposed to visible light, its upper part stops elongating because of the seedling light receptors and new leaves expand. The study of seedling heterotrophic growth thus requires mimicking dark conditions. This is afforded by thermal imaging, which operates with infrared radiation and no need of visible light. Belin et al. (2011) illustrated the possibility of transposing the computer vision tool developed for seedling growth with visible imaging to the domain of infrared thermal imaging. This is possible because the thermal contrast between the seedling and the background is strong enough (typically 0.3°C). An additional contrast was also shown in thermal images, as a thermal discontinuity was visible along the seedlings themselves. The location of this thermal discontinuity is precisely lying at the frontier between the radicle and the hypocotyl. The hypocotyl part is hotter than the radicle part with an average difference of 0.5°C.

Before pushing further the investigation of thermal imaging applied to seedling elongation, we first consider the impact of environmental parameters on the thermal discontinuity. The imaging system of Figure 8.1a is equipped with an opening of the sliding cover synchronized with image acquisition. The opening of the cover is likely to bring convection. It is therefore important to control the impact of the opening of the cover on the thermal contrast observed on the seedlings. The entire process of cover opening (open, then remain open, then close) lasts in our experiment for 120 s. To study the influence of this opening in terms of thermal exchanges, we have made a continuous recording during the 120 s of the opening of the cover. The results are presented in Figures 8.3 and 8.4. As visible in Figure 8.3, the temperature of the blotter slowly decreases during cover opening. The variation is around 0.2°C between the beginning and end. During cover opening, temperatures for seedlings almost remain constant, taking into account standard deviation. Concerning the thermal discontinuity along the seedlings themselves, as visible in Figure 8.4 for three *M. truncatula* seedlings, despite some thermal variations, the difference of temperature ($T_{hyp} - T_{rad}$) remains positive during the 120 s. This shows that the thermal discontinuity (with the hypocotyl part hotter than the radicle part) between seedling parts is well preserved during all the duration of cover opening of 120 s used in our procedure.

Another question is the impact of lightening on the thermal discontinuity. It is possible to work with the imaging system of Figure 8.1a in dark conditions or to add light. Photoreceptors influencing germination and seedling growth are sensitive in blue and red/far-red wavelengths (Smith,

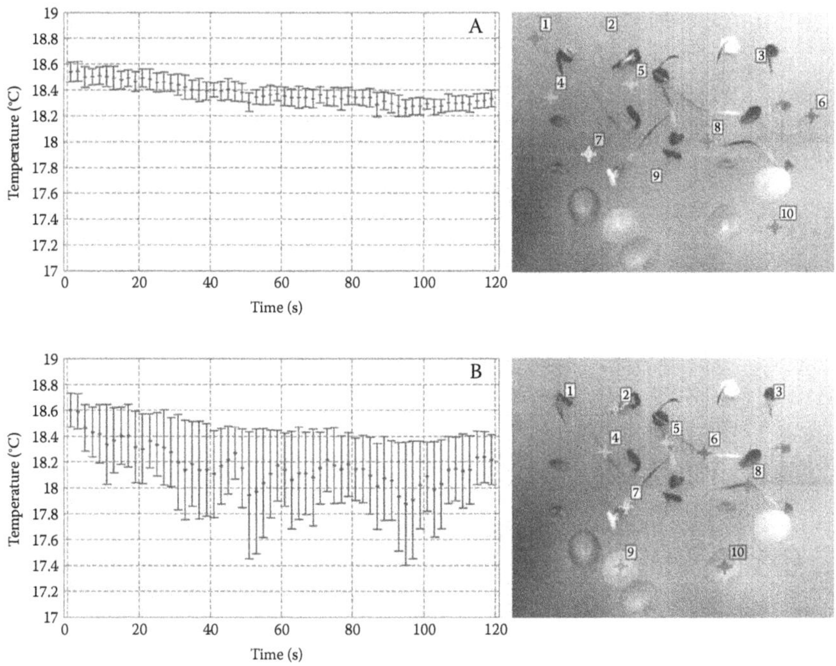

Figure 8.3 Thermal impact of cover opening. (a) Mean evolution (± standard deviation) of temperature of blotter during cover opening (duration: 120 s). Measurements are done in 10 different locations (color crosses on right image) on the blotter. (b) Mean evolution (± standard deviation) of temperature of seedling during cover opening (duration: 120 s). Measurements are done in 10 different locations (color crosses on right image) on seedlings (cotyledons, root system).

2000). One can wonder if the light has an influence on the thermal discontinuity observed. As visible in Figure 8.5 and Table 8.1, the thermal discontinuities are of the same order of magnitude in the presence of white light-emitting diode (LED) and in dark conditions. We observe no impact of such LED light on the thermal discontinuity. However, light will impact the elongation speed, and it is interesting to note that thermal imaging operating in dark and light conditions would be a potential tool to study the influence of light on seedling elongation.

Having estimated the absence of impact of environmental parameters on the thermal contrasts we measured, we push forward our investigation on the interest of thermal imaging for seedling elongation monitoring on seedlings from other species with the same procedure used for *M. truncatula* (Belin et al., 2011). Seedlings of nine different species have been placed under the imaging system of Figure 8.1a. In Figure 8.6, a thermal difference is visible between the seedlings and the background and along the seedlings themselves. The thermal discontinuity along the seedlings

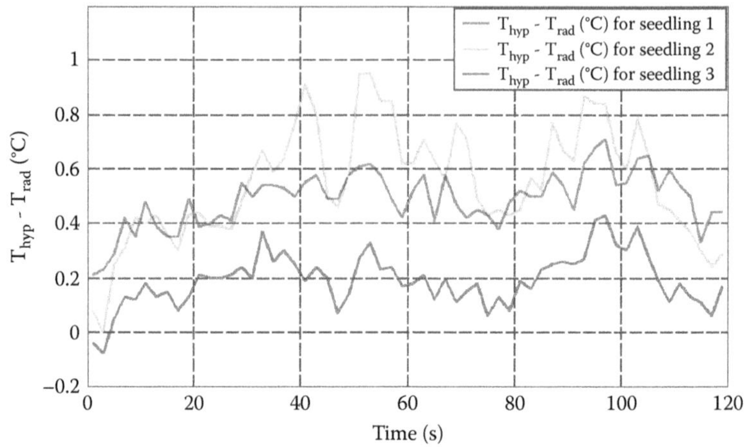

Figure 8.4 Time evolution of the difference of temperature (T_{hyp} − T_{rad}) between seedling organs (radicle and hypocotyl) during cover opening for three different seedlings of *M. truncatula*.

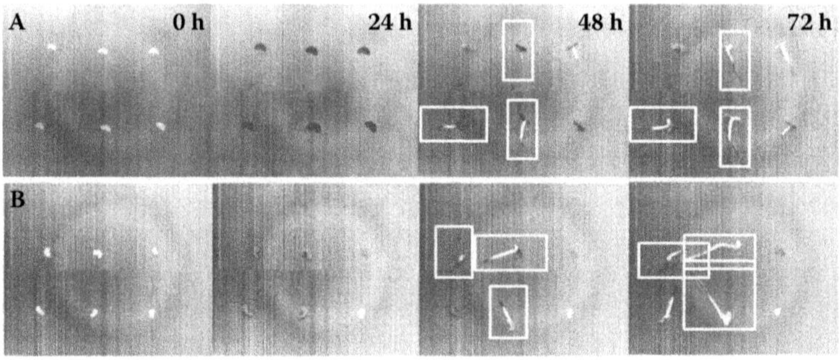

Figure 8.5 Influence of light on organs' temperature during elongation of seedlings of *M. truncatula*. Seedlings are illuminated in white light LED (a) during 1 min every 15 min, starting after sowing (0 h) and ending 3 days after (72 h). (b) Reference, as seedlings are kept in dark conditions for the entire duration of the experiment. The study of light influence on organ temperatures is taken at two measurements (48 h and 72 h), and it takes into account an average result of three seedlings (white boxes) per condition. Measurement points on seedlings are chosen manually in areas where contrasts are easily visible.

Table 8.1 Mean Value of the Difference of Temperatures ($T_{hyp} - T_{rad}$) between Organs (Hypocotyl and Radicle) Taken into Account for Three Seedlings and Two Conditions as Described in Figure 8.5

	White light	Dark conditions
Mean value of the difference of temperature at 48 h (°C)	0.900	0.945
Mean value of the difference of temperature at 72 h (°C)	0.896	0.662

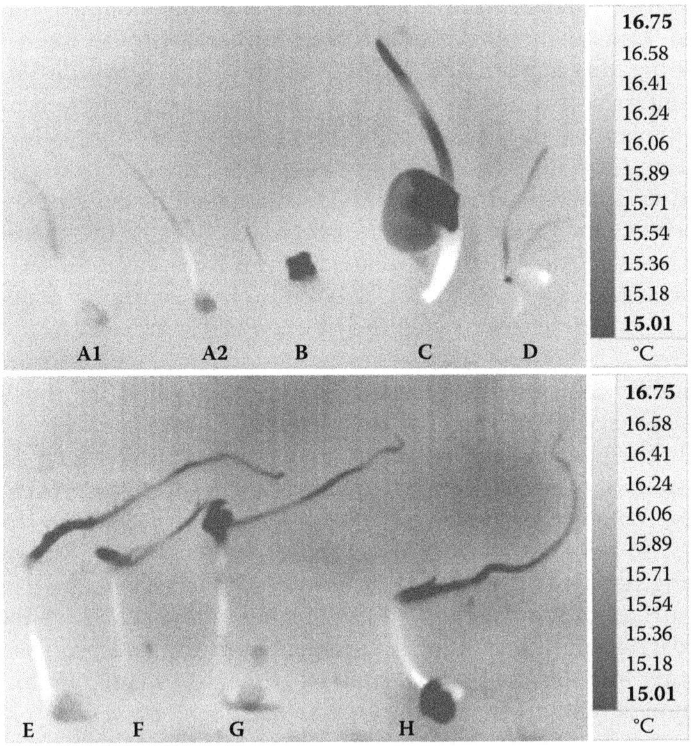

Figure 8.6 Thermal images of seedlings of various species during elongation phase. Seedlings are positioned with the entire root system (radicle + hypocotyl) in physical contact with the wet blotter. Seedlings are placed on the wet blotter 2 h before image acquisition, so that observed contrasts cannot be related to the thermal changing of the seedlings' environment. A1, *M. truncatula* (genotype A17); A2, *M. truncatula* (genotype F83); B, sugar beet; C, pea; D, wheat; E, mustard; F, ray grass; G, turnip rape; H, radish.

is located at the frontier between the radicle and the hypocotyl just as observed with *Medicago*. Illustrations of Figure 8.6 demonstrate the robustness and genericity of the presence of a thermal discontinuity observed between seedling parts of various species.

A first interpretation of the origin of this thermal discontinuity is based on functional hypotheses. During seedling heterotrophic growth, large fluxes of metabolites move from the storage organs to the elongating zones. The elongating radicle receives metabolites from seed reserve hydrolysis, and absorbs water and minerals from the growth support, as well as the seedling upper part growing toward the soil surface. Osmolyte molecules, like soluble sugars, are stored in the large cell vacuoles (Elamrani et al., 1992; Verscht, 2006). Contrasts in seedling parts in thermal images could therefore be due to such differences in fluxes and composition. Measurements of thermal discontinuities proved to be efficient for investigating sap streams on adult plants (Weibel and de Vos, 1994). Such contrasts obtained by heating a specific point in a plant part and linked to plant metabolism could be investigated in seedlings. It would be interesting to analyze how the amplitude of the thermal contrast during elongation is correlated with the chemical content of the seedling parts to investigate the validity of this functional hypothesis. The thermal contrasts observed for the nine species, varying from around 0.1°C for the lowest contrasts to around 0.5°C for the highest contrasts, may be correlated with differences in seedling biochemical compositions according to species.

Another complementary interpretation to account for the thermal contrast observed between the radicle and the hypocotyl is more anatomical. The epiderm of the radicle develops specifically long cells corresponding to very small root hairs, increasing the surface of exchange enabling the absorption of water. These long cells, measuring up to 100 microns, do not exist in the hypocotyl part. In addition to increasing the absorbing surface of the radicle, the absorbing cells change the roughness of the surface of the radicle. This could constitute a change of the emissivity, decreasing the perceived radiation from the radicle. It would be interesting to test if the thermal contrast still exists when monitoring the elongation of mutants that do not contain these long absorbing cells. Such experiments to interpret the thermal contrast observed during elongation of seedlings could be interesting to constitute thermal contrast as a new potential phenotypic trait to discriminate between different populations in a nondestructive and noninvasive way.

8.5 Conclusion

In this chapter, we have presented thermal imaging as a performance tool for the monitoring of seedling development, including germination and elongation phases. During imbibition, thermal imaging can predict whether a quiescent seed is dead or will germinate upon water uptake (Kranner et al., 2010). During elongation, thermal imaging serves to distinguish between radicle and hypocotyl (Belin et al., 2011). We have demonstrated the reproducibility of these recent results, especially by investigating the impact of environmental parameters and of the experimental setup. We have also demonstrated that similar thermal discontinuities are also observable with other seed species. The physiological interpretation of the observed thermal discontinuity is still open for further investigations.

Acknowledgments

Landry Benoit acknowledges financial support for the preparation of his PhD from Angers Loire Métropole and GEVES-SNES. This work received support from the French government supervised by the Agence Nationale de la Recherche in the framework of the program Investissements d'Avenir under reference ANR-11-BTBR-0007.

References

Armengaud, P., Zambaux, K., Hills, A., Sulpice, R., Pattison, R., Blatt, M., and Antmann, A., Ez-rhizo: integrated software for the fast and accurate measurement of root system architecture, *Plant J.*, 57, 945–956, 2009.

Baranowski, P., Mazurek, W., and Walczak, R.T., The use of thermography for pre-sowing evaluation of seed germination capacity, in *Proceedings of the International Conference on Quality Chains: An Integrated View on Fruit and Vegetable Quality*, Wageningen, Netherlands, 2003, vols. 1 and 2.

Basu, P., Pal, A., Lynch, J., and Brown, K., A novel image-analysis technique for kinematic study of growth and curvature, *Plant Physiol.*, 145, 305–316, 2007.

Belin, E., Rousseau, D., Boureau, T., and Caffier, V., Thermography versus chlorophyll fluorescence imaging for detection and quantification of apple scab, *Comput. Electron. Agric.*, 90, 159–163, 2013.

Belin, E., Rousseau, D., Rojas-Varela, J., Demilly, D., Wagner, M.H., Cathala, M.H., and Durr, C., Thermography as a non-invasive imaging for monitoring seedling growth, *Comput. Electron. Agric.*, 79, 236–240, 2011.

Breitenstein, O., Warta, W., and Langenkamp, M., *Lock-in thermography*, Springer, New York, 2003.

Chaerle, L., Leinonen, I., Jones, H.G., and der Straeten, D.V., Monitoring and screening plant populations with combined thermal and chlorophyll fluorescence imaging, *J. Exp. Bot.*, 58, 773–784, 2007.

Chaerle, L., and der Straeten, D.V., Seeing is believing: imaging techniques to monitor plant health, *Biochim. Biophys. Acta*, 1519, 153–166, 2001.

Cole, B., Kay, S.A., and Chory, J., Automated analysis of hypocotyl growth dynamics during shade avoidance in *Arabidopsis*, *Plant J.*, 65, 991–1000, 2011.

Dell'Aquila, A., Development of novel techniques in conditioning, testing and sorting seed physiological quality, *Seed Sci. Technol.*, 37, 608–624, 2009.

Elamrani, A., Gaudillere, J.P., and Raymond, P., Nature and utilization of seed reserves during germination and heterotrophic growth of young sugar beet seedlings, *Seed Sci. Res.*, 2, 1–8, 1992.

Fankhauser, C., and Staiger, D., Photoreceptors in *Arabidopsis thaliana*: light perception, signal transduction and entrainment of the endogeneous clock, *Planta*, 216, 1–16, 2002.

Fiorani, F., Rascher, U., Jahnke, S., and Schurr, U., Imaging plants dynamics in heterogenic environments, *Curr. Opin. Biotechnol.*, 23, 227, 2012.

French, A., Ubeda-Thomas, S., Holman, T.J., Bennett, M.J., and Pridmore, T., High-throughput quantification of root growth using a novel image-analysis tool, *Plant Physiol.*, 150, 1784–1795, 2009.

Kana, R., and Vass, I., Thermo-imaging as a tool for studying light-induced heating of leaves correlation of heat dissipation with the efficiency of photosystem II photochemistry and non-photochemical quenching, *Environ. Exp. Bot.*, 57, 90–96, 2008.

Kimura, K., and Yamasaki, S., Accurate root length and diameter measurement using NIH image: use of Pythagorean distance for diameter estimation, *Plant Soil*, 254, 305–315, 2003.

Kranner, I., Kastberger, G., Hartbauer, M., and Pritchard, H.W., Non-invasive diagnosis of seed viability using infrared thermography, *Proc. Natl. Acad. Sci. USA*, 107, 3912–3917, 2010.

Kümmerlen, B., Dauwe, S., Schmundt, D., and Schurr, U., Thermography to measure water relations of plant leaves, in *Handbook of computer vision and applications 3*, ed. B. Jähne, H. Haussecker, and P. Geissler, 763–781, Academic Press, Boston, 1999.

Leinonen, I., Grant, O.M., Tagliavia, C.P.P., Chaves, M.M., and Jones, H.G., Estimating stomatal conductance with thermal imagery, *Plant Cell Environ.*, 29, 1508–1518, 2006.

Naeem, A., French, A., Wells, D., and Pridmore, T., High-throughput feature counting and measurement of roots, *Bioinformatics*, 27, 1337–1338, 2011.

Nagy, F., and Schafer, E., Phytochromes control photomorphogenesis by differentially regulated, interaction signaling pathways in higher plants, *Annu. Rev. Plant Biol.*, 53, 329–355, 2002.

Smith, H., Phytochromes and light signal by plants—an emerging synthesis, *Nature*, 407, 585–591, 2000.

Verscht, J., Sugar concentrations along and across the *Ricinus communis* L. hypocotyls measured by single cell sampling analysis, *Planta*, 224, 1303–1317, 2006.

Vollmer, M., and Mollmann, K., *Infrared thermal imaging: fundamentals, research and applications*, Wiley, Weinheim, 2010.

Wang, L., Uilecan, I.V., Assadi, A.H., Kozmik, C.A., and Spalding, E.P., Hypotrace: image analysis software for measuring hypocotyl growth and shape demonstrated on *Arabidopsis* seedlings undergoing photomorphogenesis, *Plant Physiol.*, 149, 1632–1637, 2009.

Wang, X.Z., Yang, W.P., Wheaton, A., Cooley, N., and Moran, B., Automated canopy temperature estimation via infrared thermography: a first step towards automated plant water stress monitoring, *Comput. Electron. Agric.*, 73, 74–83, 2010a.

Wang, X.Z., Yang, W.P., Wheaton, A., Cooley, N., and Moran, B., Efficient registration of optical and IR images for automatic plant water stress assessment, *Comput. Electron. Agric.*, 74, 230–237, 2010b.

Weibel, F.P., and de Vos, J.A., Transpiration measurements on apple trees with an improved stem heat balance method, *Plant Soil*, 166, 203–209, 1994.
Wisniewski, M., Glenn, D.M., Gusta, L., and Fuller, M.P., Using infrared thermography to study freezing in plants, *HortScience*, 43, 1648–1651, 2008.
Yazdanbakhsh, N., and Fisahn, J., Analysis of *Arabidopsis thaliana* root growth kinetics with high temporal and spatial resolution, *Ann. Bot.*, 105, 783–791, 2010.

chapter nine

Anatomofunctional bimodality imaging for plant phenotyping: An insight through depth imaging coupled to thermal imaging

Yann Chéné, Étienne Belin, François Chapeau-Blondeau, Valérie Caffier, Tristan Boureau, and David Rousseau

Contents

9.1 Introduction

The shoots of the plants constitute challenging scenes for computer vision. Leaves, distributed with multiple orientations and various sizes, are spatially arranged along the three-dimensional (3D) networked architecture of the plants. When observed with imaging systems, the shoots of the plants form complex 3D textures with various illuminations, depths, or color details. These have been shown, for instance, to display fractal properties with scale invariance in 2D projections (Ruderman and Bialek, 1994) or in 3D (Boudon et al., 2013). As a result, the detection, or the segmentation, of items in such complex scenes raises specific questions that may require selection, adaptation, or even the developments of new imaging systems or of new image processing algorithms. These rather fundamental informational concerns also meet the need for plant scientists to perform noninvasive, rapid, automatic observations on large populations of plants to confront phenotypic data and genomic data now available at high throughput. Among applications requiring high-throughput phenotyping of plants with computer vision is the segmentation of individual leaves on the entire plant (Biskup et al., 2007; Omasa et al., 2007; Klose et al., 2009; Kraft et al., 2010; Fiorani et al., 2012; Chéné et al., 2012). Different techniques of depth cameras have been used in the researches with a variety of observation scales, including small plants (Klose et al., 2009; Kraft et al., 2010; Fiorani et al., 2012), more structured shoots of entire plants (Chéné et al., 2012), and canopy (Biskup et al., 2007; Omasa et al., 2007). It would therefore be interesting to push forward the proofs of feasibility given in the works (Biskup et al., 2007; Omasa et al., 2007; Klose et al., 2009; Kraft et al., 2010; Fiorani et al., 2012; Chéné et al., 2012) by comparing how the depth cameras compete for leaf segmentation on the same plants. We propose to contribute in this direction in this chapter. Leaf segmentation from depth information gives access to shape measurement as illustrated in several reports (Biskup et al., 2007; Omasa et al., 2007; Klose et al., 2009; Kraft et al., 2010; Fiorani et al., 2012; Chéné et al., 2012), but as underlined in Fiorani et al. (2012) and Chéné et al. (2012), it is also possible to couple the anatomic information extracted from depth imaging with functional information like water content, chlorophyll efficiency, nutrient content, or pathogen presence at the scale of the leaf. Anatomofunctional imaging providing registered bimodal information has been developed for biomedical imaging with the anatomic information of, for instance, computed tomography coupled with the functional imaging of magnetic resonance imaging or positron emitted tomography. Such anatomofunctional imaging modalities have been applied to plant imaging, but their large-scale application is rather limited by their cost and low throughput (Jahnke et al., 2009; Fiorani et al., 2012). In this report we demonstrate how the association of depth imaging with an existing functional imaging system could constitute a new direction of instrumentation development for a variety of low-cost anatomofunctional bimodal imaging in plant phenotyping. The chapter is organized as follows. Section 9.2 gives a classification and a comparison of depth cameras. We then review in Section 9.3 the existing functional imaging available in plant science and illustrate with a study case the monitoring of the development of a pathogen, apple scab, on an entire plant at the scale of the leaf with functional information extracted from thermal imaging and anatomic information extracted from depth imaging. Section 9.4 details a registration approach between the anatomic and functional images with examples on apple scab. Section 9.5 concludes and gives a set of perspective opened by this work.

9.2 Contactless optical depth cameras

3D scanning methods can be classified in terms of physical principles such as contact or contactless, destructive or nondestructive, reflection or transmission, and optical or nonoptical, as organized in Figure 9.1. In this report, we focus on depth cameras based on contactless optical 3D scanning methods. These cameras use active or passive methods. In passive methods, the depth is estimated from scene properties, sensor position, and sensor geometry in ambient natural light (see Ballan et al., 2009, for a recent overview). In active methods, a controlled light source illuminates the scene and depth is estimated from information included in the scene reflectance (Blais, 2004, for a review with some manufacturers links). The choice of a given depth camera depends on various parameters (Sansoni et al., 2009, for a methodological approach), including depth accuracy, depth range, experimental conditions (indoor or outdoor, with or without natural light), acquisition time (object in movement or not), object properties (textured or not), and the allocated budget. In this chapter, we propose to review the physical principles at work for the depth estimation and the resulting limitations for each method highlighted in Figure 9.1, and discuss their practical interest for leaf segmentation on plants.

9.2.1 Passive depth cameras

9.2.1.1 Multiview

9.2.1.1.1 Stereovision methods. Stereovision, also named computational stereovision, reconstructs scenes in depth from two or more different points of view taken with standard cameras. Figure 9.2a provides the scheme of a stereovision system with two cameras. For one real point *p* in the scene, each camera gives a projection of this point in its image plane (pixels *pl* and *pr* in Figure 9.2a). The projected points are conjugated. Extrinsic parameters of the cameras (relative positions of the projected points in camera images, global system geometry) and intrinsic parameters of the cameras (focal length, distortion) allow us to determine the real 3D position of the projected points. Intrinsic and extrinsic parameters can be determined thanks to a calibration step (Heikkila, 2000; Sturm and Maybank, 1999; Tsai, 1987; Zhang et al., 1999; Bouguet, 2004). The matching of the conjugated points in different images (two or more), as explained in Nalpantidis et al. (2007) and Brown et al. (2003), can be solved with

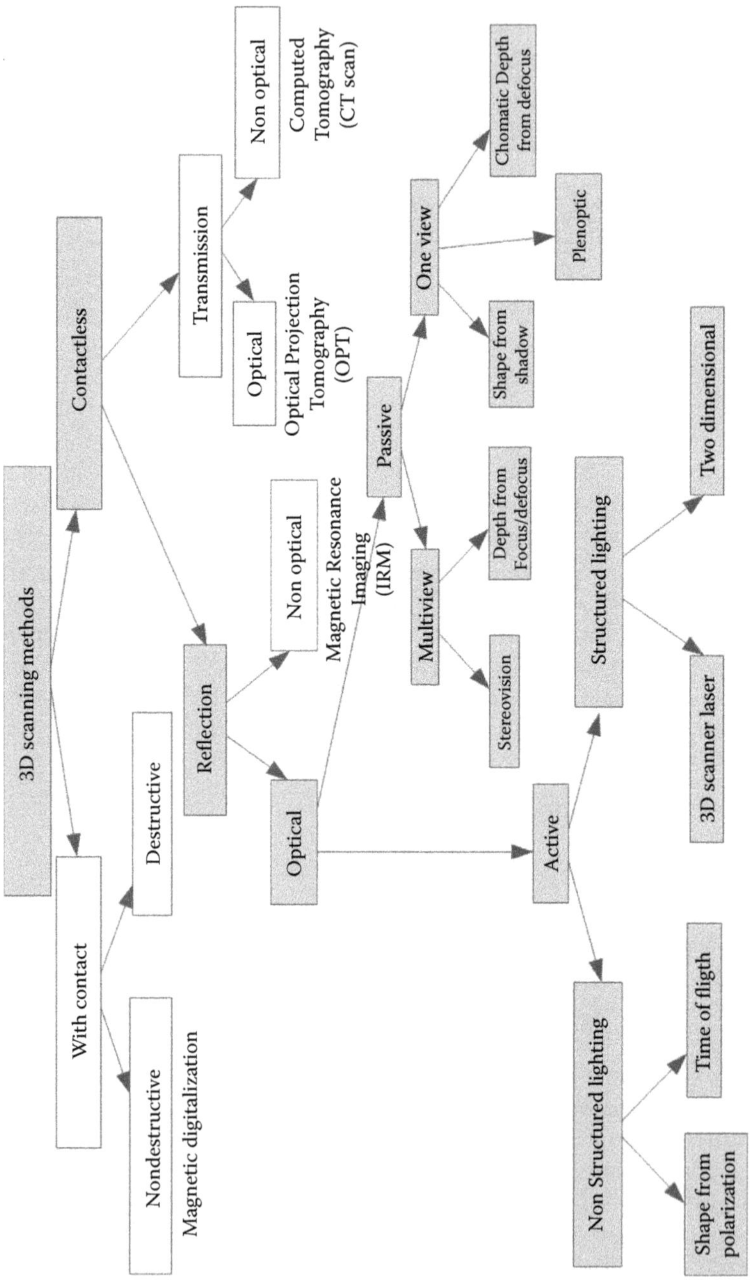

Figure 9.1 Nonexhaustive classification of depth imaging methods.

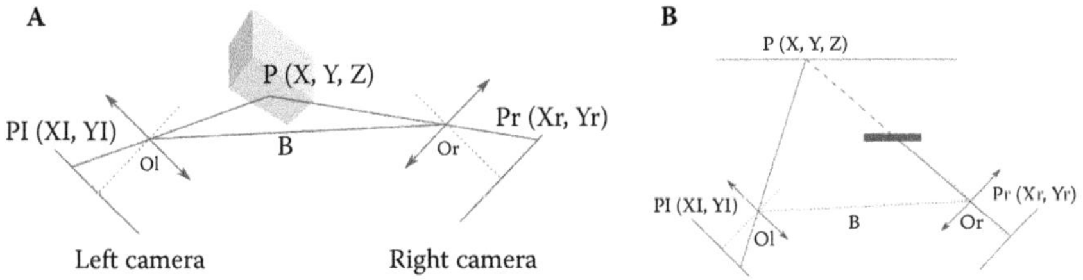

Figure 9.2 The stereovision system and its possible occlusions. (a) Principle scheme of a stereovision system with two cameras. (b) Illustration of occlusions.

local methods or global methods. Local methods analyze the subwindow around the considered point (Bhat and Nayar, 1998; Birchfield and Tomasi, 1999; Kluth, et al., 1992). These methods are fast but not very accurate. Global methods (Birchfield and Tomasi, 1999; Tomasi and Manduchi, 1998; Boykov et al., 2001; Scharstein and Szeliski, 1998; Sun et al., 2003; Kutulakos and Steiz, 2000) are more accurate but slower than local methods. Certain scene points are visible with one camera but not with the other camera, as shown in Figure 9.2b. The matching algorithms alone cannot solve such occlusion problems. Occlusion points can be detected by different algorithms either before or after matching of nonoccluded points (Wildes, 1991; Fua, 1993; Silva and Santos Victor, 2000; Birchfield and Tomasi, 1999). The use of more than two cameras improves accuracy and reduces occlusion problems (Satoh and Ohta, 1996; Nakamura et al., 1996). Stereovision systems can be composed of low-cost cameras. The depth accuracy of a stereovision system is a function of the distance between the cameras (the baseline *b*) and of the camera resolutions, but can be very high when matching and occlusion problems are solved. Stereovision systems can be used in a field or greenhouse. The relative position of cameras must not be modified after the calibration steps. The frame rate of stereovision systems depends on the chosen algorithms, but it is currently rare to access real-time acquisition and display of the depth with stereovision.

9.2.1.1.2 Depth from focus/defocus methods. Depth from focus methods and depth from defocus methods are based on a similar principle. As shown in Figure 9.3, an object point P of the scene is conjugated with the focus point P_0 across a lens. If the sensor plane is not superimposed with the focus plane, the image of P in the sensor plane is a disk of radius r. This translates in a blurry appearance in the image. The size of the disk depends on the focal length of the lens and on the distance d between the object plane and the lens. This blur can be modeled as a convolution by the point spread function (PSF). Depth from focus methods uses a measure of sharpness with translations of the object to determine its 3D shape. The object is set on a motorized translational stage (e.g., Minhas et al., 2009;

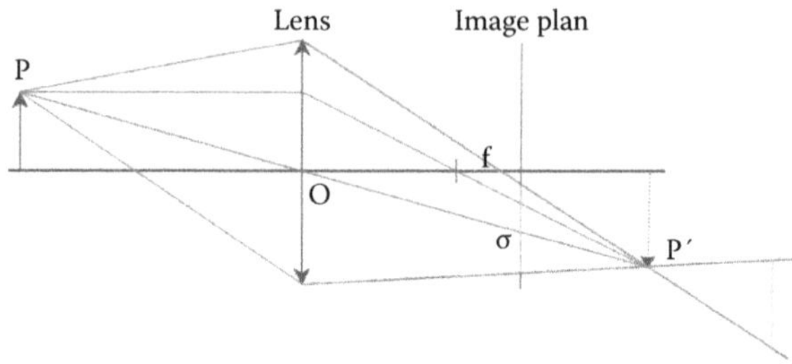

Figure 9.3 Principle scheme of blur in camera.

Nayar and Nakagawa, 1994). At each increment of the translation stage, the vision sensor acquires an image that is blurred for points out of the focused plane and sharp for points in the focus plane. In each point of the object, blur appearance is like the application of a low-pass filter on the focused image. The quantification of focus on each point of the object in an image can be done by quantifying the amount of high spatial frequencies (Xiong and Shafer, 1993; Nayar and Nakagawa, 1994; Subbarao and Choi, 1995; Martinez Baena et al., 1997; Choi and Yun, 2000; Helmli and Scherer, 2001; Ahmad and Choi, 2005, 2007; Malik and Choi, 2007, 2008; Minhas et al., 2009). All these focus measures work on highly textured objects. In each object point, the evolution of focus is computed as a function of the position of the translation stage. The position of the translational stage corresponding to the maxima of the focus measure gives the depth of the considered point. In depth from focus methods, the vision sensor can be a low-cost webcam. Due to the translation stage, depth from focus systems are cumbersome and hardly usable in the field. In depth from defocus methods the lens parameters (aperture, focal) are changed so that the scene does not have to be translated. The defocus measure consists in approximating the PSF in each point of the scene. The PSF can be approximated only for textured subwindows and for objects out of field depth. For each scene point, the PSF approximation is done locally in subwindows with a statistical framework (Rajagopalan and Chaudhuri, 1999; Schechner and Kiryati, 1999; Farid and Simoncelli, 1998) or with deterministic optimization (Xiong and Shafer, 1995; Gokstorp, 1994; Favaro and Soatto, 2000; Favaro et al., 2003; Trouvé et al., 2011). Depth from defocus methods must use vision sensors with well-known parameters (aperture, focal). The use of low-cost webcams for these methods is therefore prohibited. The depth from defocus systems are suitable for use in greenhouse on highly textured and nonmovable plants.

9.2.1.2 *One view*

9.2.1.2.1 Chromatic depth from defocus. The chromatic depth from the defocus method (Trouvé et al., 2012) uses depth from defocus principles, except that the different images required to estimate depth are acquired at the same time without changing lens parameters. It uses the association of color vision sensor and a chromatic lens. A color vision sensor uses a Bayer filter to have the red, green, and blue (RGB) values in each pixel. Each channel (red, green, and blue) of a color image represents the image of a definite wavelength interval. The focal length of a chromatic lens varies with wavelength. The association of a color vision sensor with a chromatic lens leads to three different blurred images, one for each channel. Then, for each pixel of each channel, the PSF can be locally approximated with depth from defocus methods. So, in each pixel there is a unique triplet for the estimation of the radius r. Finally, this triplet, obtained with only one acquisition, is used to estimate the depth of the scene at each pixel. Chromatic depth from defocus methods use a vision sensor associated with a chromatic lens, which is not low cost. The chromatic depth from defocus systems can be used in a field or greenhouse with plants contrasted in color and in texture.

9.2.1.2.2 Shape from shading methods. The shape of a continuous object influences the reflectance of light. Shapes from shading methods are based on the perfect knowledge of the position of the camera. The relative intensity in each camera pixel allows us to retrieve the value of the shape gradient norm. The shape gradient is equivalent to normal coordinates in each object point. The object shape is the integration of normal coordinates. To solve shape from shading problems, several approaches can be used. Minimization approaches solve the problem with an optimization process based on minimization of several constraints (Szeliski, 1991; Courteille et al., 2004; Crouzil et al., 2003). Constraints can be on brightness, smoothness, and gradient intensity. There are also propagation approaches (Bichsel and Pentland, 1992; Dupuis and Oliensis, 1992; Rouy and Tourin, 1992). These methods use reference points with a known shape or with unique solutions, like singular points (points with intensity maximum). Then, there are the local approaches where the type of shape expected is known (spherical, cylindrical) (Lee and Rosenfeld, 1985; Hayakawa et al., 1994). The normal direction to the shape at each pixel is determined with derivatives. Finally, there are the linear approaches where the nonlinear problem is solved as a linear problem (Pentland, 1990; Ping Sing and Shah, 1994). These methods linearize the luminance image. Zhang et al. (1999) and Durou et al. (2008) give an overview of shape from shading methods with comparisons. Shape from shading methods can be implemented with a low-cost webcam. However, its depth accuracy is the lowest among all presented depth cameras. These methods are disrupted for textured objects and by additional light. Objects must be continuous, and measures can be done in a field or greenhouse.

9.2.1.2.3 Plenoptic. The plenoptic (Ng et al., 2005; Lumsdaine and Georgiev, 2009) camera uses, like conventional vision system, a lens and a vision sensor. An additional array of microlenses is fixed before the vision sensor plane at a distance f corresponding to the focal length of each microlens of the sensor plane. The $n \times n$ sensor pixels matrix beyond each microlens allows us to capture the subimage provided by the microlens. In one acquisition of a scene, a plenoptic camera captures a stack of images for each sensor position of a conventional camera. This stack of focused images can be obtained with the Fourier slice photography (Ng, 2005) or the discrete focal stack transform from the global image (Nava et al., 2008). The scene depth can be processed, from the stack of focused images, with methods of depth from focus or depth from defocus (Bishop and Favaro, 2009; Lüke et al., 2009). Plenoptic systems are very costly. The depth accuracy depends on the level of texture of the scene. Plenoptic cameras can be used in a field or greenhouse.

9.2.2 *Active depth cameras*

9.2.2.1 *Structured lighting methods*

9.2.2.1.1 Three-dimensional laser scanner. Three-dimensional laser scanners (Amann et al., 2001; Pfeifer and Briese, 2007) use a laser as an optical source. The most widespread methods are based on triangulation, time of flight (TOF), phase shift, and frequency modulation. Similar to stereovision principle, the laser scanner triangulation methods (Veatch and Davis, 1990; Fu et al., 2012; Dorsch et al., 1994; Buzinski et al., 1992) rely on geometric properties. One camera of Figure 9.2a is substituted by the laser. The laser point is emitted toward the scene and the camera visualizes its reflection position. The relative position between the laser and the camera is known. Thus, the distance between the laser scanner reference and the scene can be computed thanks to the position of the reflection on the plane of the camera. With this method, the wider the baseline, the smaller the error on z. The error on z also depends on the knowledge accuracy of the relative position between the laser and the camera. Instead of the laser point, this method also works with laser stripes. Each range of the sampling stripe is defined by its position on the image sensor. The TOF method (Hiskett et al., 2008; Myllylä et al., 1999) uses the time between the emission and the reception of a laser pulse. The laser pulse is emitted at a known time. Then, it is scattered by the scene and captured by the photodiode. Knowing the speed of light, the distance between the laser scanner and the point scene can be computed. The minimum measurable depth is limited by the length of the emitted pulse. The depth error depends on the accuracy of the time calculator. The phase shift method (Nakamura et al., 2000; Poujouly and Journet, 2000) uses the phase shift between the emitted and the reflected laser wave. The emitted wave is continuous and its amplitude is modulated at a given frequency. The photodiode senses the reflected wave. Then, the phase difference is computed. The distance between laser scanner and the scene point is processed from this phase shift. The phase ambiguity, with a value more than 2π, limits the maximum measurable depth. The frequency modulation method (Piracha et al., 2010; Agishev et al., 2006) uses, like the phase shift method, a continuous laser wave with amplitude modulation, but its frequency is also modulated. The evolution of the frequency in time is used to determine depth. The reflected wave has the same frequency profile as the emitted wave, except that it is shifted in time. This shift in time corresponds to the time of flight of the wave, so that the depth can be estimated. In all these methods, to obtain a depth map, the laser source must scan the scene. This scanning is, in most cases, done with the association of a mirror fixed on a mechanical system that changes its orientation. The depth sensor space resolution depends on this system. Three-dimensional laser scanners give the highest depth accuracy among all the depth cameras presented here. However, they are also very high-cost 3D sensors. They can be used in a field or greenhouse. Objects must be fixed. Objects' color or texture did not disrupt depth measurement. Also, the time required for the mechanical scan limits the rate of image acquisition, which cannot be performed at video rates.

9.2.2.1.2 2D structured lighting. The use of a 2D structured lighting (see Geng, 2011; Salvi et al., 2010, for overviews) pattern allows us to scan the entire scene at the same time. In each case, the source of light is associated with a vision sensor. This sensor visualizes the deformation of the 2D structured lighting pattern emitted in the scene. The specific form of this pattern allows us to link each measure point of the emitted pattern with its conjugate in the reflected pattern. Finally, each scene depth is calculated by triangulation. One camera of Figure 9.2a is substituted by the 2D structured lighting pattern. The spatial resolution depends on the resolution of the emitted pattern. When scenes are static, it is possible to use sequential projections of 2D structured patterns. A method relies on the binary code principle. The association of all patterns allows us to code each stripe with a unique binary code (Ishii et al., 2007; Valkenburg and McIvor, 1998). However, a large number of patterns is needed to code a wide surface with a good spatial resolution. Thus, to decrease the number of required patterns, it is possible to add gray-level stripes in the code (Krattenthaler et al., 1994). Another method with the sequential projections is the phase shift method (Huang and Zhang, 2006). All patterns have a space sinusoidal amplitude modulation with a given frequency, but they are phase shifted. The 3D shape of the scene changes the phase of each pattern. The association of at least three patterns allows us to retrieve the phase deformation due to the arc tangent function. Finally, the 3D shape is retrieved by comparison with a reference. The use of arc tangent induces the need for an unwrapping step and forbids scanning of an the surface with wide depth skip. To improve the frame rate of the depth sensor and follow the moved object, continuous stripe indexing patterns or grid indexing patterns can be used. In continuous patterns, each stripe is coded by one gray level (Durdle et al., 1998) or color. The use of colors allows us to improve the stripe number and the stripes' segmentation. The rainbow 3D camera uses a continuous color pattern (Geng, 1996). Each stripe is defined by one color of the rainbow. Otherwise, stripes can be indexed by their composition. All stripes have the same color, but each stripe is segmented in several substripes, which have different lengths (Maruyama and Abe, 1993). For each stripe, the segmentation is unique. Moreover, stripes can be indexed in terms of their neighbors (Zhang et al., 2002; Monks and Carter, 1993). This indexation is based on the De Bruijn sequence. A De Bruijn sequence of rank n with k different intensity levels or colors allows us to code k^n stripes. In a De Bruijn sequence, each stripe is at the center of a unique n-length word. For stripe detection, two consecutive stripes cannot have the same color, so the number of possible stripes is less than k^n. A last method is the grid indexing. A grid is filled with a pseudorandom law. So each subwindow of size $n \times m$ is unique. This uniqueness allows us to link emitted points and reflected points, and there are 2^{nm} different possibilities with black and white dots (Le Moigne and Waxman, 1988; Ulusoy et al., 2009). To improve coding strength, each dot can be substituted by a codeword pattern. Codeword patterns can be coded in color (Desjardins and Payeur, 2007) or shape (Griffin et al., 1992). However, it is difficult to ensure uniqueness of each subwindow. As mentioned above, there are several types of 2D structured lighting methods. Each method has its own advantages and disadvantages. In this report, we consider a low-cost 2D lighting structured system that uses a near-infrared random dot grid. Its depth accuracy of typically 1 cm is good enough for certain plants, as illustrated in Chéné et al. (2012) and later in this chapter. This system can be disrupted by external infrared light. So its use in a greenhouse is prohibited. However, measures in the field can be done with object isolation relative to sunlight. Objects' color and texture do not disturb depth measurement, and objects can move.

9.2.2.2 *Nonstructured lighting*

9.2.2.2.1 Time of flight (TOF) depth camera. Time of flight (TOF) depth cameras (Kolb et al., 2009; Lange and Steiz, 2001) allow us to obtain an intensity map and a depth map at the same time. To our knowledge, two methods are currently used to determine the scene depth: the continuous modulation wave method and the optical shutter method. Both use near-infrared LEDs as a light source and a light sensor. In the continuous modulation wave method (Foix et al., 2011; Gokturk et al., 2004; Lange et al., 2000), the emitted wave is continuous and modulated in time. When light is reflected by the scene, the wave is disrupted and phase shifted in terms of the scene shape. The reflected wave is sensed, in each sensor pixel, with a phase shift due to the depth. Then, to find the phase shift in each pixel, the sensor computes a correlation between the emitted wave and the reflected wave. The distance between the TOF depth camera and the scene in each pixel results from the phase shift. The maximal measurable distance depends on the modulation frequency. In the optical shutter method (Iddan and Yahav, 2001; Yahav et al., 2007), the emitted wave is a short light pulse. The time of flight is limited by a faster shutter. This shutter is in front of the light sensor and is opened during a given time interval. So, the duration of this time interval sets the measurable depth interval. During the time interval, the reflected light comes in each pixel of the sensor. Thus, in each sensor pixel, the scene depth depends on the amount of light received by the pixel and the total amount of light received by all pixels. These systems can be used in a field or greenhouse. Objects' color and texture do not disturb depth measurement and objects can move.

9.2.3 *Comparison*

In this chapter, we use active depth cameras to constitute anatomofunctional imaging of plants. For illustration, we have considered the two plants of Figure 9.4 that we observed with four different depth cameras based on structured lighting methods. In this chapter the anatomical information of interest is the segmentation of the upper leaves of plants, accessible from the top view of the plant. As visible in Figures 9.5 and 9.6, the depth maps are similar, demonstrating that various choices of technologies can produce similar results in terms of anatomical information, the choice of the technologies depending on the cost or the

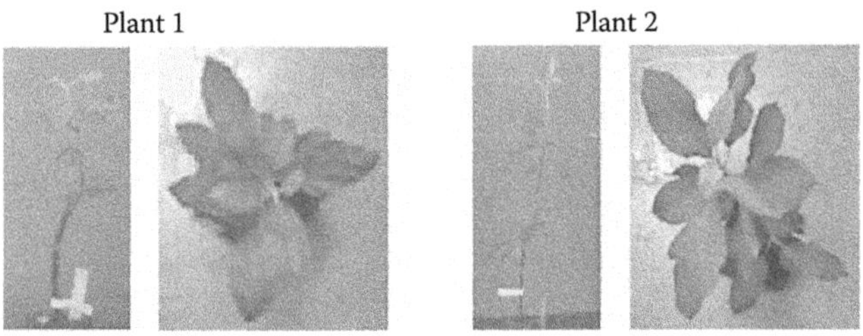

Figure 9.4 Lateral and top views of the two plants used for depth camera comparisons.

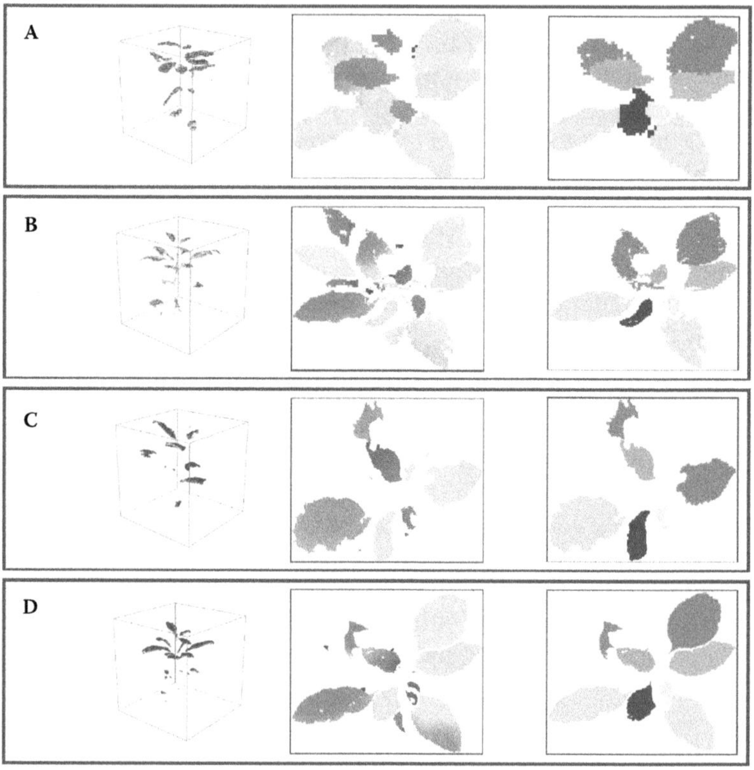

Figure 9.5 (**See color insert.**) Comparison of four depth cameras with the plant 1 of Figure 9.4. Each figure line is composed of the point cloud (left), the depth image (center), and the segmented image (right). In the image segmented with the algorithm of Chéné et al. (2012), only leaves presented in all depth images are represented with the same color for each depth camera. (a) Depth camera is a Microsoft Kinect that uses a pseudorandom dot grid as the structured lighting pattern. (b) Depth camera is a laser scanner that uses triangulation. (c) Depth camera is a hybrid depth camera that uses 2D structured lighting and stereovision (structured lighting pattern allows correspondences between two cameras). (d) Depth camera is a 2D structured lighting depth camera that uses a binary stripe sequential projection as the structured lighting pattern.

environmental constraints. A comparison of the four depth cameras used to produce Figures 9.5 and 9.6 is given in Table 9.1.

9.3 Functional imaging

We are now ready to consider the association of a functional imaging with the anatomical imaging reviewed in the previous section. In this report we illustrate anatomophysiological bimodality imaging with

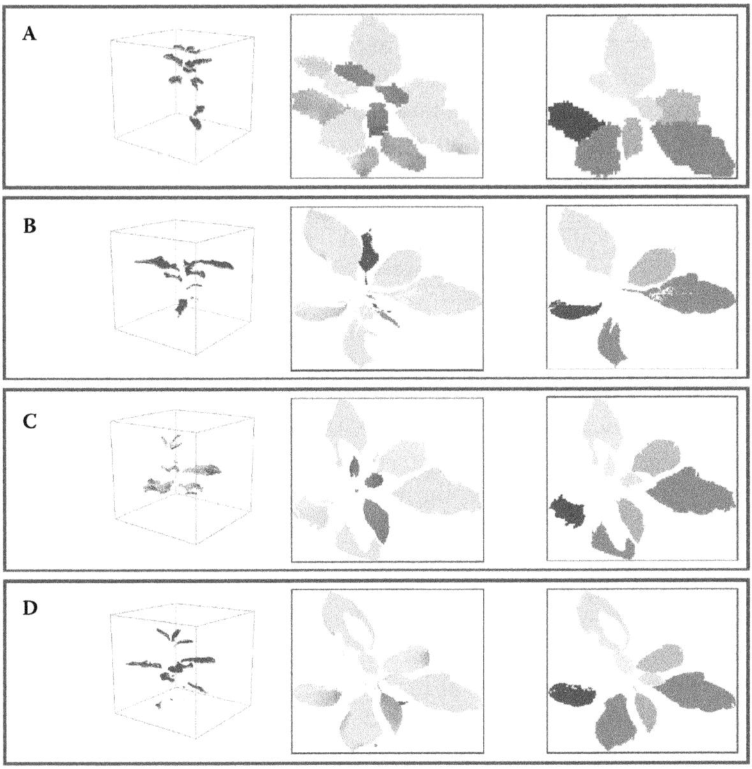

Figure 9.6 **(See color insert.)** Same as in Figure 9.5 with plant 2 of Figure 9.4.

Table 9.1 Characteristics of the Four Depth Cameras Used in Figures 9.5 and 9.6

	Kinect	Laser scanner	Hybrid method	2D structured lighting method
Price (€)	150	50,000	30,000	50,000
Lighting type	Infrared	Visible	Visible	Visible
Depth resolution (cm)	1	0.01	0.01	0.035
Points number in cloud	307,200	307,200	500,000	1,400,000
Field of view (°)	57 × 23	Unknown	23 × 21	Unknown
Depth interval (cm)	50–600	60–250	32.5–47.5	80 to unknown
Acquisition time (s)	0.03	3	0.01	3.5
Portability	+	–	+	–
Greenhouse used	No	Yes	Yes	Yes

a problem of host-pathogen interaction detection on the entire plant. There are various existing imaging modalities providing functional information on plants. Several review articles have been proposed for general purposes (see, for instance, Fiorani et al., 2012, for recent state of the art) or for some specific purposes, like host-pathogen interaction (Chaerle and Straeten, 2001; Sankaran et al., 2010). Pathogens (virus, bacteria, fungus) can be indirectly detected by their impact on the plant physiology. The main functional imaging modalities reported for plants are thermography, chlorophyll fluorescence imaging, and hyperspectral imaging. These imaging modalities operate at distinct optical wavelengths and provide distinct information. Thermography, gazing around 10 μm, is linked to many structural and functional characteristics, such as leaf orientation, heat capacity, surface properties, infrared absorption, and transpiration rate. Chlorophyll fluorescence imaging (Papageorgiou, 2004) indirectly correlates to the amount of excited electrons in photosynthetic light reactions and is thus related to the photochemical and nonphotochemical charge separation in photosystems. Hyperspectral imaging gives access to various meaningful spectral regions, such as the transition from red to near infrared, the peak of green reflectance, and the water absorption bands around 970, 1600, and 2100 nm. For a given biological question, a comparison of the contrast given by each functional imaging is a preliminary task. We focus our attention in this report on the detection of apple scab. Apple scab, which is caused by the fungus *Venturia inaequalis*, requires more than 10 fungicide treatments per year to be controlled and can be considered the most serious disease for apple (Bowen et al., 2011). Quantifying the development of apple scab at the leaf scale is of major importance for studying the apple-scab interaction, as well as for analyzing the evolution of pathogenicity in *Venturia inaequalis* populations, and for breeding scab-resistant apple cultivars. Recent studies have undertaken the comparison of the detectability of apple scab with thermography, hyperspectral, and fluorescence chlorophyll (Delalieux et al., 2009; Oerke et al., 2011; Belin et al., 2013). Thermography has been demonstrated to provide a better detectability, i.e., higher contrast at earlier dates. Briefly, in the acquisition conditions described in Belin et al. (2013), a max-min difference of temperature greater than 0.21°C on a leaf gives a probability of good detection higher than 0.9 and a probability of false alarm less than 0.15. Such preliminary detection measurement can be done on a single leaf. To translate this detection diagnostic to an entire plant, it is then necessary to apply the detection test on images separating each of the individual leaves. In the following we therefore propose to associate a depth camera with thermal imaging to perform apple scab detection on an entire plant.

9.4 Depth imaging coupled to thermal imaging

The use of several imaging systems allows the combination of different scene information. To combine information, relative positions (extrinsic parameters) and internal properties of systems (intrinsic parameters) must be known. Intrinsic and extrinsic parameters of an imaging system can be found with a calibration method. The calibration of multiple color cameras has been widely studied (Heikkila, 2000; Zhang et al., 1999; Sturm and Maybank, 1999; Tsai, 1987; Bouguet, 2004). The same calibration procedure can be used to calibrate thermography (Lagüela et al., 2011; Luhmann et al., 2010), chlorophyll fluorescence imaging (Bellasio et al., 2012), and hyperspectral imaging. The calibration of a depth camera depends on the physical principle used to find the depth of the scene (Underwood et al., 2007; Sansoni et al., 2000; Krotkov, 1991; Antone and Friedman, 2007) for laser scanner calibration (Fuchs and Hirzinger, 2008; Kahlmann et al., 2006; Lichti, 2008; Lindner and Kolb, 2006), for TOF depth camera calibration (Chen et al., 2009; Li et al., 2008; Zhang et al., 2002; Zhou and Zhang, 2005), or for 2D structured lighting system calibration. Once the calibration of each imaging system is known, the association of both imaging systems can be done. The most widespread associations found in the literature are a depth camera with color camera (Zhang and Pless, 2004; Unnikrishnan and Hebert, 2005) for laser scanner depth cameras (Lindner et al., 2007; Kim et al., 2008), for TOF depth cameras (Herrera et al., 2012; Zhang and Zhang, 2011; Geiger et al., 2012; Fortenbury and Guerra Filho, 2012), and for a 2D structured lighting depth camera. The association of a depth camera and a physiological imaging has already been reported. In a biomedical context, Nagatani et al. (2012) and Skala et al. (2011) show the association of a thermal imaging respectively with a scanner laser and with a 2D structured lighting depth camera. In a plant science context, Bellasio et al. (2012) performs the association of a chlorophyll fluorescence imaging and a 2D structured lighting depth camera.

To perform the association of a 2D structured lighting depth camera with thermal imaging, we use the method presented in Herrera et al. (2012). The thermal imaging system and the depth camera can both be modeled by the pinhole model, also named the perspective projection model. Figure 9.7 gives the principle scheme of the pinhole model. Then, the model, described in Heikkila (2000), is used to fix intrinsic and extrinsic parameters of each imaging system. In the pinhole model, the transformation of a 3D real point P_w in a 2D sensor pixel p follows a homography with the projection center O_s. Extrinsic parameters allow us to pass from the world coordinates reference $\{W\}$ to the system coordinates reference $\{S\}$. A point P_w with 3D coordinates $[X_w, Y_w, Z_w]^T$ in the world coordinates

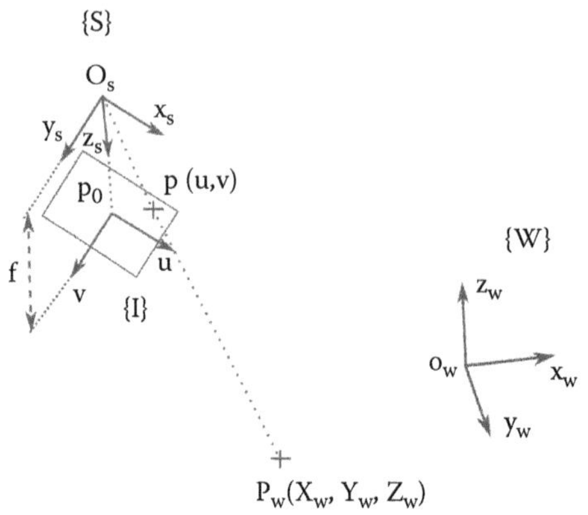

Figure 9.7 The principle scheme of the pinhole model.

reference $\{W\}$ becomes P_s with the coordinates $[X_s, Y_s, Z_s]^T$ in the system coordinates reference $\{S\}$ following

$$P_s = RP_w + t \tag{9.1}$$

where R is the 3×3 rotation matrix and t is the 3×1 translation matrix. In the pinhole model, the extrinsic parameters are the rotation matrix R and the translation matrix t.

Intrinsic parameters allow the transformation of P_s in p with 2D coordinates $[u, v]^T$ in the image coordinates reference $\{I\}$. First, the 3D point P_s is transformed in 2D point P_n by normalization, so coordinates of P_n are

$$P_n = \begin{bmatrix} X_n \\ Y_n \end{bmatrix} = \begin{bmatrix} \dfrac{X_s}{Z_s} \\ \dfrac{Y_s}{Z_s} \end{bmatrix} \tag{9.2}$$

Next, owing to optical system distortions (radial and tangential), P_n becomes P_d such as

$$P_d = (1 + k_1 r^2 + k_2 r^4 + k_5 r^6)P_n + P_g \tag{9.3}$$

where $P_g = [2k_3X_nY_n + k_4(r^2 + 2X_n^2), k_3(r^2 + 2Y_n^2) + 2k_4X_nY_n]^T$, $r^2 = X_n^2 + Y_n^2$, and k_i, with $i = 1, \ldots, 5$, are the distortion coefficients. The pixel image of the 3D real point P_w in the image coordinates reference I is

$$p = \begin{bmatrix} f_{cx} & 0 \\ 0 & f_{cy} \end{bmatrix} P_d + P_0 \tag{9.4}$$

where f_{cx} and f_{cy} are the focal length in pixels in directions x and y and $P_0 = [u_0, v_0]^T$ is named the principal point (intersection between the optical axis z_s and the image plane). In a pinhole model, intrinsic parameters are the focal length $f_c = [f_{cx}, f_{cy}]$, the principal point P_0, and the distortion vector $k = [k_1, \ldots, k_5]$. For depth cameras, depth calculation and depth distortion parameters can be added to intrinsic parameters. In the case of our 2D structured lighting depth camera, the depth is obtained from a disparity map. From Herrera et al. (2012), in each pixel, the disparity is distorted as follows:

$$d = d_k + D_\delta(u, v)\exp(\alpha_0 - \alpha_1 d) \tag{9.5}$$

with d the undistorted disparity, d_k the distorted disparity, $D_\delta(u, v)$ the spatial distortion pattern, and $\alpha = [\alpha_0, \alpha_1]$, which models the decay of the distortion effect. Then, the depth z comes as

$$z = \frac{1}{C_1 d - C_0} \tag{9.6}$$

where $c = [c_0, c_1]$ is the depth calculation matrix. Thus, α, $D_\delta(u, v)$, and c are added to intrinsic parameters of the depth camera. Once intrinsic and extrinsic parameters are defined, the calibration of the anatomofunctional imaging system can be done. Figure 9.8 gives the principle scheme of coupling two imaging systems modeled by the pinhole model. The calibration step allows us to know the relation between imaging systems. Each calibration could be done individually, but the common calibration of both imaging systems gives better results. First, intrinsic and extrinsic parameters are initialized by use of independent calibrations. The physiological imaging individual calibration is done with Zhang et al. (1999). This calibration uses a specific pattern, a chessboard with known dimensions. The chessboard must be well contrasted in terms of the information delivered by the physiological imaging system. For the depth camera, intrinsic parameters are initialized with well-known manufacturer values. Next, the relative position between imaging systems is approximated by comparing the chessboard plane equation in both calibration images. Finally, intrinsic and extrinsic parameters are optimized by minimization of the measurement reprojection errors with all the parameters. A standard Levenberg-Marquardt algorithm was used to solve the nonlinear minimization problem.

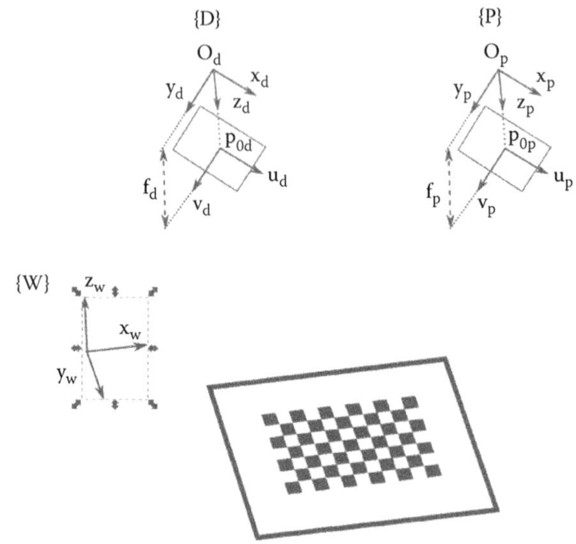

Figure 9.8 The principle scheme of coupling two imaging systems demonstrated by the pinhole model.

An example of an anatomofunctional imaging system is the coupling of our depth camera with a thermal imaging. Figure 9.9 (left) shows the rigid coupling. The whole system is calibrated with the method of Herrera et al. (2012). Figure 9.10 gives an example of the calibration of a scene. The calibration chessboard is in paper and sticks on a metallic plan. Use of a metallic plan allows us to have a well-contrasted calibration chessboard for thermal imaging, due to the strong difference in thermal reflectance between paper and metal. After calibration, the calibration file allows us

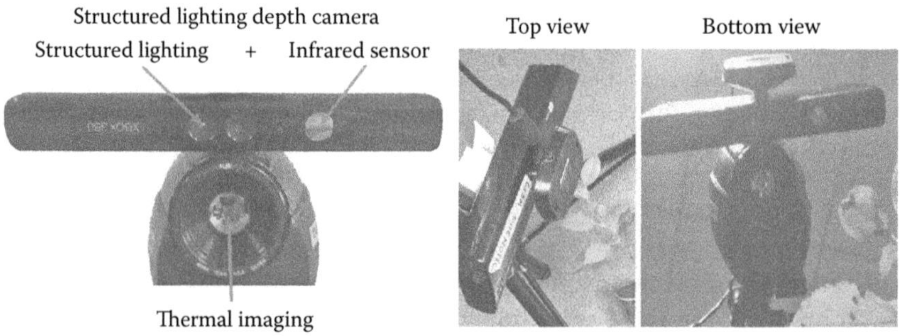

Figure 9.9 **(See color insert.)** Left: The anatomofunctional imaging system in this chapter is composed of a depth camera using the 2D structured lighting (binary dot pseudorandom grid) method and a thermal imaging. Right: An acquisition on a plant is done by one shot taken in plant top view.

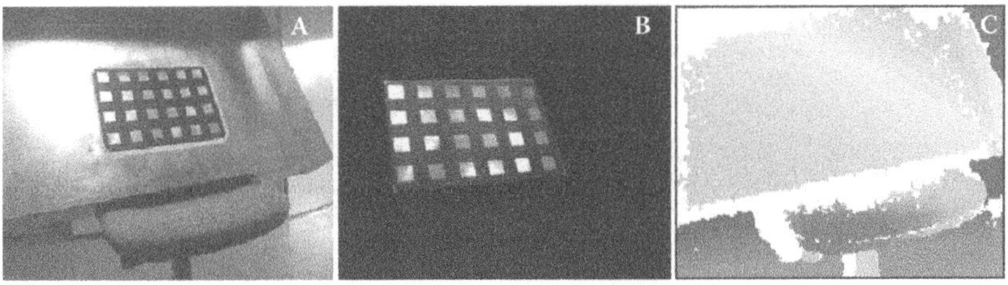

Figure 9.10 **(See color insert.)** A calibration scene. (a) The RGB image of the calibration chessboard on the metallic plane. (b) The thermal image of the calibration chessboard on the metallic plan. (c) The disparity map of the calibration chessboard plan.

to convert disparity in depth expressed in meters and to register the depth information on the thermal image. We illustrate the anatomofunctional imaging system with the example of plant 2 of Figure 9.4. As shown in Figure 9.9 (right), an acquisition on a plant is done by one shot taken from the plant top view. Figure 9.11 shows the thermal image and the registered depth map of the plant top view. The segmented image of leaves, visible in Figure 9.12, is obtained by application of the depth segmentation algorithm of Chéné et al. (2012). Application of the edges of the leaves of the plant on the thermal image is given to visualize calibration results. In this image, each leaf is associated with a number. This number allows the creation of a mask for each plant leaf. Figure 9.13 shows the application of each leaf mask on the thermal imaging. Figure 9.14 gives individual thermal images of each segmented leaf of a plant. Each image is the result of application of the leaf mask on the thermal image. The calibration step gives a good registration of depth image on the thermal image. There

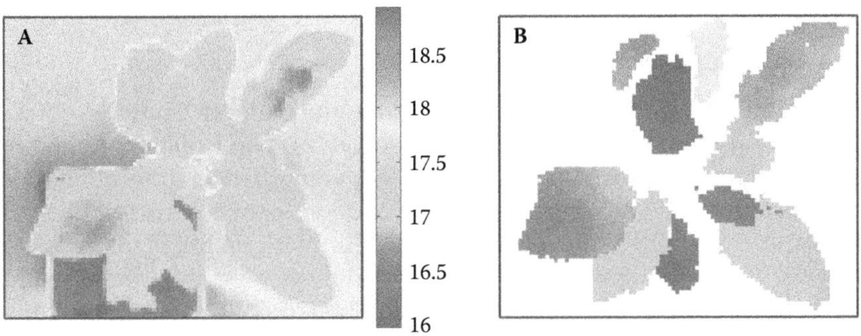

Figure 9.11 **(See color insert.)** An anatomofunctional imaging composed of depth camera and a functional imaging allows us to have two registered images: the depth map and the functional image. In this chapter, the anatomofunctional imaging system is composed of (a) a thermal imaging providing a thermal image and (b) a depth camera providing a depth map.

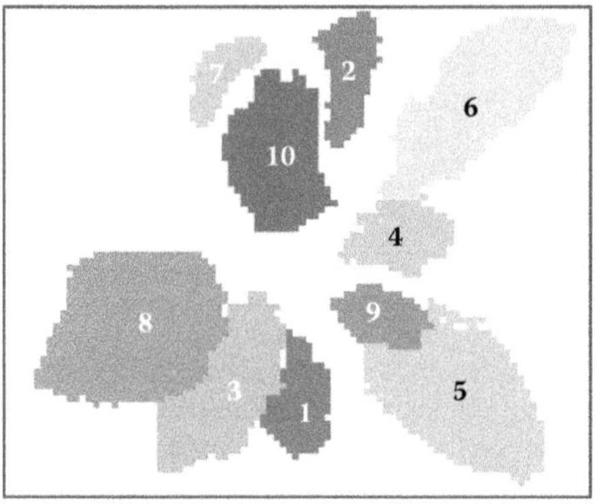

Figure 9.12 (**See color insert.**) Plant leaves are segmented by application of the depth segmentation algorithm of Chéné et al. (2012). Like the depth map, the segmented image is registered with the thermal image. In this image, each leaf is associated with a number. These numbers allow us to create a mask for each individual leaf.

Figure 9.13 (**See color insert.**) The edge of each leaf mask is applied on the thermal image. Colors of these edges are equal to the color of the leaf-segmented image of Figure 9.12. In this case only leaf 6 is hosting the pathogen.

are few errors on leaves' contour, visible in Figure 9.14, which are due to Kinect and calibration precision.

9.5 Conclusion

In this chapter, we have presented the potential of depth cameras for anatomic imaging modality of plants in association with functional imaging to deliver anatomofunctional imaging for plant science applications. We have detailed the various technologies of depth cameras currently available and discussed their advantages and limitations for plant imaging. Overall, a simple but useful message is that there is no single solution for plant phenotyping with depth cameras. The suitability of each technology depends on the biological question raised. We have illustrated this fact with a host-pathogen interaction where a thermal camera is associated with a depth camera to detect the presence or absence of a pathogen from a single top view image on each visible leaf of the upper part of the plant. We have provided practical information concerning the image registration procedure necessary to couple both the cameras. We also

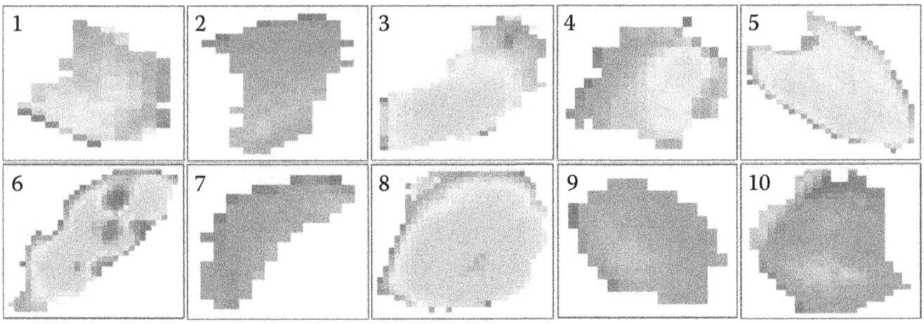

Figure 9.14 **(See color insert.)** Thermal images of each individual leaf segmented from the plant. Numbers correspond to numbers in Figure 9.12. The image resolution is the functional image resolution.

demonstrated that for such a biological question, low-cost depth cameras can be well adapted if the image acquisition is under controlled light conditions. Anatomofunctional imaging similar to the one illustrated here could also serve in extracting information from the branching structure. However, it would require higher depth resolution and possibly multiple views to allow a full reconstruction of the branching structure of bushy plants. It is likely that the grail of an ultimate phenotyping system for such more demanding applications may not exist, and that there will be an opportunity for multiple technological approaches depending on the measurement constraints and the biological question raised.

Acknowledgments

Yann Cheéné acknowledges financial support for the preparation of his PhD from Région Pays de la Loire in France. The authors thank the members of the laboratory Le2i in France (Olivier Aubreton, Vincent Daval, and Alban Bajard) for lending of the 3D scanner laser and the hybrid depth camera and Company Vecteo for the structured lighting depth camera.

References

Agishev, R., Gross, B., Moshary, F., Gilerson, A., and Ahmed, S., Range-resolved pulsed and CWFM lidars: potential capabilities comparison, *Appl. Phys. B Lasers Optics*, 85, 149–162, 2006.

Ahmad, M., and Choi, T., A heuristic approach for finding best focused shape, *IEEE Trans. Circ. Syst. Video Technol.*, 15, 566–574, 2005.

Ahmad, M., and Choi, T., Application of three dimensional shape from image focus in LCD/TFT displays manufacturing, *IEEE Trans. Consum. Electron.*, 53, 1–4, 2007.

Amann, M., Bosch, T., Lescure, M., Myllyla, R., and Rioux, M., Laser ranging: a critical review of usual techniques for distance measurement, *Opt. Eng.*, 40, 10–19, 2001.

Antone, M., and Friedman, Y., Fully automated laser range calibration, in *Proceedings of British Machine Vision Conference*, Warwick, UK, 2007, pp. 1–11.

Ballan, L., Brusco, N., and Cortelazzo, G., 3D content creation by passive optical methods, in *3D online multimedia and games: processing, visualization and transmission*, ed. I. Cheng, G.M. Cortelazzo, A. Basu, and S.K. Tripathi, 231–270, World Scientific Publishing, 2009.

Belin, E., Rousseau, D., Bourreau, T., and Caffier, V., Thermography versus chlorophyll fluorescence imaging for detection and quantification of apple scab, *Comput. Electron. Agric.*, 90, 159–163, 2013.

Bellasio, C., Olejníčková, J., Tesař, R., Šebela, D., and Nedbal, L., Computer reconstruction of plant growth and chlorophyll fluorescence emission in three spatial dimensions, *Sensors*, 12, 1052–1071, 2012.

Bhat, D., and Nayar, S., Ordinal measures for image correspondence, *IEEE Trans. Pattern Anal. Mach. Intell.*, 20, 415–423, 1998.

Bichsel, M., and Pentland, A., A simple algorithm for shape from shading, in *IEEE Computer Society Conference on Computer Vision and Pattern Recognition*, 1992, pp. 459–465.

Birchfield, S., and Tomasi, C., Depth discontinuities by pixel-to-pixel stereo, *Int. J. Comput. Vision*, 35, 269–293, 1999a.

Birchfield, S., and Tomasi, C., Multiway cut for stereo and motion with slanted surfaces, in *Proceedings of the 7th IEEE International Conference on Computer Vision*, Kerkyra, Greece, 1999b, pp. 489–495.

Bishop, T., and Favaro, P., Plenoptic depth estimation from multiple aliased views, in *IEEE 12th International Conference on Computer Vision*, 2009, pp. 1622–1629.

Biskup, B., Scharr, H., Schurr, U., and Rascher, U., A stereo imaging system for measuring structural parameters of plant canopies, *Plant Cell Environ.*, 30, 1299–1308, 2007.

Blais, F., Review of 20 years of range sensor development, *J. Electron. Imaging*, 13, 231–240, 2004.

Boudon, F., Godin, C., Puech, O., Pradal, C., and Sinoquet, H., Estimating the fractal dimension of plants using the two-surface method: an analysis based on 3D-digitized tree foliage, *Fractals*, 14, 149–163, 2006.

Bouguet, J., Camera calibration toolbox for MatLab, http://www.vision.caltech.edu/bouguetj/calib_doc/, 2004 (accessed August 27, 2013).

Bowen, J., Mesarich, C., Bus, V., Beresford, R., Plummer, K., and Templeton, M., *Venturia inaequalis*: the causal agent of apple scab, *Mol. Plant Pathol.*, 12, 105–122, 2011.

Boykov, Y., Veksler, O., and Zabih, R., Fast approximate energy minimization via graph cuts, *IEEE Trans. Pattern Anal. Mach. Intell.*, 23, 1222–1239, 2001.

Brown, M., Burschka, D., and Hager, G., Advances in computational stereo, *IEEE Trans. Pattern Anal. Mach. Intell.*, 25, 993–1008, 2003.

Buzinski, M., Levine, A., and Stevenson, W., Laser triangulation range sensors: a study of performance limitations, *J. Laser Appl.*, 4, 29–36, 1992.

Chaerle, L., and Straeten, D., *Venturia inaequalis*: the causal agent of apple scab, *Biochim. Biophys. Acta*, 1519, 153–166, 2001.

Chen, X., Xi, J., Jin, Y., and Sun, J., Accurate calibration for a camera-projector measurement system based on structured light projection, *Opt. Lasers Eng.*, 47, 310–319, 2009.

Chéné, Y., Belin, E., Rousseau, D., and Chapeau-Blondeau, F., Multiscale analysis of depth images from natural scenes: scaling in the depth of the woods, *Chaos Solit. Fractals*, 54, 134–149, 2013.

Chéné, Y., Rousseau, D., Lucidarme, P., Bertheloot, J., Caffier, V., Morel, P., Belin, E., and Chapeau-Blondeau, F., On the use of depth camera for 3D phenotyping of entire plants, *Comput. Electron. Agric.*, 82, 122–127, 2012.

Choi, T., and Yun, J., Three-dimensional shape recovery from the focused-image surface, *Opt. Eng.*, 39, 1321–1326, 2000.

Courteille, F., Crouzil, A., Durou, J., and Gurdjos, P., Towards shape from shading under realistic photographic conditions, in *Proceedings of the 17th International Conference on Pattern Recognition*, 277–280, 2004.

Crouzil, A., Descombes, X., and Durou, J., A multiresolution approach for shape from shading coupling deterministic and stochastic optimization, *IEEE Trans. Pattern Anal. Mach. Intell.*, 25, 1416–1421, 2003.

Delalieux, S., Auwerkerken, A., Verstraeten, W., Somers, B., Valcke, R., Lhermitte, J., and Coppin, P., Hyperspectral reflectance and fluorescence imaging to detect scab induced stress in apple leaves, *Remote Sensing*, 1, 858–874, 2009.

Desjardins, D., and Payeur, P., Dense stereo range sensing with marching pseudo-random patterns, in *4th Canadian Conference on Computer and Robot Vision*, 2007, pp. 216–226.

Dorsch, R., Häusler, G., and Herrmann, J., Laser triangulation: fundamental uncertainty in distance measurement, *Appl. Opt.*, 33, 1306–1314, 1994.

Dupuis, P., and Oliensis, J., Direct method for reconstructing shape from shading, in *IEEE Computer Society Conference on Computer Vision and Pattern Recognition*, 1992, pp. 453–458.

Durdle, N., Thayyoor, J., and Raso, V., An improved structured light technique for surface reconstruction of the human trunk, in *IEEE Canadian Conference on Electronics and Computer Engineering*, 1998, pp. 874–877.

Durou, J., Falcone, M., and Sagona, M., Numerical methods for shape-from-shading: a new survey with benchmarks, *Comput. Vision Image Underst.*, 109, 22–43, 2008.

Farid, H., and Simoncelli, E., Range estimation by optical differentiation. *J. Opt. Soc. Am. A*, 15, 1777–1786, 1998.

Favaro, P., Mennucci, A., and Soatto, S., Observing shape from defocused images, *Int. J. Comput. Vision*, 52, 25–43, 2003.

Favaro, P., and Soatto, S., Shape and radiance estimation from the information divergence of blurred images, in *European Conference on Computer Vision*, 2000, pp. 153–158.

Fiorani, F., Rascher, U., and Jahnke, S., Imaging plants dynamics in heterogenic environments, *Curr. Opin. Biotechnol.*, 23, 227–235, 2012.

Foix, S., Alenya, G., and Torras, C., Lock-in time-of-flight (ToF) cameras: a survey, *IEEE Sens. J.*, 11, 1917–1926, 2011.

Fortenbury, B., and Guerra Filho, G., Robust 2D/3D calibration using RANSAC registration, *Adv. Visual Comput.*, 7431, 179–188, 2012.

Fu, G., Menciassi, A., and Dario, P., Development of a low-cost active 3D triangulation laser scanner for indoor navigation of miniature mobile robots, *Robotics Auton. Syst.*, 60, 1317–1326, 2012.

Fua, P., A parallel stereo algorithm that produces dense depth maps and preserves image features, *Machine Vision Appl.*, 6, 35–49, 1993.

Fuchs, S., and Hirzinger, G., Extrinsic and depth calibration of ToF-cameras, in *IEEE Conference on Computer Vision and Pattern Recognition*, Anchorage, AK, 2008, pp. 1–6.

Geiger, A., Moosmann, F., Car, O., and Schuster, B., Automatic camera and range sensor calibration using a single shot, in *IEEE International Conference on Robotics and Automation*, St. Paul, MN, 2012, pp. 3936–3943.

Geng, J., Structured-light 3D surface imaging: a tutorial, *Adv. Opt. Photon.*, 3, 128–160, 2011.

Geng, Z., Rainbow three-dimensional camera: new concept of high-speed three-dimensional vision systems, *Opt. Eng.*, 35, 376–383, 1996.

Gokstorp, M., Computing depth from out-of-focus blur using a local frequency representation, in *Proceedings of the 12th IAPR International Conference on Pattern Recognition, Conference A: Computer Vision and Image Processing*, 1994, pp. 153–158.

Gokturk, S., Yalcin, H., and Bamji, C., A time-of-flight depth sensor—system description, issues and solutions, in *Conference on Computer Vision and Pattern Recognition*, 2004, pp. 35–35.

Griffin, P., Narasimhan, L., and Yee, S., Generation of uniquely encoded light patterns for range data acquisition, *Pattern Recogn.*, 25, 609–616, 1992.

Hayakawa, H., Nishida, S., Wada, Y., and Kawato, M., A computational model for shape estimation by integration of shading and edge information, *Neural Networks*, 7, 1193–1209, 1994.

Heikkila, J., Geometric camera calibration using circular control points, *IEEE Trans. Pattern Anal. Mach. Intell.*, 22, 1066–1077, 2000.

Helmli, F., and Scherer, S., Adaptive shape from focus with an error estimation in light microscopy, in *Proceedings of the 2nd International Symposium on Image and Signal Processing and Analysis*, 2001, pp. 188–193.

Herrera, C., Kannala, J., and Heikkilä, J., Joint depth and color camera calibration with distortion correction, *IEEE Trans. Pattern Anal. Mach. Intell.*, 34, 2058–2064, 2012.

Hiskett, P., Parry, C., McCarthy, A., and Buller, G., A photon-counting time-of-flight ranging technique developed for the avoidance of range ambiguity at gigahertz clock rates, *Opt. Expr.*, 16, 13685–13698, 2008.

Huang, P., and Zhang, S., Fast three-step phase-shifting algorithm, *Appl. Opt.*, 45, 5086–5091, 2006.

Iddan, G., and Yahav, G., Three-dimensional imaging in the studio and elsewhere, in *Proceedings of the SPIE Three-Dimensional Image Capture and Applications IV*, 2001, pp. 48–55.

Ishii, I., Yamamoto, K., and Tsuji, T., High-speed 3D image acquisition using coded structured light projection, in *IEEE/RSJ International Conference on Intelligent Robots and Systems*, 2007, pp. 925–930.

Jahnke, S., Menzel, M., Van Dusschoten, D., Roeb, G., Buhler, J., Minwuyelet, S., Blumler, P., Temperton, V., Hombach, T., and Streun, M., Combined MRI–PET dissects dynamic changes in plant structures and functions, *Plant J.*, 59, 634–644, 2009.

Kahlmann, T., Remondino, F., and Ingensand, H. Calibration for increased accuracy of the range imaging camera SwissRanger™, in *International Society for Photogrammetry and Remote Sensing*, Dresden, Germany, 2006, pp. 136–141.

Kim, Y., Chan, D., Theobalt, C., and Thrun, S., Design and calibration of a multi-view TOF sensor fusion system, in *IEEE Conference on Computer Vision and Pattern Recognition*, Anchorage, AK, 2008, pp. 1–7.

Klose, R., Penlington, J., and Ruckelshausen, A., Usability study of 3D time-of-flight cameras for automatic plant phenotyping, *Bornimer Agrartechnische Berichte*, 69, 93–105, 2009.
Kluth, V., Kunkel, G., and Rauhala, U., Global least squares matching, in *Geoscience and Remote Sensing Symposium*, Houston, TX, 1992, pp. 1615–1618.
Kolb, A., Barth, E., Koch, R., and Larsen, R., Time-of-flight sensors in computer graphics, *Comput. Graph. Forum*, 29, 141–159, 2010.
Kraft, M., Saloma De Freitag, N., and Munack, A., Test of a 3D time of flight camera for shape measurements of plants, in *CIGR Workshop on Image Analysis in Agriculture*, 2010, pp. 108–116.
Krattenthaler, W., Mayer, K.J., and Duwe, H.P., 3D-surface measurement with coded light approach, in *Proceedings of the 17th Meeting of the Austrian Association for Pattern Recognition on Image Analysis and Synthesis*, Munich, Germany, 1993, pp. 103–114.
Krotkov, E., Laser rangefinder calibration for a walking robot, *in IEEE International Conference on Robotics and Automation*, Sacramento, CA, 1991, pp. 2568–2573.
Kutulakos, K., and Seitz, S., A theory of shape by space carving, *Int. J. Comput. Vision*, 38, 199–218, 2000.
Lagüela, S., González Jorge, H., Armesto, J., and Herráez, J., High performance grid for the metric calibration of thermographic cameras, *Meas. Sci. Technol.*, 23, 15402–15410, 2011.
Lange, R., and Seitz, P., Solid-state time-of-flight range camera, *IEEE J. Quantum Electron.*, 37, 390–397, 2001.
Lange, R., Seitz, P., Biber, A., and Lauxtermann, S., Demodulation pixels in CCD and CMOS technologies for time-of-flight ranging, in *Proceedings of SPIE*, 2001, pp. 177–188.
Le Moigne, J., and Waxman, A., Structured light patterns for robot mobility, *IEEE J. Robot. Autom.*, 4, 541–548, 1988.
Lee, C., and Rosenfeld, A., Improved methods of estimating shape from shading using the light source coordinate system, *Art. Intell.*, 26, 125–143, 1985.
Li, Z., Shi, Y., Wang, C., and Wang, Y., Accurate calibration method for a structured light system, *Opt. Eng.*, 47, 053604, 1–9, 2008.
Lichti, D., Self-calibration of a 3D range camera, *Int. Arch. Photogram Rem. Sens. Spatial Inform. Sci.*, 37, 927–932, 2008.
Lindner, M., and Kolb, A., Lateral and depth calibration of PMD-distance sensors, in *Advances in Visual Computing, 2nd International Symposium, ISVC*, Lake Tahoe, CA, 2006, pp. 524–533.
Lindner, M., Kolb, A., and Hartmann, K., Data-fusion of PMD-based distance information and high-resolution RGB-images, in *International Symposium on Signals, Circuits and Systems*, Lasi, Romania, 2007, pp. 1–4.
Luhmann, T., Ohm, J., Piechel, J., and Roelfs, T., Geometric calibration of thermographic cameras, *Int. Arch. Photogram Rem. Sens. Spatial Inform. Sci.*, 38, 411–416, 2010.
Lüke, J., Pérez Nava, F., Marichal Hernández, J., Rodríguez Ramos, J., and Rosa, F., Near real-time estimation of super-resolved depth and all-in-focus images from a plenoptic camera using graphics processing units, *Int. J. Digit. Multimed. Broadcast.*, 2010, 12–15, 2009.
Lumsdaine, A., and Georgiev, T., The focused plenoptic camera, in *IEEE International Conference on Computational Photography*, 2009, pp. 1–8.

Malik, A., and Choi, T., Application of passive techniques for three dimensional cameras, *IEEE Trans. Cons. Electron.*, 53, 258–264, 2007.

Malik, A., and Choi, T., A novel algorithm for estimation of depth map using image focus for 3D shape recovery in the presence of noise, *Pattern Recogn.*, 41, 2200–2225, 2008.

Martinez Baena, J., Fdez Valdivia, J., and Garcia, J., A multi-channel autofocusing scheme for gray-level shape scale detection, *Pattern Recogn.*, 30, 1769–1786, 1997.

Maruyama, M., and Abe, S., Range sensing by projecting multiple slits with random cuts, *IEEE Trans. Pattern Analysis Mach. Intell.*, 15, 647–651, 1993.

Minhas, R., Mohammed, A., Wu, Q., and Sid Ahmed, M., 3D shape from focus and depth map computation using steerable filters, in *Proceedings of the 6th International Conference on Image Analysis and Recognition*, 2009, pp. 573–583.

Monks, T., and Carter, J., Improved stripe matching for color encoded structured light, in *5th International Conference on Computer Analysis of Images and Patterns*, 476, 1993.

Myllylä, R., Marszalec, J., Kostamovaara, J., M¨antyniemi, A., and Ulbrich, G., Imaging distance measurements using TOF lidar, *J. Opt.*, 29, 188, 1999.

Nagatani, K., Otake, K., and Yoshida, K., Three-dimensional thermography mapping for mobile rescue robots, presented at *8th International Conference on Field and Service Robotics*, Matsushimas, Japan, 2012.

Nakamura, K., Hara, T., Yoshida, M., Miyahara, T., and Ito, H., Optical frequency domain ranging by a frequency-shifted feedback laser, *IEEE J. Quantum Electron.*, 36, 305–316, 2000.

Nakamura, Y., Matsuura, T., Satoh, K., and Ohta, Y., Occlusion detectable stereo-occlusion patterns in camera matrix, in *Proceedings of IEEE Computer Society Conference on Computer Vision and Pattern Recognition*, San Francisco, CA, 1996, pp. 371–378.

Nalpantidis, L., Sirakoulis, G., and Gasteratos, A., Review of stereo matching algorithms for 3D vision, in *Proceedings of the 16th International Symposium on Measurement and Control in Robotics*, Warsaw, Poland, 2007, pp. 116–124.

Nava, F., Marichal Hernández, J., and Rodríguez Ramos, J., The discrete focal stack transform, in *Proceedings of the 16th European Signal Proceedings Conference*, 25–106, 2008.

Nayar, S., and Nakagawa, Y., Shape from focus, *IEEE Trans. Pattern Anal. Mach. Intell.*, 16, 824–831, 1994.

Ng, R., Fourier slice photography, *ACM Trans. Graph.*, 24, 735–744, 2005.

Ng, R., Levoy, M., Brédif, M., Duval, G., Horowitz, M., and Hanrahan, P., Light field photography with a hand-held plenoptic camera, *Comput. Sci. Tech. Rep. CSTR*, 2, 1–11, 2005.

Oerke, E., Fröhling, P., and Steiner U., Thermographic assessment of scab disease on apple leaves, *Prec. Agric.*, 12, 699–715, 2011.

Omasa, K., Hosoi, F., and Konishi, A., 3D lidar imaging for detecting and understanding plant responses and canopy structure, *J. Exp. Bot.*, 158, 881–898, 2007.

Papageorgiou, G.G., *Chlorophyll a fluorescence: a probe of photosynthesis*, Kluwer Academic, Dordrecht, The Netherlands, 2004.

Pentland, A., Linear shape from shading, *Int. J. Comput. Vision*, 4, 153–162, 1990.

Pfeifer, N., and Briese, C., Laser scanning—principles and applications, presented at *3rd International Exhibition and Scientific Congress on Geodesy, Mapping, Geology, Geophisics*, Cadaster GEO-SIBERIA, 2007.

Ping Sing, T., and Shah, M., Shape from shading using linear approximation, *Image Vision Comput.*, 12, 487–498, 1994.
Piracha, M., Nguyen, D., Mandridis, D., Yilmaz, T., Ozdur, I., Ozharar, S., and Delfyett, P., Range resolved lidar for long distance ranging with submillimeter resolution, *Opt. Expr.*, 18, 7184–7189, 2010.
Poujouly, S., and Journet, B., Laser range finding by phase-shift measurement: moving towards smart systems, in *Proceedings of the SPIE Machine Vision and Three-Dimensional Imaging Systems for Inspection and Metrology*, 2000, pp. 152–160.
Rajagopalan, A., and Chaudhuri, S., Simultaneous depth recovery and image restoration from defocused images, in *IEEE Conference on Computer Vision and Pattern Recognition*, 1999, pp. 1348–1353.
Rouy, E., and Tourin, A., A viscosity solutions approach to shape-from-shading, *SIAM J. Numer. Anal.*, 29, 867–884, 1992.
Ruderman, D.L., and Bialek, W., Statistics of natural images: scaling in the woods, *Phys. Rev. Lett.*, 73, 814–817, 1994.
Salvi, J., Fernandez, S., Pribanic, T., and Llado, X., A state of the art in structured light patterns for surface profilometry, *Pattern Recogn.*, 43, 2666–2680, 2010.
Sankaran, S., Mishra, A., Ehsani, R., and Davis, C., A review of advanced techniques for detecting plant diseases, *Comput. Electron. Agric.*, 59, 634–644, 2010.
Sansoni, G., Carocci, M., and Rodella, R., Calibration and performance evaluation of a 3D imaging sensor based on the projection of structured light, *IEEE Trans. Instrum. Meas.*, 49, 628–636, 2000.
Sansoni, G., Trebeschi, M., and Docchio, F., State-of-the-art and applications of 3D imaging sensors in industry, cultural heritage, medicine, and criminal investigation, *Sensors*, 9, 568–601, 2009.
Satoh, K., and Ohta, Y., Occlusion detectable stereo-systematic comparison of detection algorithms, in *Proceedings of the 13th International Conference on Pattern Recognition*, Vienna, Austria, 1996, pp. 280–286.
Scharstein, D., and Szeliski, R., Stereo matching with nonlinear diffusion, *Int. J. Comput. Vision*, 28, 155–174, 1998.
Schechner, Y., and Kiryati, N., The optimal axial interval in estimating depth from defocus, in *Proceedings of the 7th IEEE International Conference on Computer Vision*, Kerkyra, Greece, 1999, pp. 843–848.
Silva, C., and Santos Victor, J., Intrinsic images for dense stereo matching with occlusions, *Comput. Vision*, 1842, 100–114, 2000.
Skala, K., Lipić, T., Sović, I., Gjenero, L., and Grubišić, I., 4D thermal imaging system for medical applications, *Period. Biol.*, 113, 407–416, 2011.
Sturm, P., and Maybank, S., On plane-based camera calibration: a general algorithm, singularities, applications, in *IEEE Computer Society Conference on Computer Vision and Pattern Recognition*, Fort Collins, CO, 1999, pp. 432–437.
Subbarao, M., and Choi, T., Accurate recovery of three-dimensional shape from image focus, *IEEE Trans. Pattern Anal. Mach. Intell.*, 17, 266–274, 1995.
Sun, J., Zheng, N., and Shum, H., Stereo matching using belief propagation, *IEEE Trans. Pattern Anal. Mach. Intell.*, 25, 787–800, 2003.
Szeliski, R., Fast shape from shading, *Comput. Vision Graphics Image Proc.*, 53, 129–153, 1991.
Tomasi, C., and Manduchi, R., Stereo matching as a nearest-neighbor problem, *IEEE Trans. Pattern Anal. Mach. Intell.*, 20, 333–340, 1998.

Trouvé, P., Champagnat, F., Le Besnerais, G., Druart, G., and Idier, J., Chromatic depth from defocus: a theoretical and experimental performance study, in *Computational Optical Sensing and Imaging*, CM3B.3, 2012.

Trouvé, P., Champagnat, F., Le Besnerais, G., and Idier, J., Single image local blur identification, in *18th IEEE International Conference on Image Processing*, 2011, pp. 613–616.

Tsai, R., A versatile camera calibration technique for high-accuracy 3D machine vision metrology using off-the-shelf TV cameras and lenses, *IEEE J. Robotics Autom.*, 3, 323–344, 1987.

Ulusoy, A., Calakli, F., and Taubin, G., One-shot scanning using de Bruijn spaced grids, in *IEEE 12th International Conference on Computer Vision Workshops*, 2009, pp. 1786–1792.

Underwood, J., Hill, A., and Scheding, S., Calibration of range sensor pose on mobile platforms, in *IEEE/RSJ International Conference on Intelligent Robots and Systems*, San Diego, CA, 2007, pp. 3866–3871.

Unnikrishnan, R., and Hebert, M., *Fast extrinsic calibration of a laser rangefinder to a camera*, Robotics Institute Technical Report, 2005, http://repository.cmu.edu/robotics/339/ (accessed August 27, 2013).

Valkenburg, R., and McIvor, A., Accurate 3D measurement using a structured light system, *Image Vision Comput.*, 16, 99–110, 1998.

Veatch, P., and Davis, L., Efficient algorithms for obstacle detection using range data, *Comput. Vision Graphics Image Proc.*, 50, 50–74, 1990.

Wildes, R., Direct recovery of three-dimensional scene geometry from binocular stereo disparity, *IEEE Trans. Pattern Anal. Mach. Intell.*, 13, 761–774, 1991.

Xiong, Y., and Shafer, S., Depth from focusing and defocusing, in *IEEE Computer Society Conference on Computer Vision and Pattern Recognition*, 1993, pp. 68–73.

Xiong, Y., and Shafer, S., Moment filters for high precision computation of focus and stereo, in *IEEE/RSJ International Conference on Intelligent Robots and Systems, Human Robot Interaction and Cooperative Robots*, 1995, pp. 108–113.

Yahav, G., Iddan, G., and Mandelboum, D., 3D imaging camera for gaming application, in *International Conference on Consumer Electronics, Digest of Technical Papers*, 2007, pp. 1–2.

Zhang, C., and Zhang, Z., Calibration between depth and color sensors for commodity depth cameras, in *IEEE International Conference on Multimedia and Expo*, 2011, pp. 1–6.

Zhang, L., Curless, B., and Seitz, S., Rapid shape acquisition using color structured light and multi-pass dynamic programming, in *Proceedings of the 1st International Symposium on 3D Data Processing, Visualization, and Transmission*, 2002, pp. 24–36.

Zhang, Q., and Pless, R., Extrinsic calibration of a camera and laser range finder (improves camera calibration), in *IEEE/RSJ International Conference on Intelligent Robots and Systems*, 2301–2306, 2004.

Zhang, R., Tsai, P., Cryer, J., and Shah, M., Shape-from-shading: a survey, *IEEE Trans. Pattern. Anal. Mach. Intell.*, 21, 690–706, 1999.

Zhou, F., and Zhang, G., Complete calibration of a structured light stripe vision sensor through planar target of unknown orientations, *Image Vision Comput.*, 23, 59–67, 2005.

chapter ten

Chlorophyll fluorescence imaging for plant health monitoring

Kotaro Takayama

Contents

10.1 Introduction

Chlorophyll fluorescence imaging provides information on photosynthetic performance without destruction or contact with the living plant. The chlorophyll fluorescence involves an emission of red light from chlorophyll *a* pigments (Krause and Weis, 1991; Govindjee, 1995) that can be used to assess photosynthetic functions, thereby allowing for plant health monitoring (Maxwell and Johnson, 2000; Takayama and Nishina, 2009). The imaging technique of chlorophyll fluorescence, originally developed by Omasa et al. (1987) and Daley et al. (1989), has been used to evaluate the heterogeneous distribution of photosynthetic activities over a leaf surface, and thus to detect photosynthetic dysfunctions caused by biotic and abiotic stress factors (Daley et al., 1989; Omasa and Shimazaki, 1990; Omasa et al., 1991; Balachandran et al., 1994; Daley, 1995; Siebke and Weis, 1995; Rolfe and Scholes, 1995; Genty and Meyer, 1995; Osmond et al., 1998; Takayama et al., 2003). Recently, the objects of chlorophyll fluorescence imaging have been scaled up to the levels of a whole plant (Takayama et al., 2010), a tree canopy (Nichol et al., 2012), and tomato crops cultivated in a large-scale greenhouse (Takayama et al., 2011a, 2011b, 2011c).

The following sections describe the basics of chlorophyll fluorescence imaging, that is, chlorophyll fluorescence emission, optical settings of the light source and optical filters, performance of the charge-coupled device (CCD) camera, and calibration of fluorescence images, with special attention to the technical and engineering aspects of chlorophyll fluorescence imaging for researchers and students intending to develop a handmade imaging system customized for their own studies. Then, two popular methods, the saturation pulse method and the induction method, are reviewed. These methods are used to monitor plant health by applications of chlorophyll fluorescence imaging, such as detection of invisible photosynthetic dysfunction at the levels of leaf surface, individual plant, and crop canopies due to environmental stress factors such as herbicides, drought stress, and strong light stress. The physiological aspects of chlorophyll fluorescence imaging are reviewed in detail by Oxborough (2004).

10.2 Basics of chlorophyll fluorescence

Figure 10.1 shows a schematic presentation of the light energy distribution in the chlorophyll *a* pigment, which is the main photosynthetic pigment in green leaves. Plant leaves utilize light energy absorbed by photosynthetic pigments for photosynthetic photochemical reactions. However, plant leaves cannot use all the absorbed light energy (L in Figure 10.1) for photosynthetic reactions (P in Figure 10.1). The residual light energy unutilized for photosynthetic reactions is dissipated as heat (H in Figure 10.1)

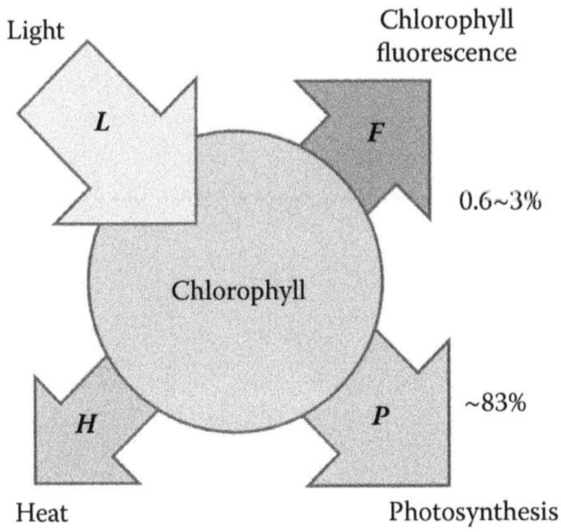

Figure 10.1 Energy distribution in chlorophyll *a* pigment, which governs in vivo chlorophyll fluorescence emission. *L*, *F*, *P*, and *H* represent absorbed light energy, energy used for chlorophyll fluorescence emission, photosynthetic reaction, and heat dissipation, respectively.

or reemitted as red light (*F* in Figure 10.1). This red light emission is the chlorophyll fluorescence (Krause and Weis, 1991). Chlorophyll fluorescence intensity thus varies with the photosynthetic reactions (*P*) and heat dissipation process (*H*), even if the excitation light intensity (*L*) is stable. The maximum energy distribution for photosynthesis (*P*/*L*) is 83%, and the energy distribution for chlorophyll fluorescence emission (*F*/*L*) varies from 0.6 to 3%. Hence, accurate measurement of chlorophyll fluorescence emission allows the evaluation of photosynthetic functions, namely, both photosynthetic photochemical reactions and the status of heat dissipation processes, without physical contact with the plant (Maxwell and Johnson, 2000; Takayama and Nishina, 2009).

In the case that the chlorophyll fluorescence intensity is expressed as *F*, the chlorophyll fluorescence yield should be expressed as *F*/*L* (Figure 10.1). So, an increase in absorbed light energy increases the chlorophyll fluorescence intensity but does not affect the chlorophyll fluorescence yield if the energy distribution ratio (*F*:*P*:*H* in Figure 10.1) is stable.

10.3 Settings for chlorophyll fluorescence imaging

10.3.1 Light source and optical filters

The chlorophyll fluorescence is visible red light, so imaging of chlorophyll fluorescence emission is possible with a commercially available CCD. As seen in the excitation and emission matrix of steady-state fluorescence of a healthy cucumber leaf (Figure 10.2), the emission peaks (660 to 770 nm)

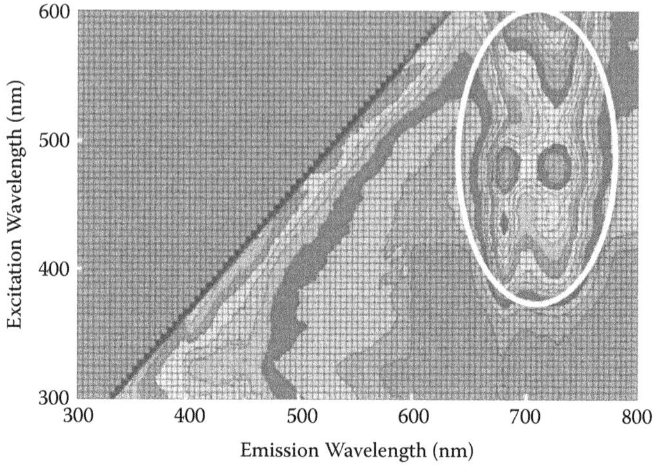

Figure 10.2 Excitation and emission matrix of steady-state fluorescence of a healthy cucumber leaf measured with a spectrofluorometer. The peaks surrounded by a circle represent strong chlorophyll fluorescence emission excited by visible light. (Reproduced from Omasa and Takayama, in *Air Pollution and Plant Biotechnology*, ed. Omasa et al., 287–308, Springer-Verlag, Berlin, 2002.)

are the chlorophyll fluorescence emission induced by the excitation light in the visible range (380 to 600 nm) (Omasa and Takayama, 2002).

Two examples (Takayama and Nishina, 2009) of optical settings for the chlorophyll fluorescence imaging system are shown in Figure 10.3. The chlorophyll fluorescence induced by the blue excitation light, which is passed through a short-pass filter ($\lambda < 600$ nm) (Figure 10.3a) or directly provided by a blue light-emitting diode (LED) panel (Figure 10.3b), is

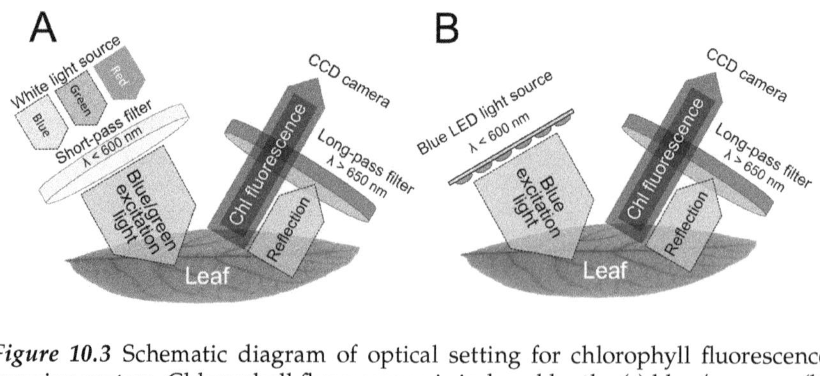

Figure 10.3 Schematic diagram of optical setting for chlorophyll fluorescence imaging system. Chlorophyll fluorescence is induced by the (a) blue/green or (b) pure blue excitation light and (a, b) captured by a CCD camera filtered with a long-pass filter.

captured by a CCD camera equipped with a long-pass filter ($\lambda > 650$ nm) that prevents reflective images of the blue excitation light.

10.3.2 Performance of CCD camera

The CCD camera used for the chlorophyll fluorescence imaging system must have adequate sensitivity to far-red and near-infrared lights (660 to 770 nm), as shown in Figure 10.2. Figure 10.4 shows the spectral response of a commercially available black and white CCD camera (Stingray F-145, Allied Vision Technologies), which was used for our chlorophyll fluorescence imaging system (Takayama et al., 2011b, 2011c).

USB cameras are also available for chlorophyll fluorescence imaging, but in most cases, the built-in infrared cutoff filter should be removed to improve the sensitivity to far-red and near-infrared light. Furthermore, in practical use, the gamma correction of the CCD camera module that compensates human vision should be switched off to keep the linear response to fluorescence intensity, and the shutter speed and gain of the CCD camera should be regulated to capture the appropriate chlorophyll fluorescence intensity images (Takayama et al., 2003). The frame rate of the CCD camera should be higher than 10 fps for common use.

10.3.3 Calibration of fluorescence images

If the above-mentioned functions of gamma correction, shutter speed, and gain are automatically controlled by the CCD camera module, a

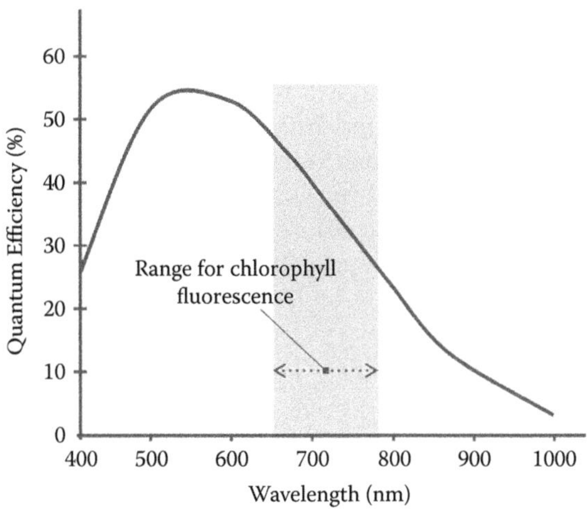

Figure 10.4 Spectral response of a CCD camera for a chlorophyll fluorescence imaging system. (Based on datasheet of Stingray F-145.)

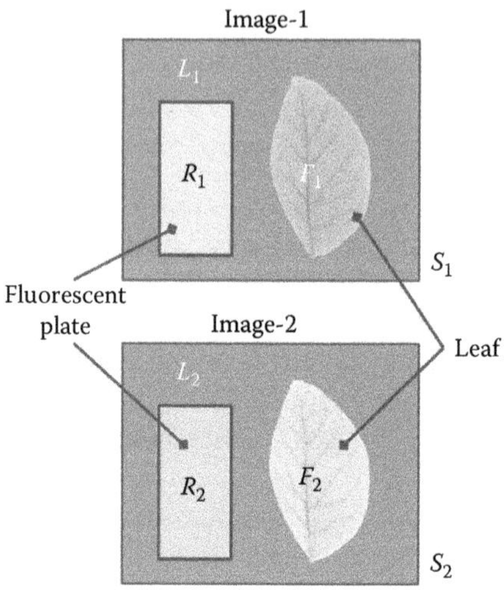

Figure 10.5 Schematic presentation of a set of two fluorescence images captured under different light intensities, L_1 for Image-1 and L_2 for Image-2. R_1 and R_2 are the pixel values of the fluorescent plate, and F_1 and F_2 are the pixel values of the leaf in Image-1 and Image-2, respectively. S_1 and S_2 are the sensitivities of the CCD for each image.

fluorescent plate/paper/sheet of the stable fluorescence yield should be imaged together with the experimental leaf. The fluorescence intensity of the fluorescent plate can be used as a reference to obtain the relative chlorophyll fluorescence intensity of the leaf. Figure 10.5 shows a set of two fluorescence images captured under different light intensities, L_1 for Image-1 and L_2 for Image-2. R_1 and R_2 are the pixel values of the fluorescent plate in Image-1 and Image-2, respectively. F_1 and F_2 are the pixel values of the leaf in Image-1 and Image-2, respectively. S_1 and S_2 are the sensitivities of the CCD for each image. The sensitivity is automatically controlled by the CCD camera module and an unmeasurable parameter. For accurate comparison of the chlorophyll fluorescence intensities of the leaf between Image-1 and Image-2, the ratio of S_1 to S_2 is determined by the following equation:

$$\frac{S_1}{S_2} = \frac{L_2 \times R_1}{L_1 \times R_2}$$

By using S_1/S_2, the relative chlorophyll fluorescence intensity of the leaf under the light intensity of L_2 can be derived by $F_2 \times S_1/S_2$. In addition, the relative chlorophyll fluorescence yields, ΦF_1 in Image-1 and ΦF_2 in Image-2, are obtained by the following equations:

$$\Phi F_1 = \frac{F_1}{L_1}$$

$$\Phi F_2 = \frac{F_2 \times \dfrac{S_1}{S_2}}{L_2}$$

10.4 *Simple comparison between saturation pulse method and induction method*

The following two methods are implemented for plant health monitoring by chlorophyll fluorescence imaging: saturation pulse method (Daley et al., 1989) and induction method (Omasa et al., 1987). The saturation pulse method is the most common for measuring photosynthetic activity, as it incorporates widely used chlorophyll fluorescence parameters, such as Φ_{PSII} (quantum yield of PSII electron transport) and NPQ (non-photochemical quenching) (Genty et al., 1989; Bilger and Björkman, 1990; Krause and Weis, 1991). To measure these parameters, the plant body should be illuminated with a pulsed strong light, i.e., saturation light pulse (Genty et al., 1989; Bilger and Björkman, 1990). In contrast, a weak excitation light is enough to induce chlorophyll fluorescence emission in the induction method (Govindjee, 1995; Omasa and Takayama, 2002; Takayama and Nishina, 2009). A comparison between the saturation pulse method (SP-M) and the chlorophyll fluorescence induction method (CFI-M) is summarized in Table 10.1. In comparison with the induction method, the advantages of the saturation pulse method are characterized by the measurement under a light/dark condition, well-defined parameters, and availability of dynamic/static photosynthetic information. On the other hand, the disadvantages of the saturation pulse method are the strict control of light sources, small measureable area, and difficulties of a self-produced imaging system.

10.5 *Saturation pulse method*

10.5.1 *Parameters derived by saturation pulse method*

The saturation pulse method provides common chlorophyll fluorescence parameters such as Φ_{PSII} (Genty et al., 1989) and NPQ (Bilger and Björkman, 1990; Krause and Weis, 1991). These parameters are derived from the relative chlorophyll fluorescence yield images (ΦF_m, $\Phi F'_m$, ΦF_s) obtained from the chlorophyll fluorescence intensity images (F_m, F'_m, F_s) by

Table 10.1 Comparison between the Saturation Pulse Method (SP-M) and the Chlorophyll Fluorescence Induction Method (CFI-M)

	SP-M	SFI-M
Generality for research	General	Conventional Additive to SPM
Intensities of light for measurement	Two Saturation light/actinic light	Single < PPFD 100 < mol m^{-2} s^{-1}
Uniformity of light for measurement	Severe	Not so severe
Ambient light condition	Light/dark	Dark
Available parameters	Well-defined	Not defined well
Available photosynthetic information	Dynamic/static Photosynthetic activity/ability	Static Photosynthetic ability
Measurable plant area	Small leaf level	Large canopy level
Easiness of self-produced imaging system	Not easy	Relatively easy

$$\Phi F_m \cong \frac{F_m}{L_{SL}}$$

$$\Phi F_s \cong \frac{F_s}{L_{AL}}$$

$$\Phi F'_m \cong \frac{F'_m}{L_{SL}}$$

where F_m is the maximal chlorophyll fluorescence intensity measured during the saturation pulse light under the dark condition, F_s is the steady-state chlorophyll fluorescence intensity under actinic light, F'_m is the maximal chlorophyll fluorescence intensity measured during the saturation pulse light under the light condition (normally, $F'_m < F_m$), L_{SL} is the intensity of saturation pulse light, and L_{AL} is the intensity of actinic light (Omasa and Takayama, 2002). The saturation pulse light is a pulsed (for 1 s) strong light (> PPFD 2000 μmol m^{-2}s^{-1}, where PPFD is photosynthetic photon flux density) used for the full reduction of primary electron acceptors, known as Q_A, in the photosynthetic electron transport chain (Krause and Weis, 1991). During the saturation pulse light, the energy distribution to photosynthesis (*P* in Figure 10.1) can be assumed to be zero. Hence, the chlorophyll fluorescence yields during the saturation pulse light (ΦF_m and $\Phi F'_m$) are expressed as $F/(F + H)$ in contrast to the steady-state chlorophyll fluorescence yield expressed as $F/(F + H + P)$.

By using these chlorophyll fluorescence yields, images of the photochemical yield of photosystem II (PSII) (Φ_{PSII}) and NPQ were computed. Images of Φ_{PSII} obtained under actinic light represent the activity of linear electron transport in PSII and ranged from 0 to 1 (Genty et al., 1989; Genty and Meyer, 1995; Bro et al., 1996; Meyer and Genty, 1999). Images of Φ_{PSII} can be used as relative maps of CO_2 fixation in the nonphotorespiratory condition and were computed for each corresponding pixel by the following formula (Genty et al., 1989; Genty and Meyer, 1995):

$$\Phi_{PSII} = \frac{(\Phi F'_m - \Phi F_s)}{\Phi F'_m}$$

NPQ, ranging from 0 to infinity, represents the extent of the intrathylakoid pH gradient and the ability of chloroplasts to dissipate excess excitation energy as heat (Bilger and Björkman, 1990; Osmond et al., 1998). Thereby, NPQ has been used as a sensitive indicator of regulation and protection of photosynthesis in environments where light energy absorption exceeds the capacity for light utilization (Daley et al., 1989; Osmond et al., 1998). Images of NPQ were calculated for each corresponding pixel by the following formula (Bilger and Björkman, 1990; Maxwell and Johnson, 2000; Omasa and Takayama, 2002; Takayama et al., 2003):

$$NPQ = \frac{(\Phi F_m - \Phi F'_m)}{\Phi F'_m}$$

10.5.2 Imaging system for saturation pulse method

Figure 10.6a shows a schematic diagram of the chlorophyll fluorescence imaging system (Takayama et al., 2003). The actinic light at a PPFD of 350 $\mu mol\ m^{-2}s^{-1}$ for photosynthesis was provided by a 180 W metal halide lamp (LS-M180, Sumita Optical Glass, Inc.), and the saturation light pulse at a PPFD of 2700 $\mu mol\ m^{-2}s^{-1}$ for 1 second was provided by three 180 W metal halide lamps (LS-M180, Sumita Optical Glass, Inc.). The actinic or saturation light pulse was provided through an optical fiber equipped with a short-pass filter (blue filter, $\lambda < 570$ nm) (Sumita Optical Glass, Inc.). The variation in the distribution of the actinic or saturation light within the measurement area ($5 \times 5\ cm^2$) was kept within ±5%. The spectral properties of the short-pass and long-pass filters are shown in Figure 10.6b.

10.5.3 Detection of invisible photosynthetic dysfunction caused by an herbicide

Figure 10.7 shows changes in the distribution of chlorophyll fluorescence intensity (F_s and Fm'), Φ_{PSII}, and NPQ on a kidney bean (*Phaseolus vulgaris*

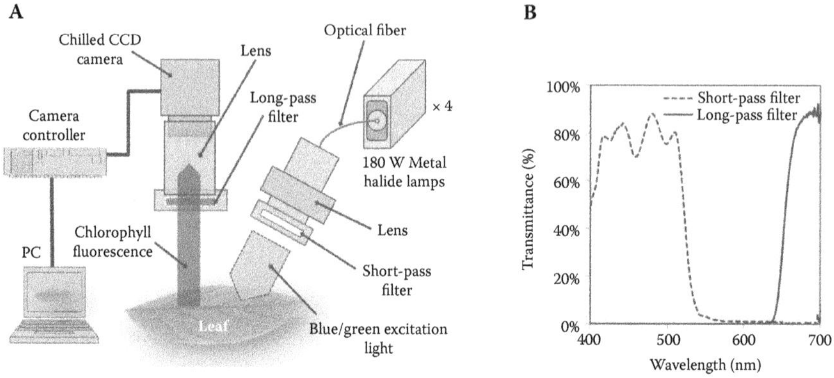

Figure 10.6 (a) Schematic diagram of chlorophyll fluorescence imaging system for a single leaf. An attached plant leaf is placed on the fixing apparatus. The 5 ´ 5 cm^2 measurement area is illuminated by actinic or saturation pulse light sources. (b) The spectral properties of the short-pass and long-pass filters are represented as the dashed and solid lines, respectively.

L.) leaf surface, as well as the visual appearance during 24 h after the Basta treatment (control, 30 min, 3 h, and 24 h). The active ingredient of Basta is glufosinate. Glufosinate inhibits the activity of glutamine synthetase, which is essential for the removal of toxic ammonia produced in the metabolism of living plant systems (Lea and Ridley, 1989). Therefore, application of glufosinate leads to increased levels of ammonia in plant tissues. This causes photosynthesis to stop and the plant dies within a few days (Lea and Ridley, 1989). A 1:100 dilution of Basta with pure water was used according to the label directions for killing annual plants. The upper surface area on the left side of the leaf was covered with gauze soaked in the Basta solution for 10 min. An image of F_m was captured before treatment and used in the calculation for the NPQ image.

Before treatment, chlorophyll fluorescence intensity (F and F_m'), Φ_{PSII}, and NPQ were uniformly distributed and the mean values of Φ_{PSII} and NPQ were 0.44 and 0.96, respectively. At 30 min after treatment, the patchy decrease in chlorophyll fluorescence intensity (F and F_m') within the treated area was confirmed. In those areas, Φ_{PSII} decreased to 0.25 and NPQ increased to 2.26. The time after treatment suggested that Basta rapidly moved from the leaf surface into the mesophyll cells. Afterwards, it inhibited the photosynthetic electron transport activity and activated the heat dissipation pathway of absorbed light energy (Genty and Meyer, 1995; Osmond et al., 1998). The patchiness may be caused by heterogeneous absorption of herbicide through the stomata during the herbicide treatment. At 3 h after treatment, the injured areas were connected with each other and expanded to the adjacent untreated area. This showed

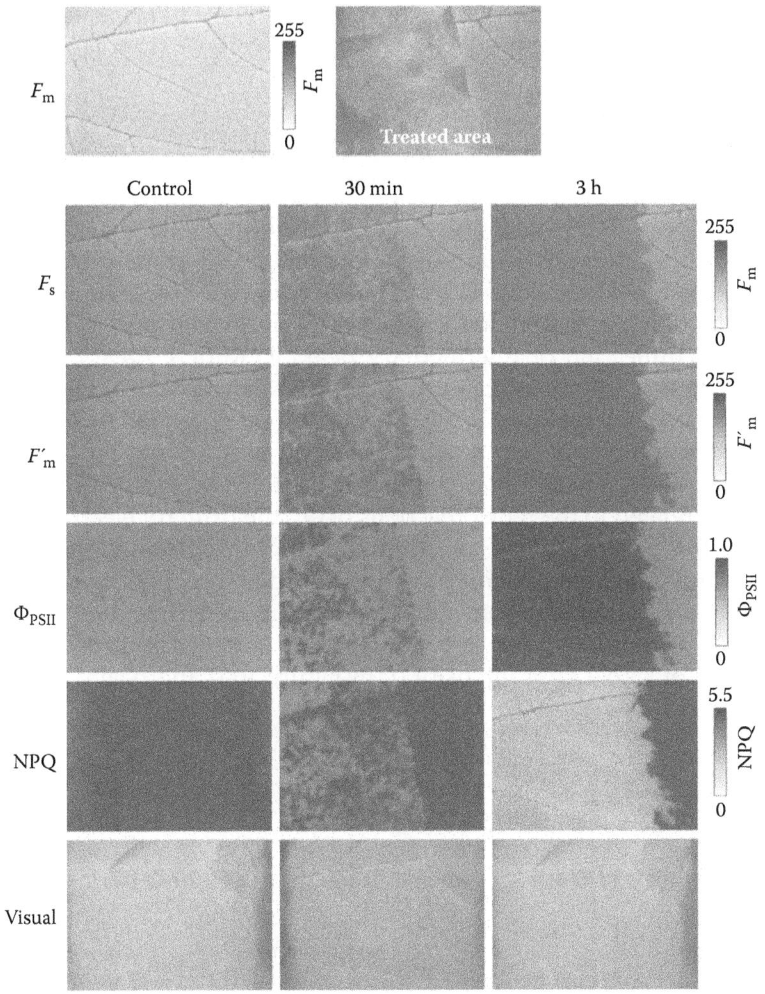

Figure 10.7 **(See color insert.)** Effects of Basta treatment on a *Phaseolus vulgaris* L. leaf. The fluorescence image titled "Treated area" represents the Basta treatment. Chlorophyll fluorescence intensity images (F_s and F'_m) were captured before (control), 30 min, and 3 h after treatment, and Φ_{PSII} and NPQ images were calculated each time.

that Basta absorbed in the leaf diffused in the leaf apoplast first and then was transported along veins and veinlets. The mean values of Φ_{PSII} and NPQ in the injured areas were 0.11 and 3.47, respectively. The values demonstrated that photosynthetic electron transport activity was severely inhibited, and the heat dissipation pathway of absorbed light energy was significantly activated in these areas (Krause and Weis, 1991; Genty and

Meyer, 1995; Osmond et al. 1998). However, no visible damage was detectable at 3 h after treatment.

10.6 Induction method

10.6.1 Chlorophyll fluorescence induction

Figure 10.8 shows a typical time course of chlorophyll fluorescence intensity during a chlorophyll fluorescence induction (Takayama and Nishina, 2009; Stirbet and Govindjee, 2011). The curve of this time course is called the induction curve and has characteristic inflection points named O (origin), I (inflection or intermediary peak), D (dip), P (peak), S (quasi-steady state), M (maximum), and T (terminal steady state) (Govindjee, 1995; Omasa and Takayama, 2002; Papageorgiou et al., 2007). The induction curve consists of two phases. The fast phase (from O to P) occurs during the few seconds after commencement of the excitation light and is closely related to the redox reactions of Q_A, the primary electron acceptors in the photosynthetic electron transport chain (Govindjee, 1995; Stirbet and Govindjee, 2011). Basically, the reduction of Q_A decreases the energy distribution for photosynthesis (*P* in Figure 10.1) and results in an increase in chlorophyll fluorescence intensity (*F* in Figure 10.1) under the condition of stable light intensity (*L* in Figure 10.1) and heat dissipation

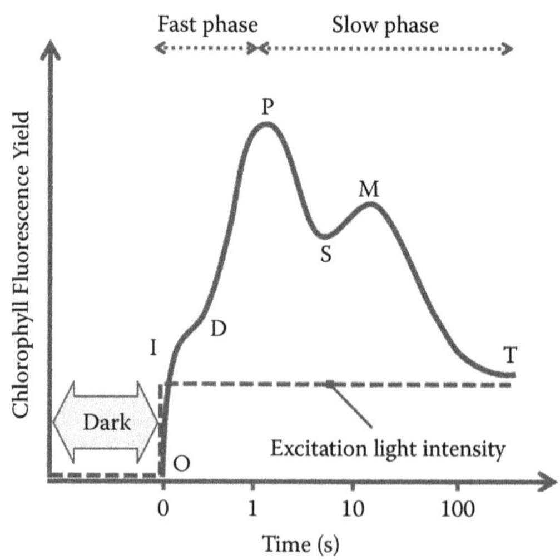

Figure 10.8 Schematic diagram of changes in chlorophyll fluorescence yield during a chlorophyll fluorescence induction plotted against on a logarithmic timescale. The characteristic inflection points are named O, I, D, P, S, M, and T. The O-P and P-T transients are classified as the fast and slow phases, respectively. The dotted line represents the intensity of the excitation light.

(*H* in Figure 10.1). The transient O-I represents the photoreduction of Q_A by the PSII reaction center, I-D represents the rapid oxidation of Q_A by the plastoquinone (PQ) pool and PSI, and D-P indicates the photoreduction of Q_A by the PSII reaction linked to the water-splitting enzyme system. The slow phase (from P to T) requires a couple of minutes, and the physiological interpretation of this phase is far more difficult than that of the first phase because a number of different processes are involved (Papageorgiou et al., 2007; Stirbet and Govindjee, 2011). These processes that lower the chlorophyll fluorescence intensity in the slow phase are called quenching and are classified into two major categories: photochemical quenching and nonphotochemical quenching.

10.6.2 *Physiological interpretation of induction curve*

The contributions of photochemical and nonphotochemical quenching on the slow phase of chlorophyll fluorescence induction can be determined by consecutive measurements of Φ_{PSII} and NPQ during induction. Figure 10.9a shows an example of the changes in Φ_{PSII} and NPQ during the chlorophyll fluorescence induction of a mature tomato leaf (Takayama et al., 2012). The spike-like increase in the chlorophyll fluorescence yield during the chlorophyll fluorescence induction is due to the measurement of the saturation pulse method for the determination of Φ_{PSII} and NPQ

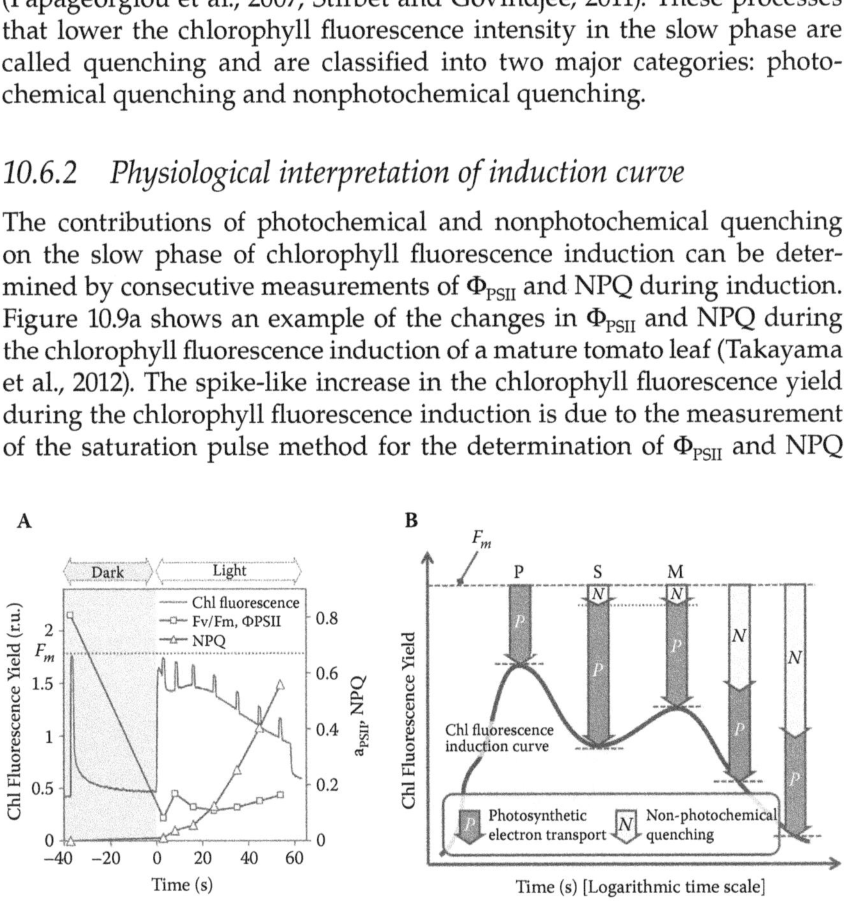

Figure 10.9 (a) Time courses of chlorophyll fluorescence yield, Φ_{PSII} (F_v/F_m), and NPQ of a mature tomato leaf before and during a chlorophyll fluorescence induction with repetitive applications of the saturation light pulse. These data are plotted on a linear timescale. The horizontal dotted line indicates the chlorophyll fluorescence yield of F_m. (b) Schematic illustration of the contributions of photosynthetic electron transport activity and nonphotochemical quenching, i.e., energization of the thylakoid membrane, state transition, and xanthophyll cycle on the slow phase of the chlorophyll fluorescence induction curve of tomato leaves.

at each time point. This example proves that the P-S transient is mainly caused by a temporal increase in Φ_{PSII} accompanied by a slight increase in NPQ, and the following S-M transient is caused by prompt depression of Φ_{PSII} (Figure 10.9b). In such a case, a parameter such as P/S (the chlorophyll fluorescence intensity at P divided by the chlorophyll fluorescence intensity at S) can be used as an index of the initial activation of photosynthetic electron transport.

10.6.3 Single leaf measurement—photosynthetic dysfunction caused by an herbicide

Chlorophyll fluorescence induction images of a kidney bean (*Phaseolus vulgaris* L.) leaf before and after herbicide treatment are shown in Figure 10.10. In this treatment, the petiole was soaked in a 1/1000 diluted solution of Nekosogi-ace (Rainbow Chemicals) for 20 min (Takayama et al., 2003). The active ingredient of Nekosogi-ace is 3-(3,4-dichlorophenyl)-1,1-dimethylurea (DCMU). The chlorophyll fluorescence imaging was conducted by the chlorophyll fluorescence imaging system shown in Figure 10.6. Although the chlorophyll fluorescence intensity was relatively uniform all over the

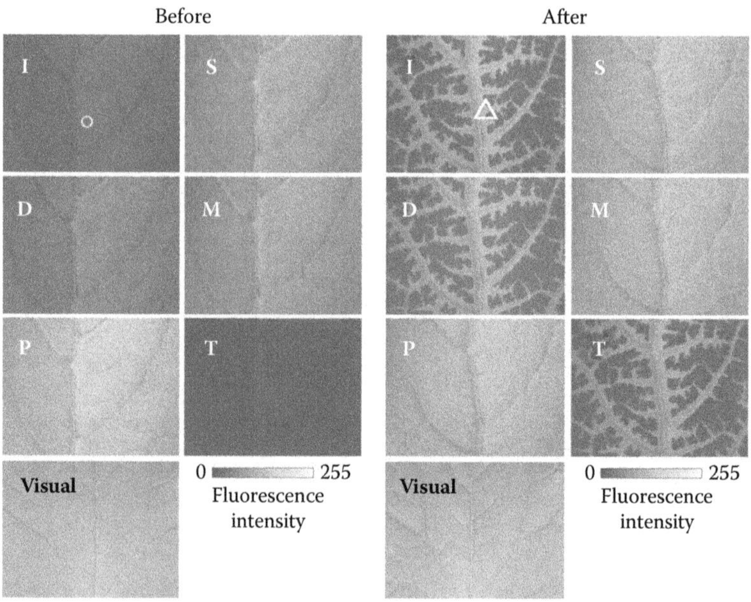

Figure 10.10 **(See color insert.)** Changes in chlorophyll fluorescence induction images of a kidney bean (*Phaseolus vulgaris* L.) leaf before and after herbicide treatment. The petiole was soaked in 1/1000 diluted solution of Nekosogi-ace (Rainbow Chemicals) for 20 min. The symbols ○ and Δ indicate the leaf areas used for the chlorophyll fluorescence induction analysis shown in Figure 10.11.

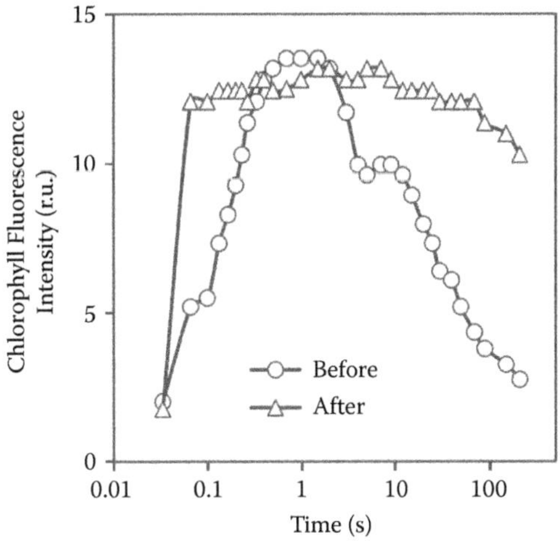

Figure 10.11 Changes in chlorophyll fluorescence induction curve of a kidney bean (*Phaseolus vulgaris* L.) leaf before (○ shown in Figure 10.10) and after (Δ shown in Figure 10.10) herbicide (DCMU) treatment.

leaf surface before treatment, it was remarkably increased at the mesophyll cells around the major veins (see the images of I, D, and T after treatment).

Figure 10.11 shows the chlorophyll fluorescence induction at sites around the major vein marked as ○ (before) and Δ (after) in Figure 10.10. As DCMU blocks photosynthetic electron transport from Q_A to PQ, probably by binding to the Q_B site of the D1 protein, the increase in chlorophyll fluorescence intensity from O to P is much faster in the presence of DCMU than in the absence of an inhibitor (Krause and Weis, 1991; Govindjee, 1995). In addition, the inhibition of photosynthetic electron transport by DCMU causes a low intrathylakoid pH gradient and results in lower heat dissipation. Therefore, we can diagnose the rapid increase in chlorophyll fluorescence intensity from O to P at sites around major veins (Δ in Figure 10.11) as a photosynthetic dysfunction caused by the DCMU. In contrast to such a clear observation in chlorophyll fluorescence induction imaging, no visible injury occurred on the leaf during the measurement.

10.6.4 *Whole tomato plant measurement—detection of light stress*

The chlorophyll fluorescence imaging system used to measure chlorophyll fluorescence induction of whole tomato plants is illustrated in Figure 10.12a (Takayama and Nishina, 2009; Takayama et al., 2010). An optical fiber equipped with a short-pass filter was used to condense light from four 180 W metal halide lamps into a blue excitation light (LS-M180, Sumita Optical Glass, Inc.) ($\lambda < 620$ nm) (Corning 4-96). The blue excitation light could

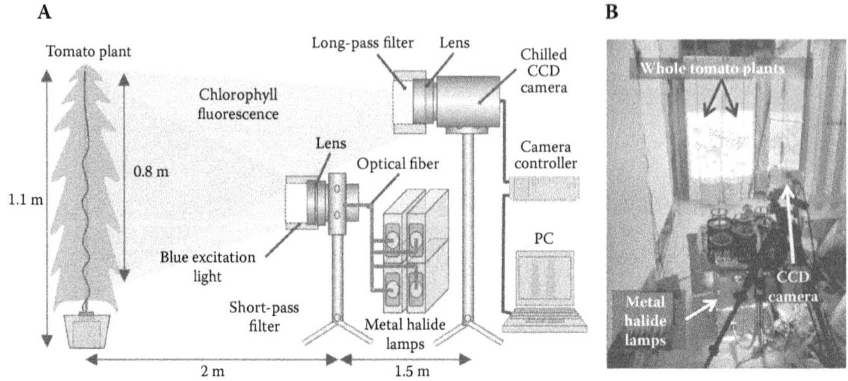

Figure 10.12 (a) Schematic diagram of the chlorophyll fluorescence imaging system for whole tomato plants. The intensity of the blue excitation light at the plant surface was approximately PPFD 10 μmol $m^{-2}s^{-1}$. (b) The laboratory setup for chlorophyll fluorescence imaging of the two whole tomato plants.

illuminate an area of 0.8 × 0.8 m^2, covering all the leaves from two 1.1 m high tomato plants located 2 m away from the excitation light sources. Induction of chlorophyll fluorescence was captured with a CCD camera (C5985-02, Hamamatsu Photonics) equipped with a long-pass filter ($\lambda > 680$ nm) (Corning 2-64), located 3.5 m from the plants. All images were recorded by a computer. The intensity of the blue excitation light was approximately PPFD 10 μmol $m^{-2}s^{-1}$ on the plant surface. Actual imaging of the chlorophyll fluorescence induction of two whole tomato plants is presented in Figure 10.12b.

Changes in chlorophyll fluorescence intensity at the inflection point of P measured before and just after a light stress treatment are shown in Figure 10.13 (Takayama and Nishina, 2009; Takayama et al., 2010). One plant (stressed) was exposed to full sunlight (approximately PPFD 1500 μmol $m^{-2}s^{-1}$) for 2.5 h, and a second plant (control) was kept in the dark throughout the experiment. Although the chlorophyll fluorescence intensities of both tomato plants were almost the same before treatment, the intensity of the stressed plant after treatment was clearly lower than that of the control plant. The decrease in chlorophyll fluorescence intensity at the inflection point of P after light stress treatment is probably due to activation of the heat dissipation process (see Figure 10.1) and photoinhibition (Krause and Weis, 1991; Müller et al., 2001).

10.6.5 *Tomato plants grown in a greenhouse—detection of drought stress*

A schematic presentation of the chlorophyll fluorescence induction imaging system for tomato plants cultivated in a greenhouse is depicted in

Figure 10.13 Effect of light stress on chlorophyll fluorescence intensity at the inflection point P of the induction curve. The control plant on the left was kept in the dark throughout the experiment. The experimental plant on the right was exposed to full sunlight for 2.5 h. (Reproduced from Takayama and Nishina, *Environ. Control Biol.*, 47, 101–109, 2009.)

Figure 10.14a (Takayama and Nishina, 2009; Takayama et al., 2011c). Figure 10.14b shows the actual imaging system in the greenhouse. A 60 × 60 cm^2 blue light-emitting diode (LED) panel provided excitation light ($\lambda < 520$ nm) irradiating the leaves around the shoot apex of the tomato plant. The duration of excitation light irradiation was set at 30 s. The PPFD of the excitation light at the plant surface ranged from 30 to 100 $\mu mol\ m^{-2}s^{-1}$. A CCD camera with a long-pass filter ($\lambda > 640$ nm) (Figure 10.14c), fixed in the middle of the LED panel, captured the chlorophyll fluorescence emission induced by the excitation light. The imaging system was approximately 0.6 m from the target plant. Temporal resolution of the imaging was 15 fps, and the captured image was recorded on the hard disk of a laptop computer in BMP format at 1024 (W) × 768 (H) pixels. The captured images were analyzed using Visual Basic 6.0 software developed by us. The software selected an image of maximum chlorophyll fluorescence intensity from the 450 captured images and determined the plant area. Average pixel values within the determined plant area were calculated for all captured images, and a chlorophyll fluorescence induction curve was made. All imaging was performed under dark conditions at night.

The set of chlorophyll fluorescence intensity images at inflection points I-D, P, and S-M and the resulting chlorophyll fluorescence induction curve are shown in Figure 10.15. The inflection points of I-D and P are clearly recognized in the induction curve, but inflection points S and M are unclear because the decrease in chlorophyll fluorescence intensity from P to S-M was very gentle. Such a chlorophyll fluorescence induction curve, with no clear inflection points for S and M, is usually observed in tomato plants during winter.

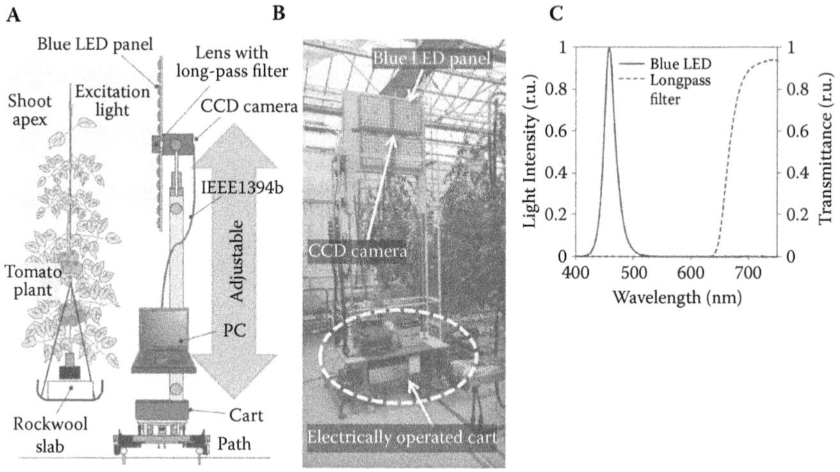

Figure 10.14 (a) Chlorophyll fluorescence imaging system for tomato crops grown in a greenhouse. The LED panel provides the excitation light, and the induced chlorophyll fluorescence is captured by the CCD camera filtered with a long-pass filter. (b) The semicommercial greenhouse in which chlorophyll fluorescence imaging was performed at night. (c) The emission spectrum of the excitation light provided by the blue LED panel (solid line) and the spectral property of the long-pass filter (dashed line).

Figure 10.15 A set of chlorophyll fluorescence intensity images, including several leaves around the shoot apex at the inflection points I-D, P, and S-M and the resulting chlorophyll fluorescence induction curve.

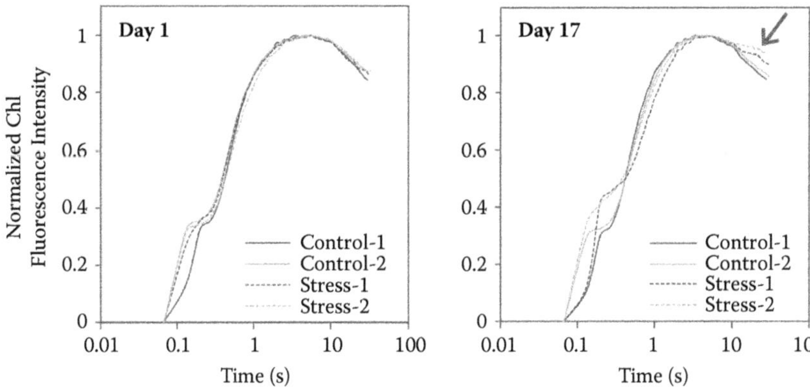

Figure 10.16 Changes in shapes of chlorophyll fluorescence induction curves between day 1 and day 17 of drought stress treatment (the arrow marks a clear difference between control and stress plants). The chlorophyll fluorescence intensity of an induction curve was normalized to the maximum chlorophyll fluorescence intensity of the induction curve.

Four healthy tomato plants grown in a semicommercial greenhouse were used for the next experiment. Two plants (Stress-1 and Stress-2) were exposed to drought stress by stopping the nutrient water supply for 17 days. The other two plants (Control-1 and Control-2) were cultivated normally during the 17 days as controls. Figure 10.16 shows the chlorophyll fluorescence induction curves of the control (Control-1 and Control-2) and stressed (Stress-1 and Stress-2) plants on days 1 and 17 of drought stress treatment. The chlorophyll fluorescence intensities of the induction curves were normalized to their maximum chlorophyll fluorescence intensity (P) to enhance the differences in shapes of the chlorophyll fluorescence induction curves. At the beginning of the drought stress treatment, the shapes of all induction curves were similar, but a clear difference in the shapes between the control (Control-1 and Control-2) and the stressed (Stress-1 and Stress-2) plants was recognized on day 17 of drought stress treatment. In particular, the inflection points of S-M for Stress-1 and Stress-2 were relatively higher; that is, a small decline was seen from P to S-M, compared with those for Control-1 and Control-2 (see arrow in Figure 10.16) (Takayama et al., 2011c).

10.7 Conclusion

In this chapter, the basics of chlorophyll fluorescence imaging were first explained with special attention to the technical and engineering aspects, and then some examples of plant health monitoring with chlorophyll fluorescence imaging techniques were introduced. The examples included the detection of invisible photosynthetic dysfunctions of plants caused by stress factors such as herbicides, drought stress, and strong light stress for a leaf surface, a whole plant, and a crop canopy.

In detail, the saturation pulse method and the induction method were mentioned as popular chlorophyll fluorescence imaging techniques to evaluate photosynthetic functions. The differences between these two techniques were emphasized. The saturation pulse method is the most common method and provides widely used chlorophyll fluorescence parameters, not only NPQ and Φ_{PSII}, but also F_v/F_m (maximum quantum yield of PSII), qP (proportion of open PSII), and so on (Genty et al., 1989; Bilger and Björkman, 1990; Krause and Weis, 1991; Maxwell and Johnson, 2000; Osmond et al., 1998; Omasa and Takayama, 2002). However, the saturation pulse method requires keeping a strong saturation light intensity to reduce all the primary electron acceptors; otherwise, F'_m can be underestimated, and this results in the overestimation of NPQ and the underestimation of Φ_{PSII}. This characteristic suggests that the saturation pulse method may not be applicable for plant health monitoring of whole plants and plant canopies. In contrast, a weak excitation light is enough to induce chlorophyll fluorescence in the induction method (Govindjee, 1995; Omasa and Takayama, 2002). Although widely used parameters are not derived from the induction method, it has potential as a tool for plant health monitoring of whole plants and plant canopies (Omasa and Takayama, 2002; Takayama and Nishina, 2009).

Because of recent remarkable increases in the availability of high-performance yet affordable instruments such as blue LED panels and high spatial and temporal resolution CCD cameras with a USB connection, various kinds of high-throughput phenotyping applications with chlorophyll fluorescence imaging techniques seem more achievable (Nichol et al., 2012; Takayama et al., 2011c). In our studies, a chlorophyll fluorescence imaging robot that can be installed in a commercial tomato production greenhouse was developed for plant health monitoring at the crop canopy level (Takayama et al., 2011a, 2011b, 2011c). In the near future, chlorophyll fluorescence imaging is expected to become a common method for plant health monitoring in agricultural production.

References

Balachandran, S., Osmond, C.B., and Daley, P.F., Diagnosis of the earliest strain-specific interactions between tobacco mosaic virus and chloroplasts of tobacco leaves in vivo by means of chlorophyll fluorescence imaging, *Plant Physiol.*, 104, 1059–1065, 1994.

Bilger, W., and Björkman, O., Role of the xanthophyll cycle in photoprotection elucidated by measurements of light-induced absorbance changes, fluorescence and photosynthesis in leaves of *Hedera canarienisis*, *Photosynth. Res.*, 25, 173–185, 1990.

Bro, E., Meyer, S., and Genty, B., Heterogeneity of leaf CO_2 assimilation during photosynthetic induction, *Plant Cell Environ.*, 19, 1349–1358, 1996.

Daley, P.F., Chlorophyll fluorescence analysis and imaging in plant stress and disease, *Can. J. Plant Pathol.*, 17, 167–173, 1995.

Daley, P.F., Raschke, K., Ball, J.T., and Berry, J.A., Topography of photosynthetic activity of leaves obtained from video images of chlorophyll fluorescence, *Plant Physiol.*, 90, 1233–1238, 1989.

Genty, B., Briantais, J.M., and Baker, N.R., The relationship between the quantum yield of photosynthetic electron transport and quenching of chlorophyll fluorescence, *Biochim. Biophys. Acta*, 990, 87–92, 1989.

Genty, B., and Meyer, S., Quantitative mapping of leaf photosynthesis using chlorophyll fluorescence imaging, *Aust. J. Plant Physiol.*, 22, 277–284, 1995.

Govindjee, Sixty-three years since Kautsky: chlorophyll a fluorescence, *Aust. J. Plant Physiol.*, 22, 131–160, 1995.

Krause, G.H., and Weis, E., Chlorophyll fluorescence and photosynthesis: the basics, *Annu. Rev. Plant Physiol. Plant Mol. Biol.*, 42, 313–349, 1991.

Lea, P.J., and Ridley, S.M., Glutamine synthetase and its inhibition, in *Herbicides and plant metabolism (seminar series/Society for Experimental Biology; 38)*, ed. A.D. Dodge, 137–170, Cambridge University Press, Cambridge, 1989.

Maxwell, K., and Johnson, G.N., Chlorophyll fluorescence: a practical guide, *J. Exp. Bot.*, 51, 659–668, 2000.

Meyer, S., and Genty, B., Heterogeneous inhibition of photosynthesis over the leaf surface of *Rosa rubiginosa* L. during water stress and abscisic acid treatment: induction of a metabolic component by limitation of CO_2 diffusion, *Planta*, 210, 126–131, 1999.

Müller, P., Li, X.P., and Niyogi, K.K., Non-photochemical quenching: a response to excess light energy, *Plant Physiol.*, 125, 1558–1566, 2001.

Nichol, C.J., Pieruschka, R., Takayama, K., et al., Canopy conundrums: building on the Biosphere 2 experience to scale measurements of inner and outer canopy photoprotection from the leaf to the landscape, *Funct. Plant Biol.*, 39, 1–24, 2012.

Omasa, K., and Shimazaki, K., Image analysis of chlorophyll fluorescence in leaves, in *Measurement technique in plant science*, ed. Y. Hashimoto, P.J. Kramer, H. Nonami, and B.R. Strain, 387–401, CRC Press, Boca Raton, FL, 1990.

Omasa, K., Shimazaki, K., Aiga, I., Larcher, W., and Onoe, M., Image analysis of chlorophyll fluorescence transients for diagnosing the photosynthetic system of attached leaves, *Plant Physiol.*, 84, 748–752, 1987.

Omasa, K., and Takayama, K., Image instrumentation of chlorophyll *a* fluorescence for diagnosis photosynthetic injury, in *Air pollution and plant biotechnology*, ed. K. Omasa, H. Saji, S. Youssefian, and N. Kondo, 287–308, Springer-Verlag, Berlin, 2002.

Osmond, C.B., Daley, P.F., Badger, M.R., and Lüttge, U., Chlorophyll fluorescence quenching during photosynthetic induction in leaves of *Abutilon striatum* Dicks. infected with Abutilon mosaic virus, observed with a field-portable imaging system, *Bot. Acta*, 111, 390–397, 1998.

Oxborough, K., Imaging of chlorophyll *a* fluorescence: theoretical and practical aspects of an emerging technique for the monitoring of photosynthetic performance, *J. Exp. Bot.*, 55, 1195–1205, 2004.

Papageorgiou, G.C., Tsimilli-Michael, M., and Stamatakis, K., The fast and slow kinetics of chlorophyll a fluorescence induction in plants, algae and cyanobacteria: a viewpoint, *Photosynth. Res.*, 94, 275–290, 2007.

Rolfe, S.A., and Scholes, J.D., Quantitative imaging of chlorophyll fluorescence, *New Phytol.*, 131, 69–79, 1995.

Siebke, K., and Weis, E., Assimilation images of leaves of *Glechoma hederacea*: analysis of non-synchronous stomata related oscillations, *Planta*, 196, 155–165, 1995.

Stirbet, A., and Govindjee, On the relation between the Kautsky effect (chlorophyll *a* fluorescence induction) and photosystem II: basics and applications of the OJIP fluorescence transient, *J. Photochem. Photobiol. B*, 104, 236–257, 2011.

Takayama, K., Konishi, A., and Omasa, K., Diagnosis of invisible photosynthetic injury caused by a herbicide (Basta) with chlorophyll fluorescence imaging system, *CIGR J.*, 03 001, 1–12, 2003.

Takayama, K., Miguchi, Y., Manabe, Y., Takayamashi, N., and Nishina, H., Analysis of Φ_{PSII} and NPQ during the slow phase of the chlorophyll fluorescence induction phenomenon in tomato leaves, *Environ. Control Biol.*, 50, 181–187, 2012.

Takayama, K., and Nishina, H., Chlorophyll fluorescence imaging of chlorophyll fluorescence induction phenomenon for plant health monitoring, *Environ. Control Biol.*, 47, 101–109, 2009.

Takayama, K., Nishina, H., Iyoki, S., Arima, S., Hatou, K., Ueka, Y., and Miyoshi, Y., Early detection of drought stress in tomato plants with chlorophyll fluorescence imaging—practical application of the speaking plant approach in a greenhouse, in *Preprints of the 18th IFAC World Congress*, 2011c, pp. 1785–1790.

Takayama, K., Nishina, H., Mizutani, K., Arima, S., Hatou, K., and Miyoshi, Y., Chlorophyll fluorescence imaging for health condition monitoring of tomato plants in greenhouse, *Acta Hortic.*, 893, 333–339, 2011a.

Takayama, K., Nishina, H., Mizutani, K., Iyoki, S., Arima, S., Hatou, K., and Miyoshi, Y., Evaluation of photosynthetic functions of tomato plants in greenhouse with chlorophyll fluorescence induction imaging system, *Acta Hortic.*, 907, 337–342, 2011b.

Takayama, K., Sakai, Y., Oizumi, T., and Nishina, H., Assessment of photosynthetic dysfunction in a whole tomato plant with chlorophyll fluorescence induction imaging, *Environ. Control Biol.*, 48, 151–159, 2010.

chapter eleven

PRI imaging and image-based estimation of light intensity distribution on plant canopy surfaces

Yasuomi Ibaraki and S. Dutta Gupta

Contents

11.1 Introduction

Photosynthesis is a primary physiological process for plant growth, and the evaluation of photosynthetic status is important in cultivation practice and botanical research. Recently, image-based methods have been developed for nondestructively evaluating the spatial distribution of properties of photosynthesis or gas exchange controlled by stomatal regulation. The analysis of chlorophyll fluorescence is a powerful method for evaluating the functioning of an intact photosynthetic system (Jones, 1992), particularly the functioning of photosystem II (PSII) (Lichtenthaler et al., 1996). Several reports have shown the possibility of imaging chlorophyll fluorescence parameters based on a saturating light pulse method to obtain an image of the photochemical efficiency of PSII for a leaf (Genty and Meyer, 1994; Meng et al., 2001; Omasa and Takayama, 2003) or several leaves (Ibaraki and Matsumura, 2005; Guidi et al., 2007; Berger et al., 2007). Moreover, some chlorophyll fluorescence imaging systems (Walz, IMAGING-PAM, Germany; PSI, FluorCam, Czech Republic) are now commercially available.

However, these systems have some imaging limitations. Chlorophyll fluorescence imaging using a saturated light pulse requires a pulse of very high intensity (>2000 μmol m^{-2} s^{-1} of photosynthetic photon flux density (PPFD)). A high-intensity light source is required, and the image must be acquired at a short distance from the leaf surface. In addition, variations of leaf angle in a canopy or surface roughness on a single leaf affect the success of saturating pulse irradiation.

Photochemical reflectance index (PRI) has been proposed for estimating the efficiency of photosynthetic radiation use (Gamon et al., 1997), primarily in remote sensing. This index is derived from reflectance at 531 and 570 nm and is a measure of the degree of the photoprotective xanthophyll cycle pigment zeaxanthin, which is related to heat dissipation in PSII. PRI has been reported to be highly correlated with the quantum yield of PSII ($\Delta F/Fm'$) (Gamon et al., 1997; Stylinski et al., 2002; Ibaraki et al., 2010), which is determined by chlorophyll fluorescence. PRI has a potential to be used as a substitution for chlorophyll fluorescence-based estimation of PSII quantum yield (Ibaraki et al., 2010), although some caution should be observed in its use for evaluating light use efficiency (Nakaji et al., 2006; Hall et al., 2008). PRI measurement is based on reflectance and has the potential for simple and nondestructive image-based evaluation of photosynthetic properties because it does not require a saturated light pulse (very high intensity light) or a modulated lighting system for measurement of fluorescence quantum yield (Ibaraki et al., 2010).

However, the actual quantum yield of PSII under light conditions, which can be estimated from chlorophyll fluorescence or PRI measurements, depends on the level of the light irradiating a leaf (Ibaraki, 2006).

Figure 11.1 (**See color insert.**) An example of thermal image of rose canopy in a greenhouse. The values in the figure indicate photosynthetic photon flux density (PPFD) measured with a quantum sensor.

Therefore, it is essential to determine the actual light intensity irradiating a plant canopy surface for image-based evaluation of the PSII function by chlorophyll fluorescence or PRI. Moreover, a measure of the light intensity irradiating a leaf provides critical information for the evaluation of plant gas exchange (stomatal response) on the basis of thermal imaging because the light intensity irradiating the leaf is one of the main components in the energy balance of the leaf, and consequently affects leaf temperature (Figure 11.1). Therefore, in these methods, light intensity on the leaf surface provides critical information to accurately evaluate the plant status. A method for simple evaluation of light intensity distribution is desirable.

In this chapter, we introduce a low-cost system for PRI imaging along with its application in micropropagated plant leaves from outside culture vessels and discuss its feasibility. Moreover, we have introduced an image-based method for estimating light intensity on leaf surfaces, which is a critical factor in evaluating plant photosynthetic status using PRI, with application to a tomato canopy in a greenhouse.

11.2 *PRI imaging*

11.2.1 *Basic principles and methods of PRI imaging*

The xanthophyll cycle, in which the carotenoid pigment violaxanthin is converted to antheraxanthin and zeaxanthin through de-epoxidase reactions (Yamamoto, 1979), is associated with heat dissipation in PSII, and reflectance changes near 531 nm are associated with the de-epoxidation of violaxanthin to zeaxanthin through antheraxanthin (Gamon et al., 1990). PRI can be estimated using reflectance at two different wavelengths, 531 and 570 nm, by the following equation:

$$PRI = \frac{R_{531} - R_{570}}{R_{531} + R_{570}} \tag{11.1}$$

where R_{531} and R_{570} denote reflectance at 531 and 570 nm, respectively.

PRI was highly correlated with the quantum yield of PSII, $\Delta F/Fm'$, determined by chlorophyll fluorescence for 20 species representing three functional types of plants (Gamon et al., 1997). Stylinski et al. (2002) also reported a strong correlation of PRI with the chlorophyll fluorescence parameter $\Delta F/Fm'$ across species and seasons. Rashimzadeh-Bajgiran et al. (2012) reported good correlation between PRI and nonphotochemical quenching and between PRI and electron transfer rate, although they pointed out limitations in the application of PRI. The use of PRI is promising for simple evaluation of the photosynthetic properties of plants in field conditions because it does not require a saturated light pulse or a modulated lighting system.

For PRI imaging, it is necessary to acquire two images at two different wavelengths. A high-cost device such as a spectral camera (e.g., Nakaji et al., 2005) is normally required to obtain spectral reflectance information for estimation of PRI. Imaging of reflectance at specific wavelengths using band-pass filters is a promising approach to develop a low-cost imaging system of reflectance indices. However, unlike NDVI, PRI requires imaging in a relatively narrow wavelength band. Therefore, narrow band-pass filters are necessary. Thus, spectral reflectance may be imaged by appropriate use of band-pass filters. However, linearity of output (pixel value (PV)) of a camera and input is critical for success in imaging reflectance. It is worthy to note that most gamma values for digital cameras are not 1.0, and these cameras often use auto gain control.

11.2.2 Low-cost PRI imaging system and its application in micropropagated plants

11.2.2.1 Low-cost PRI imaging system

A PRI imaging system using commercially available band-pass filters was developed (Ibaraki et al., 2010). In brief, each filter had a half width of 10 nm, and both filters were doubled for acquisition of narrow band-pass images. The reflected light intensities of leaves at 530 and 570 nm were imaged with a monochrome charge-coupled device (CCD) camera (Watec, WAT-120N, Japan) with 530 and 570 nm band-pass filters (MX0530 and MX0570, respectively; Asahi Spectra, Japan). Pairs of each filter were placed in filter holders (Suruga, F53-100, Japan), and holders were manually changed. The CCD camera used was a noncooled type; however, it could accumulate up to 256 frames and was used with a constant gamma value ($\gamma = 1$) and without automatic gain control.

For imaging in an experimental room, plant leaves were illuminated from above at a fixed PPFD intensity by an artificial light (a ring-shaped white LED lamp) attached to a diffuse panel. In outdoor imaging, direct solar radiation was blocked with a fabric shade. In both indoor and outdoor imaging, the depression angle of a CCD camera was fixed at 45°. The target leaf was placed on the imaging stand almost vertically to the optical axis of the CCD camera at a distance of 50 cm from the camera. A reflectance standard (Kodak, gray card) with constant reflectance over 400–700 nm was placed beside the targeted leaf at the same angle as the leaf to allow conversion of light intensity to relative reflectance. The relative reflectance of leaves could be estimated under different light sources using the reflectance standard displaying constant reflectance over the visible wavelength region.

Because outputs of both light source and camera are susceptible to variation in applied voltage, power supply for the light source and the camera was constantly regulated using a voltage regulator (Aswan, ASA-10000, Japan). The NTSC video signal from the CCD camera was digitized and converted into 8-bit monochrome images by a video capture board. Each image was digitized twice and the average was used.

PRI image, I_{PRI}, was constructed by image calculation using the following equation:

$$I_{PRI} = \frac{kI_{531} - I_{570}}{kI_{531} + I_{570}} \tag{11.2}$$

where I_{530} and I_{570} are images acquired through 530 and 570 nm band-pass filters, respectively, and k is a coefficient used to convert reflection intensity (i.e., PV) into relative reflectance and calculated as $k = S_{570}/S_{530}$, where $S5_{30}$ and S_{570} are the PV of the reflectance standards for I_{530} and I_{530}, respectively.

First, to test the stability of the system's measurement of light intensity, the linearity of PV in images to light intensity incident to the camera and repeatability of the PRI measurement were tested. The reflectance standard was illuminated with a halogen lamp and was imaged through several neutral density (ND) filters. The relationships between PV in images of the reflectance standard and the transmissivity of ND filters were investigated (Figure 11.2). The average PV in a dark frame acquired under no lighting conditions in a dark room was 16. This value was used as an offset value. The linearity of PV to the light intensity incident to

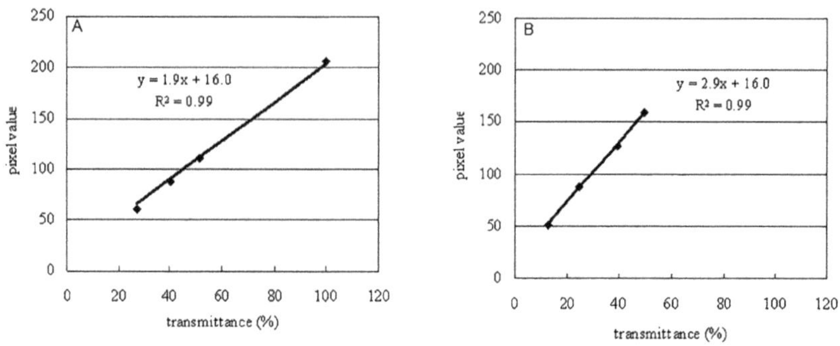

Figure 11.2 Relationships between pixel value (PV) in images of the reflectance standard and incident light intensity to the camera (transmissivity of neutral density (ND) filters) at (a) 530 and (b) 570 nm. (Reprinted from Ibaraki et al., *Comput. Electron. Agric.*, 71, 170–175, 2010. With permission of Elsevier.)

the camera was confirmed for both 530 nm (Figure 11.2a) and 570 nm (Figure 11.2b) measurements.

Repeatability of PRI measurement was tested using a pothos leaf whose photochemical efficiency, $\Delta F/Fm'$, was relatively stable under room light conditions. The leaf was placed on the imaging stand and imaged repeatedly under 10 μmol m^{-2} s^{-1} of PPFD. The variation in PRI was estimated from the resulting images. The results of 40 replications gave an average PRI value of −0.026 and a PRI standard error of 6.3×10^{-4} (2.4% of the average value) for a leaf whose $\Delta F/Fm'$ was 0.75. This result confirmed the relatively high repeatability of PRI measurement by the system.

Next, the relationships between image-based PRI and PSII quantum yield ($\Delta F/Fm'$), determined by chlorophyll fluorescence, were investigated for lettuce leaves treated with abscisic acid (ABA) outside and strawberry leaves exposed to high light intensity in an experimental room. Linear relationships between $\Delta F/Fm'$ and PRI, estimated with the developed imaging system, were obtained in both settings. The system was also applied to micropropagated plants grown in culture vessels, and images were acquired from outside the culture vessels (Figure 11.3). The time course of PSII quantum yield for a single leaf could be evaluated using PRI estimated from images of a potato plantlet (Figure 11.4). These findings suggested that the system is potentially inexpensive, simple, and efficient for estimating PRI of plant leaves or micropropagated plantlets.

11.2.2.2 *Application for evaluation of photosynthetic capacity of micropropagated plants*

As described above, the PRI imaging system developed by us has the potential to provide information on the quantum yield of PSII, normally

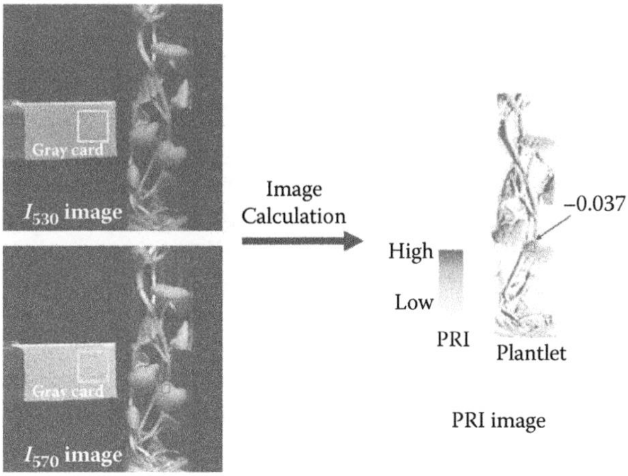

Figure 11.3 (**See color insert.**) An example of acquired images (I530, I570) and the PRI image of a potato plantlet. (Reprinted from Ibaraki et al., *Comput. Electron. Agric.*, 71, 170–175, 2010. With permission of Elsevier.)

determined as $\Delta F/Fm'$ by chlorophyll fluorescence measurement (Ibaraki and Matsumura, 2005), nondestructively from the outside of culture vessels. However, because $\Delta F/Fm'$ depends on the level of light intensity on leaves, care is required in determining photosynthetic properties from values of $\Delta F/Fm'$. In particular, in micropropagation, it is difficult to determine the exact light intensity on a plantlet leaf inside a culture vessel because

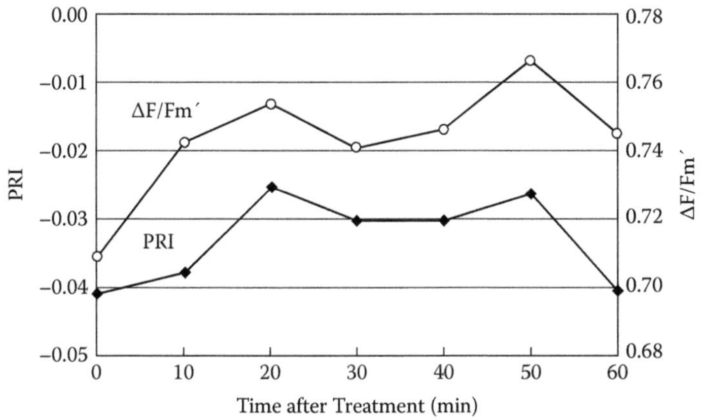

Figure 11.4 Relationship between photochemical reflectance index (PRI) and photosystem II (PSII) quantum yield ($\Delta F/Fm'$) in potato plantlet leaves treated with saturated light. (Reprinted from Ibaraki et al., *Comput. Electron. Agric.*, 71, 170–175, 2010. With permission of Elsevier.)

the light intensity inside the vessel differs from that of the outside (Ibaraki and Nozaki, 2005). Therefore, the use of maximum PSII quantum yield, *Fv/Fm*, which is determined under weak light conditions after dark adaptation, is desirable for routine estimation of PRI for micropropagated plant leaves within culture vessels. PRI is normally measured in daylight under high light conditions. Ibaraki and Dutta Gupta (2010) proposed a promising method for evaluating the photosynthetic properties of micropropagated plantlets. They nondestructively acquired PRI images of micropropagated potato leaves from outside the culture vessel using the system described here and showed that PRI under weak light conditions after dark treatment could be used as a substitute for the chlorophyll fluorescence parameter *Fv/Fm*. A linear relationship between PRI and *Fv/Fm* was observed for both plantlets exposed to short-term high-light treatment and grown under different culture conditions (Figure 11.5). The combined plot of high-light treatment and culture conditions showed a significant correlation between PRI and *Fv/Fm*, suggesting that the relationship may be independent of *Fv/Fm*-decreasing factors. They concluded that PRI estimated from reflectance images under weak light conditions may be used for rapid and noninvasive

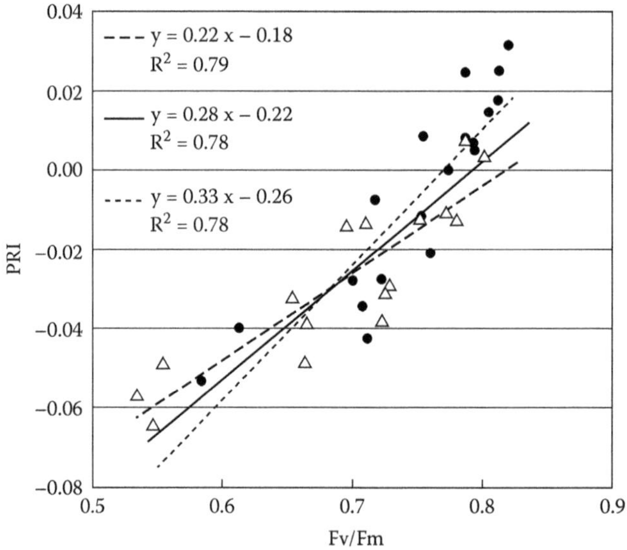

Figure 11.5 Relationship between photochemical reflectance index (PRI) after dark adaptation and *Fv/Fm* observed for plantlets exposed to both short-term high-light treatment (open triangle, broken line) and cultured under different conditions (closed circle, dotted line). The regression line for a combined plot is also shown (solid line). (Reproduction of Figure 3 in Ibaraki and Dutta Gupta, *In Vitro Cell. Dev. Biol.-Plant*, 46, 530–536, 2010. With kind permission from Springer Science + Business Media B.V.)

evaluation of the photosynthetic properties of micropropagated plantlets in a manner similar to that used for determining *Fv/Fm*.

However, it should be noted that there are some limitations for using PRI to evaluate photosynthetic properties of plant leaves. Factors reported to affect PRI measurement are as follows: chlorophyll content (Nakaji et al., 2006; Suárez et al., 2009), temperature (Nakaji et al., 2006), shadow fraction (Hall et al., 2008), viewing (imaging) direction (Suárez et al., 2009), and leaf structure (Suárez et al., 2009). In the study of micropropagated potato plants (Ibaraki and Dutta Gupta, 2010), no correlation was observed between PRI and *Fv/Fm* in leaves that turned yellow because of prolonged culture. Nakaji et al. (2006) also reported that seasonal variation of pigment contents affected PRI values. Rashimzadeh-Bajgiran et al. (2012) showed that PRI was highly correlated with content of plant pigments, particularly chlorophyll. Chlorophyll content is a critical factor for determining the validity of PRI as a substitute for chlorophyll fluorescence. Several methods for estimating chlorophyll content in micropropagated plant leaves using RGB color images have been reported (Yadav et al., 2010; Dutta Gupta et al., 2013). PRI imaging combined with these methods would be an effective tool for evaluation of photosynthetic capacity in micropropagated plants.

Another critical point is the dependence of PRI on light intensity irradiating the leaf. PSII status cannot be assessed by PRI without knowing the light intensity. In particular, in PRI imaging, it is essential to know the light intensity distribution. A simple method for estimating light intensity distribution on a canopy surface is described in the next section.

11.3 *Image-based estimation of PPFD distribution on plant canopy surface*

As described above, determining light intensity on the leaf surface is critical for accurately evaluating plant photosynthetic status, particularly in methods using remotely acquired images. Light intensity is often evaluated as PPFD in evaluations of photosynthetic status. PPFD per se can be measured by several types of quantum sensor, including point, line, and globular-type sensors. However, determining the distribution of light intensity on leaves located at the canopy surface is difficult even with these sensors because of the dynamic variation in solar radiation and large variations in leaf angle/orientation in a plant canopy (Ibaraki, et al., 2012b). The reflection image of a leaf includes information on light irradiating the leaf; therefore, it has the potential to be used for simple estimation of light intensity. Ibaraki et al. (2012a, 2012b) developed a simple method for evaluating PPFD distribution on a plant canopy surface using the relationship between light intensity on the leaf surface and PV in the reflection image. In this section, the method is introduced with an application in tomato plants in a greenhouse.

11.3.1 Methods

11.3.1.1 Acquisition of reflection images of plant canopy and analysis of the images

Reflection images of plant canopies were acquired with a monochrome 14-bit CCD camera (Bitran, BU-41L, Japan) or a digital camera (Canon, SX130, Japan) through a blue-green band-pass filter (Suruga, S76-BG28 or S76-BG7, Japan). The filter had a peak wavelength of approximately 450–500 nm and was chosen on the basis of the results of a preliminary experiment with several optical filters, including yellow, red, and far-red filters. To minimize the effect of specular reflection, images were acquired from several directions. Cameras were moved horizontally at a fixed distance from the plant material (e.g., 50 cm) and at a fixed angle of depression (e.g., 45°) to acquire images from different directions. The imaging field of view from a distance of 50 cm was 40 × 23 cm, and the pixel sizes of the acquired images were 0.54 × 0.52 mm. The linearity of output (PV in an image) and input (light intensity entering the camera) was confirmed for each camera. The gamma value was set to 1.0, if possible, and if it could not be changed, it was determined *par avance* and used for gamma correction in analyses.

Ibaraki et al. (2012b) compared actual PPFD with PV in reflection images. To determine the actual PPFD distribution on the canopy surface for comparison, after imaging, the light intensity was measured as the PPFD with a quantum sensor (LI-COR, LI-190SA, United States) placed just behind the targeted leaf at an angle similar to that of the leaf surface. PPFD was measured when the leaf was gently displaced by hand. Measurements were conducted for leaves located only on the canopy surface. Acquired images were analyzed with image processing software (Mathworks, MATLAB®, and MATLAB Image Processing Toolbox, United States; Adobe, Adobe Photoshop 6). PVs corresponding to the sensitive area of the quantum sensor on the measurement point in each reflection image were averaged and used for analysis of the relationship between the PPFD and PV.

Figure 11.6 shows an example of the relationship between the measured PPFD and PV in a reflection image. Linear relationships between these values were observed in both outdoor (under natural sunlight) and indoor (under artificial light) conditions for several plant species.

11.3.1.2 Estimation of the PPFD distribution on plant canopy

A simple method was proposed to estimate the distribution of light intensity on a plant canopy surface in a greenhouse (Ibaraki et al., 2012b). In this method, the actual light intensity was measured with a PPFD sensor at one point on the canopy surface and was used for constructing a linear regression model for the calculation of the PPFD from PV at any point in the reflection image. A small PPFD sensor (Apogee, MQSO-SUN, United

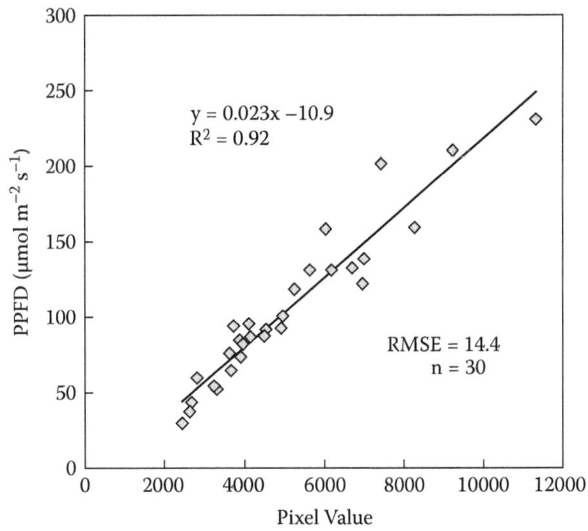

Figure 11.6 An example of the relationship between measured photosynthetic photon flux density (PPFD) and average pixel value (PV) for reflection images from three directions for a strawberry canopy under indoor conditions. (Reproduced from Ibaraki et al., *Environ. Control Biol.*, 50, 117–126, 2012b.)

States) was placed adjacent to a leaf on which the PPFD appeared to be high in the target canopy. After imaging, slope a in the linear regression model ($y = ax$) was calculated as the ratio of measured PPFD versus PV at the measurement point in the image.

Figure 11.7 shows an example of the relationship between PV and PPFD for a rose canopy in the greenhouse on clear and cloudy days. The circles in the figure denote the values of the points at which PPFD was measured with the PPFD sensor, and the lines represent the regression models for estimating PPFD. On both clear and cloudy days, the method could roughly estimate PPFD on the leaves of a rose canopy in a greenhouse, showing a root mean square error (RMSE) of 49 and 8.4 μmol m^{-2} s^{-1} for clear (PPFD range, 170–423 μmol m^{-2} s^{-1}) and cloudy (PPFD range, 17–65 μmol m^{-2} s^{-1}) days, respectively (Figure 11.7). This method, in which the regression model was determined for each imaging on the basis of actual measurement at one point, can minimize the effects of canopy type (plant species) and light characteristics (spectrum).

11.3.2 Application for tomato plants in the greenhouse

Ibaraki et al. (2012a) analyzed light intensity distribution on a greenhouse-grown tomato canopy as a PPFD histogram using the method described here. The experiment is summarized below.

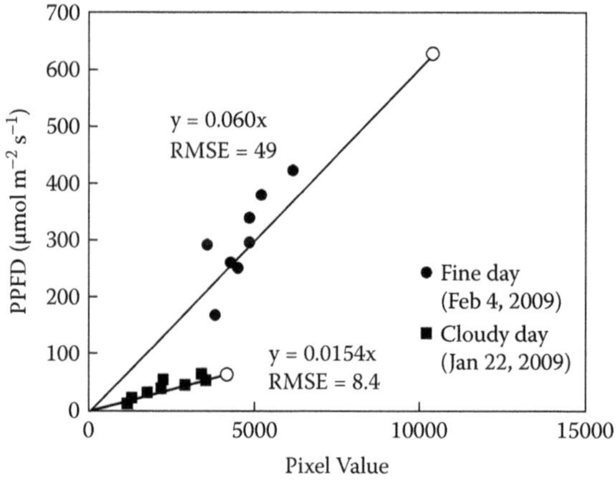

Figure 11.7 Estimation of PPFD distribution. An example for a rose canopy in the greenhouse on fine (closed circle) and cloudy (closed square) days. The slope a in the linear model ($y = ax$) was determined as a ratio of measured PPFD (open circle) to pixel value (PV) at a measuring point in the image and used for estimation of PPFD at other points. (Reproduced from Ibaraki et al., *Environ. Control Biol.*, 50, 117–126, 2012b.)

11.3.2.1 Plant materials and image acquisition

Tomato plants (*Solanum lycopersicum* L. "Momotaro") grown in a plastic greenhouse were assessed. The greenhouse, located at Yamaguchi University (Yamaguchi, Japan), was oriented toward northwest and had a width of 5.0 m and height of 2.5 m. Polyolefin films, which did not block UV radiation from sunlight, were used as the covering material.

Reflection images of the tomato plant canopy were acquired from three different directions using three digital cameras through a blue-green band-pass filter (Suruga, S76-BG7, Japan). To simultaneously acquire images from different directions, the three cameras were placed on a horizontal bar at 11 cm intervals, and their shutters were controlled with shutter release cables. The cameras were placed 50 cm from the canopy surface. The imaging field of view from the distance of 50 cm was 60 × 50 cm, and the resolution of the images was 0.75 mm per pixel in this configuration. Shutter speeds varied for different measurement times on the basis of brightness; other settings of the cameras were not changed.

11.3.2.2 Construction of PPFD histogram

Pixels corresponding to a plant canopy in the images were extracted using an automatic selection tool of the software (Adobe Photoshop 6.0), and then the images were converted into grayscale images. First, a histogram of PV before gamma correction was obtained. Then, a histogram of PV after gamma correction was converted into a PPFD histogram using a linear model based on the measured PPFD at one point on the canopy in each image. In this step, each class mark (PV) was gamma corrected and then converted into a PPFD value using the linear model described above. The number of pixels for each PPFD mark was reconstructed according to the interval of the PPFD histogram (e.g., 20 μmol or 50 μmol m^{-2} s^{-1}).

11.3.2.3 Diurnal changes in PPFD histogram on house-grown tomato canopy

Figure 11.8 shows the histograms estimated from reflection images and observations (actual measurements). To obtain a PPFD histogram based on the actual measurements, after imaging, PPFD was measured for all leaves in the imaging field. The estimated and actual PPFD histograms had similar patterns, with close mean values. However, the pattern of estimated histograms was smoother. Only one point for each leaf was measured to construct the histogram based on actual measurements, although there was variation in PPFD within a single leaf. The histogram estimated from reflection images was expected to reproduce the distribution within a leaf.

Figure 11.9 shows an example of the time course of a histogram estimated from reflection images obtained on a clear day in summer. The distribution pattern changed over time. At 10:00 and 15:00 JST, similar patterns were obtained (Figure 11.9a, c). However, at noon, PPFD was distributed more widely, ranging from 100 to 2000 μmol m^{-2} s^{-1} (Figure 11.9b). Direct radiation is dominant at this time, and a large difference was observed in light intensity between the shaded and sunny parts of the canopy surface. At 18:00 JST, there was little variation in PPFD on the leaf surface (Figure 11.9d).

The results of this study suggest that light intensity distribution on a canopy surface can be estimated as a PPFD histogram using reflection images. The method uses commercially available digital cameras and is

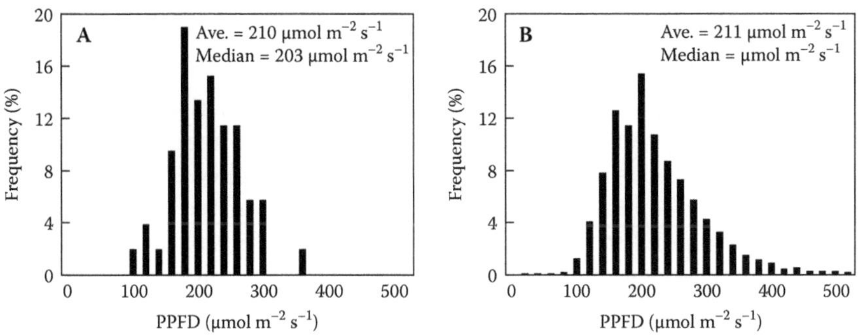

Figure 11.8 An example of comparison of histograms estimated from reflection images (right) and observations (actual measurements) (left) on a clear day. (Reproduced from Ibaraki et al., *Acta Hort. (ISHS)*, 956, 577–582, 2012a.)

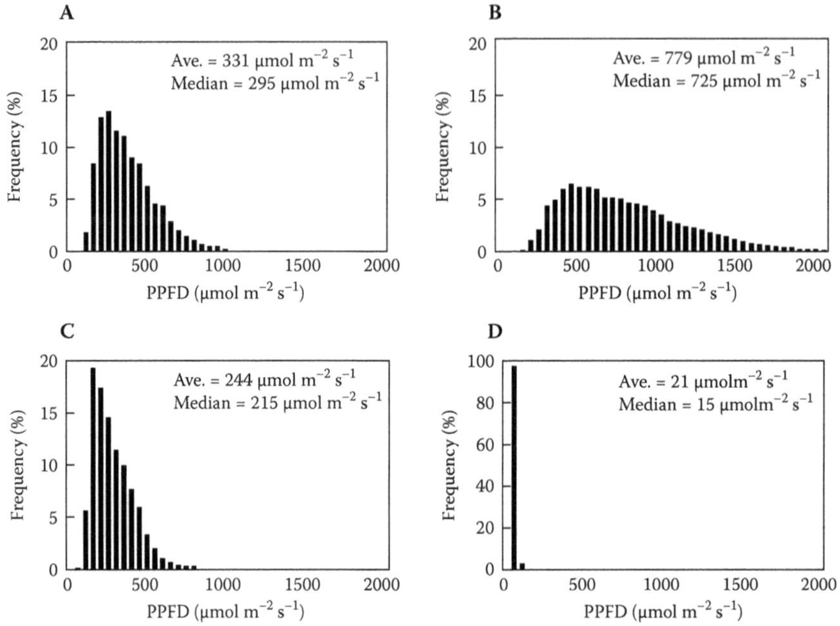

Figure 11.9 Examples of the time course of a histogram estimated from reflection images obtained at (a) 10:00, (b) 12:00, (c) 15:00, and (d) 18:00 JST on a clear day. (Reproduced from Ibaraki et al., *Acta Hort. (ISHS)*, 956, 577–582, 2012a.)

a simple and cost-effective means for estimating the actual light intensity distribution when the light environment is controlled in protected cultivation. It can also be linked to image-based plant diagnostic systems using chlorophyll fluorescence, PRI, or thermal imaging, and can thereby improve their applicability for protected cultivation.

11.4 Conclusion

PRI imaging is promising for simple and nondestructive evaluation of photosynthetic properties (PSII status), although it is necessary to understand the limitations in its application. It could be applied to assessment of photosynthetic capacity of micropropagated plants inside culture vessels.

A critical point in the use of PRI is the light intensity on the leaf surface. In image-based methods for evaluating plant photosynthetic status, including PRI imaging, this intensity is critical information for the accurate evaluation of plant status. The method described here is a simple and cost-effective means of estimating the actual light intensity distribution when the light environment is controlled for protected cultivation. As an effective use of this method, Ibaraki and Shigemoto (2013) proposed an image-based index for lighting efficiency in supplementary lighting. The method can be linked to image-based plant diagnostic systems using thermal and PRI imaging, and can thereby improve its applicability to practical cultivation or botanical research.

References

Berger, S., Benediktyová, Z., Matouš, K., et al., Visualization of dynamics of plant-pathogen interaction by novel combination of chlorophyll fluorescence imaging and statistical analysis: differential effects of virulent and avirulent strains of *P. syringae* and of oxylipins on *A. thaliana*, *J. Exp. Bot.*, 58, 797–806, 2007.

Dutta Gupta, S., Ibaraki, Y., and Pattanayak, A.K., Development of a digital image analysis method for real-time estimation of chlorophyll content in micropropagated potato plants, *Plant Biotechnol. Rep.*, 7, 91–97, 2013.

Gamon, J.A., Field, C.B., Bilger, W., Björkman, O., Fredeen, A.L., and Peñuelas, J., Remote sensing of the xanthophyll cycle and chlorophyll fluorescence in sunflower leaves and canopies, *Oecologia*, 85, 1–7, 1990.

Gamon, J.A., Serrano, L., and Surfus, J.S., The photochemical reflectance index: an optical indicator of photosynthetic radiation use efficiency across species, functional types, and nutrient levels, *Oecologia*, 112, 492–501, 1997.

Genty, B., and Meyer, S., Quantitative mapping of leaf photosynthesis using chlorophyll fluorescence imaging, *Aust. J. Plant Physiol.*, 22, 277–284, 1994.

Guidi, L., Mori, S., Degl'Innocenti, E., and Pecchia, S., Effects of ozone exposure or fungal pathogen on white lupin leaves as determined by imaging of chlorophyll a fluorescence, *Plant Physiol. Biochem.*, 45, 851–857, 2007.

Hall, F.G., Hilker, T., Coops, N.C., et al., Multi-angle remote sensing of forest light use efficiency by observing PRI variation with canopy shadow fraction, *Remote Sens. Environ.*, 112, 3201–3211, 2008.

Ibaraki, Y., Evaluation of photosynthetic capacity in micropropagated plants by image analysis, in *Plant tissue culture engineering*, ed. S. Dutta Gupta and Y. Ibaraki, 15–29, Springer, Berlin, 2006.

Ibaraki, Y., and Dutta Gupta, S., Nondestructive evaluation of the photosynthetic properties of micropropagated plantlets by imaging photochemical reflectance index under low light intensity, *In Vitro Cell. Dev. Biol.-Plant*, 46, 530–536, 2010.

Ibaraki, Y., Kishida, T., and Shigemoto, C., Image-based estimation of PPFD distribution on the canopy surface in a greenhouse, *Acta Hort. (ISHS)*, 956, 577–582, 2012a, http://www.actahort.org/books/956/956_69.htm.

Ibaraki, Y., and Matsumura, K., Non-destructive evaluation of the photosynthetic capacity of PSII in micropropagated plants, *J. Agric. Meteorol.*, 60, 1073–1076, 2005.

Ibaraki, Y., Matsumura, K., and Dutta Gupta, S., Low-cost photochemical reflectance index measurements of micropropagated plantlets using image analysis, *Comput. Electron. Agric.*, 71, 170–175, 2010.

Ibaraki, Y., and Nozaki, Y., Estimation of light intensity distribution in a culture vessel, *Plant Cell Tiss. Org. Cult.*, 80, 111–113, 2005.

Ibaraki, Y., and Shigemoto, C., Estimation of supplemental lighting efficiency based on PPFD distribution on the canopy surface, *J. Agric. Meteorol.*, 69, 47–54, 2013.

Ibaraki, Y., Yano, Y., Okuhara, H., and Tazuru, M., Estimation of light intensity distribution on a canopy surface from reflection images, *Environ. Control Biol.*, 50, 117–126, 2012b.

Jones, H.G., *Plants and microclimate*, 2nd ed., Cambridge University Press, New York, 1992.

Lichtenthaler, H.K., Lang, M., Sowinska, M., Heisel, F., and Miehe, J.A., Detection of vegetation stress via a new high resolution fluorescence imaging system, *J. Plant Physiol.*, 148, 599–612, 1996.

Meng, Q., Siebke, K., Lippert, P., et al., Sink-source transition in tobacco leaves visualized using chlorophyll fluorescence imaging, *New Phytol.*, 151, 585–595, 2001.

Nakaji, T., Oguma, H., and Fujinuma, Y., Seasonal changes in the relationship between photochemical reflectance index and photosynthetic light use efficiency of Japanese larch needles, *Int. J. Remote Sens.*, 27, 493–509, 2006.

Nakaji, T., Takeda, T., Fujimura, Y., and Oguma, H., Effect of autumn senescence on the relationship between the PRI and LUE of young Japanese larch tree, *Phyton*, 45, 535–542, 2005.

Omasa, K., and Takayama, K., Simultaneous measurement of stomatal conductance, non-photochemical quenching, and photochemical yield of photosystem II in intact leaves by thermal and chlorophyll fluorescence imaging, *Plant Cell Physiol.*, 44, 1290–1300, 2003.

Rashimzadeh-Bajgiran, P., Munehiro, M., and Omasa, K., Relationship between the photochemical reflectance index (PRI) and chlorophyll fluorescence parameters and plant pigment indices at different leaf growth stages, *Photosynth. Res.*, 113, 261–271, 2012.

Stylinski, C.D, Gamon, J.A., and Oechel, W.C., Seasonal patterns of reflectance indices, carotenoid pigments and photosynthesis of evergreen chaparral species, *Oecologia*, 131, 366–374, 2002.

Suárez, L., Zarco-Tejada, P.J., Berni, J.A.J., González-Dugo, V., and Fereres, E., Modelling PRI for water stress detection using radiative transfer models, *Remote Sens. Environ.*, 113, 730–744, 2009.

Yadav, S.P., Ibaraki, Y., and Dutta Gupta, S., Estimation of the chlorophyll content of micropropagated potato plants using RGB based image analysis, *Plant Cell Tiss. Org. Cult.*, 100, 183–188, 2010.

Yamamoto, H.Y., Biochemistry of violaxanthin cycle in higher plant, *Pure Appl. Chem.*, 51, 639–648, 1979.

chapter twelve

ROS and NOS imaging using microscopical techniques

Nieves Fernandez-Garcia and Enrique Olmos

Contents

12.1 Introduction

In this chapter we describe different microscopic approaches for making different measurements in reactive oxygen and nitrogen species (ROS and NOS). Microscopic techniques have proved increasingly to be a powerful tool in the studies of life science. The greatest advances have been made in confocal laser scanning microscopy with the development of new fluorescent sensors and the implementation of new microscopical methods such as fluorescence recovery after photobleaching (FRAP), fluorescence lifetime imaging microscopy (FLIM), and fluorescence resonance energy transfer (FRET), which have opened new ways for ROS and NOS determination and location.

12.1.1 ROS production and imaging

In green parts of plants, the most important sources of ROS are chloroplast and peroxisomes (Foyer and Noctor, 2003; Asada, 2006; Fernandez-Garcia et al., 2010). In plants grown in the darkness or nongreen parts of the plant, the mitochondrion is the most important source of ROS production (Maxwell et al., 1999; Moller, 2001). It is well known that ROS are continuously produced by the cell with different accumulation and can induce oxidative damages to proteins, DNA, and lipids (Fridovich, 1986; Wise and Naylor, 1987; Imlay and Linn, 1988; Halliwell and Gutteridge, 2001). Moreover, the production of ROS plays important roles in signal transduction in response to different biotic and abiotic factors in both plants and animals (Moller et al., 2007; Swanson et al., 2011). ROS is a key factor controlling different plant functions.

Alteration in the environmental conditions that produce any change in the redox homeostasis can induce oxidative stress in plants. Oxygen is essential for aerobic life, but accumulation of ROS, such as hydrogen peroxide, superoxide radicals, singlet oxygen, and hydroxyl radicals, during the aerobic metabolism can be dangerous for the cell metabolism (Halliwell, 2006; Kim et al., 2008). The most important ROS produced by the cell metabolism are singlet oxygens, superoxide radicals, hydrogen peroxide, and hydroxyl radicals.

12.1.1.1 Singlet oxygens

Singlet oxygen ($^1O_2$) is a nonradical, usually generated photodynamically in an energy transfer reaction between a dye molecule in a triplet excited state and molecular oxygen. The singlet oxygens ($^1O_2$) are the most reactive forms of O_2, as the oxygen is in a higher energy state. The singlet oxygens can oxidize protein, lipids, and DNA directly (Halliwell, 2006). They will rapidly oxidize molecules containing carbon-carbon double bonds to form hydroperoxides or endoperoxides. The chloroplasts produce $^1O_2$ during photosynthesis in photosystem I (Moller et al., 2007). Moreover, under conditions of photoinhibition, photosystem II is preferentially damaged, resulting in the production of $^1O_2$.

12.1.1.2 Superoxide radicals

Compared to other oxy-radical species the superoxide radicals are not toxic per se; the toxicity is due to their being the precursor of extremely reactive hydroxyl radicals ($HO^{\cdot}$). The $O_2^{\cdot-}$ is involved in several physiological processes mediated by oxygen, such as lipid peroxidation, viral biotic inactivation, membrane damage, cellular toxicity, and DNA breaking (Fridovich, 1986; Fernandez-Garcia et al., 2010). During photosynthesis, the chloroplasts produce $O_2^{\cdot-}$ at photosystems I and II (Moller et al., 2007). In mitochondria, the production of $O_2^{\cdot-}$ is in complexes I and III and between 1–5% of the oxygen consumption for isolated mitochondria results in ROS production (Moller et al., 2007).

In peroxisomes, the $O_2^{\cdot-}$ production is the consequence of their normal metabolism, with an important production in several key metabolic reactions (Del Rio et al., 2006; Moller et al., 2007). Superoxide radicals can be also produced by action of NADPH oxidases in the plasma membrane (Moller et al., 2007).

12.1.1.3 Hydrogen peroxide

Hydrogen peroxide (H_2O_2) is a relatively stable molecule; under normal conditions, the half-life of H_2O_2 is probably 1 ms, whereas other forms of ROS (superoxide radical, hydroxyl radicals, and singlet oxygen) are much shorter, around 2–4 μs (Bhattacharjee, 2005). This relative stability and high permeability through membranes makes H_2O_2 a good messenger for signals generated by means of ROS (Neil et al., 2002). Hydrogen peroxide is produced in many subcellular compartments. In the apoplast, it can be accumulated by the coordinated action of NADPH oxidases and Superoxide Dismutase (SOD) that produces H_2O_2. In the cell wall, hydrogen peroxide can be used for the lignification processes.

12.1.1.4 Hydroxyl radical

Hydroxyl radical ($OH^{\cdot}$) is considered the most harmful ROS for cellular components. Hydroxyl radical is a strongly oxidizing ROS that can react with the different components of the cell, so damaging their molecular structure. Hydroxyl radicals are generated by the Haber-Weiss reaction by the interaction of H_2O_2 and $O_2^{\cdot-}$ or via the Fenton reaction in the presence of iron or copper ions.

12.1.2 NOS production and imaging

NOS and ROS are versatile molecules that may be acting in many molecular processes. NO is a type of free radical reactive gas whose first physiological function was described in mammals (Palmer et al., 1987). The discovery that NO may also be acting in plants was made in 1998 (Delladonne et al., 1998; Durner et al., 1998). Since then, NO has been described as being involved in many processes in plants, such as germination, root growth, stomatal closure, gravitropism, cell death, flowering, and adaptation to plant stress. Strikingly, the understanding of the molecular mechanisms underlying the biosynthesis of NO in a plant is still incomplete (Gupta et al., 2010).

NO radical is a very unstable molecule. It is a transient species with a single electron in its 2p-π antibonding orbital and a half-life of around 5–10 s. When NO is in aqueous solution, this electron is lost, giving the nitrosonium cation (NO^+). This cation is moderately stable in aqueous solution and can react with the different NO-sensitive fluorochromes (Xie and Shen, 2012). It is also possible to add a new electron to the NO radical, giving a nitroxyl radical (NO^-).

NO can be removed by the cells in different ways. NO can interact with ROS as a superoxide to form compounds such as peroxynitrite ($OONO^-$). It can also react with O_2 to form nitrite and nitrate. NO can also react with glutathione (GSH) to form S-nitrosylated GSH (GSNO). The equilibrium between the synthesis and removal of the cellular NO will determine the available NO for signaling.

To understand how ROS and NOS act in so many processes, it has become critical to define their location and spatial dynamics. Therefore, ROS and NOS imaging is vital for a correct interpretation of the results.

Finally, it is important to take into account some considerations when looking for a method for the detection of reactive oxygen species (ROS) and nitrogen (NOS). In some biological structures, the ROS and NOS detection is complicated. For this reason, it is important to know how to obtain the best results, with in vivo or processed material. In the following sections, we describe the most frequently used methods used for subcellular localization of ROS and NOS.

12.2 ROS imaging at subcellular level using transmission electron microscopy (TEM) techniques

The application of TEM techniques involves an in vitro approach to the subcellular location of ROS. The major advantage of these techniques is the accurate location of ROS formation in the different subcellular compartments.

12.2.1 *Hydrogen peroxide location using the cerium precipitation technique*

Hydrogen peroxide (H_2O_2) is probably the most studied ROS component in plant tissues. It is involved in many subcellular mechanisms and in plant signaling. H_2O_2 exhibits a dual function in plant responses to stress conditions. An excess of H_2O_2 may be damaging to the cells, whereas a moderate production can help to enhance plant adaptation to different stresses. To ascertain the role of H_2O_2 in stress response, it is important not only to study its total concentration, but also to investigate the subcellular location in the different compartments. The cerium precipitation technique is probably the most useful for an accurate subcellular location of hydrogen peroxide.

The cerium precipitation technique has been used by many researchers for the specific location of hydrogen peroxide. Cerous ions are known to act as perhydroxide-trapping agents producing a well-located electron-dense reaction product (Briggs et al., 1975). This cytochemical method is based on the reaction of hydrogen peroxide with cerium chloride ($CeCl_3$) to form electron-dense, insoluble precipitates of cerium perhydroxides [$Ce(OH)]_2OOH$ and $Ce(OH)_3OOH$]. This technique was first described for cytochemical location of the NADH oxidase (Briggs et al., 1975). The technique was later adapted to localize glycolate oxidase in plants by Thomas and Trelease (1981). This technique permits the location of the specific site of the endogenous production of hydrogen peroxide by ultrastructural location of the precipitates. To our knowledge, the first study for subcellular location of endogenous production of hydrogen peroxide in plants was done by Czaninski et al. (1993) by studying the endogenous production of hydrogen peroxide in the lignifying tissues.

Cerium precipitation presents an important interference during the process of incubation. A by-product can be formed during the incubation of tissues with $CeCl_3$, cerium phosphate ($CePO_4$). Phosphate (PO_4^{3-}) is an anion highly abundant in the living cells and can be an insoluble precipitate of $CePO_4$. This cerium phosphate precipitation technique has also been used for location of plasma membrane ATPase and acid phosphatase activities (Olmos and Hellin, 1998). However, these precipitates cannot be distinguished from cerium pehydroxides by transmission electron microscopy. One way to discriminate between these precipitates is x-ray analysis of the samples. The analysis can also be used for quantification of subcellular production of H_2O_2 (Chen et al., 2009).

12.2.2 *Standard method for cerium precipitation*

For the subcellular localization of H_2O_2, the cytochemical method based on the generation of cerium perhydroxides developed by Bestwick et al. (1997) can be used. The steps include:

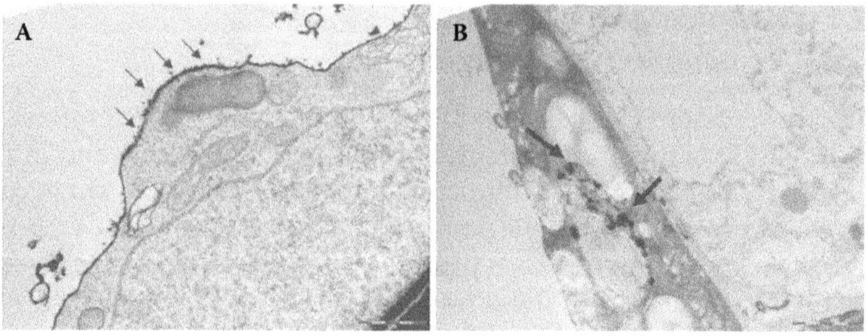

Figure 12.1 Subcellular location of hydrogen peroxide using the cerium precipitation technique in salt stressed *Nicotiana tabacum* BY-2 cell cultures. (a) Detail of cerium precipitation in the tonoplast (arrows). (b) Detail of cerium precipitation in a plastid (arrows).

Fixation: Sections are collected and incubated in freshly prepared 5 mM $CeCl_3$ in 50 mM MOPS (3-[N-morpholino] propane sulfonic acid) at pH 7.0 for 30 min. After incubation, samples are collected and fixed in a mixture of 2% (v/v) paraformaldehyde and 0.5% (v/v) glutaraldehyde in 50 mM CAB (sodium cacodylate buffer), pH 7.0, for 1 h.

Postfixation: After fixation, samples are washed twice for 10 min in CAB buffer and postfixed for 1 h in 1% (v/v) osmium tetroxide in CAB. Samples are then washed again in CAB (twice for 10 min).

Dehydration: All fixed tissues are dehydrated in a graded series of ethanol for 30 min once each (35, 50, 70, 96, and twice 100%).

Embedding: Samples need to be infiltrated, first with a propylene oxide (twice for 30 min) and then with a mixture of 50% propylene oxide and 50% Spurr's resin (45 min). The samples are then immersed in Spurr's resin overnight at 4°C. Finally, the samples are embedded in new Spurr resin and solidified at 68–70°C for 12–24 h (Fernandez-Garcia et al., 2009).

Sectioning: Blocks are sectioned on a Reichert ultramicrotome. Ultrathin sections for transmission electron microscopy are collected on copper grids. Sections are then examined using a transmission electron microscope. Figure 12.1 shows the subcellular location of H_2O_2 using the cerium technique in tobacco cell cultures.

12.3 *In vivo ROS and NOS imaging using confocal laser scanning microscopy (CLSM)*

Different imaging tools have been developed to visualize the delivery and distribution of biomolecules and process in different tissues. However, optical fluorescence imaging tools are known to have several disadvantages, such as limitations in resolution, penetration depth of fluorescence, and real-time accessibility. Alternatively, in vivo real-time confocal laser scanning microscopy (CLSM) is a powerful imaging technique to circumvent those problems. Today, the CLSM is widely utilized to visualize biological processes, such as cell trafficking and cell-cell and cell-microenvironment interactions with a high resolution. In addition, in vivo confocal microscopy has been utilized to image ROS and NOS.

CLSM is a critical technique for studying the functions within organisms in cellular biology. Fluorescence microscopy gives researchers insights into an area normally hidden with other techniques. The structures within a tissue or cell culture can be targeted in vivo with different fluorescent probes. This technique permits us to make visible the subcellular location of different elements (sodium, potassium, calcium, etc.) and other compounds. such as ROS, NOS, and pH variations.

The CLSM technique includes the possibility of finding in vivo the precise location of ROS and NOS production of cellular scavenging (Mullineaux and Karpinski, 2002). In recent years, new fluorochromes have been developed that can identify subcellular sites for ROS and NOS accumulation in the cytoplasm, mitochondria, endoplasmic reticulum (ER), and nuclear region.

In the following sections, we describe the most currently used methods to obtain in vivo plant sections and the frequently used fluorochromes that can be directly used in CLSM studies.

12.3.1 Sample preparation for in vivo techniques using CLSM

Preparation of fresh material for in vivo locations can be performed by different methods. Two options to section living tissue are to use a hand microtome or a vibratome. The advantage of these methods is that the absence of different treatments (dehydration with organic solvents, chemical fixation, etc.) preserves protein conformation, lipids, and enzyme activities.

12.3.1.1 Hand microtome

Fresh-cut sections (50–100 μm) from plant material (leaves, root, fruit, flower buds, etc.) can be obtained using a hand microtome (Figure 12.2a). The sections can be stained and observed when the research requires a quick microscopy technique of a certain tissue or organ. The section obtained from a hand microtome is thin enough for study under CLSM. Depending on the material size and shape, the sections can be done using thin stems and roots that can be directly sectioned as fresh material, although it needs some support in the microtome clamp. The support is normally cork (Figure 12.2c).

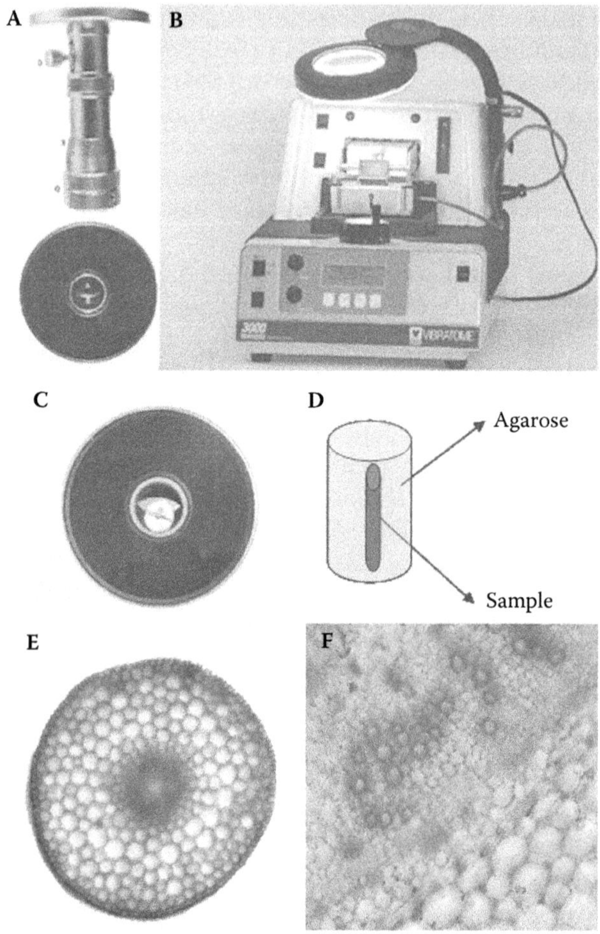

Figure 12.2 (a) Hand microtome. (b) Vibratome. (c) Clamped material using cork cylinders. (d) Agarose embedded sample. (e) Root transversal section obtained with a hand microtome. (f) Detail of stem transversal section obtained with a hand microtome.

12.3.1.2 Vibratome

Not all the tissues are suitable to be cut by the hand microtome, due to its thinness or hardness. Vibratome (Figure 12.2b) is an alternative for this kind of fresh tissue. Very thin sections of plants tissues can be thus obtained (10–40 μm). Prior to sectioning, the samples should be submerged in de-ionized water. The sample can then be sectioned directly. However, soft and very thin tissues should be embedded in a matrix of agarose (low melting point) or polyacrylamide before obtaining the sections (Figure 12.2d).

12.3.1.3 Whole organ

Very thin samples like young roots, seedlings, pollen, or cell cultures can be monitored by confocal laser microscopy without sectioning. Material needs to be incubated directly in a buffer or fresh culture medium (cell culture or plant growth in hydroponic system).

12.3.1.4 Mounting sections on glass slides

Tissue sections obtained by hand microtome or vibratome must be mounted on glass slides with a glycerol/buffer solution (1/1 v/v).

12.3.2 Fluorochromes used for ROS imaging

There are different fluorimetric techniques for assessing intracellular ROS produced in plant cells. These methods are based on the oxidation of a nonfluorescent compound to a fluorescent compound by the presence of different ROS. Table 12.1 shows the excitation/emission wavelength (nm) and some references of the fluorochromes described in the following sections.

12.3.2.1 ROS (hydrogen peroxide) imaging by 2,7-dichlorofluorescin diacetate

To our knowledge, two of the first papers to describe the use 2′,7′-dichlorofluorescin diacetate (DCFH-DA) as a fluorometric technique of hydrogen peroxide (H_2O_2) in the presence of peroxidase were LeBel et al. (1992) and Keston and Brandt (1965). The 2,7-dichlorofluorescin diacetate (DCFH-DA) is a sensitive ROS fluorochrome (a chemically reduced form of fluorescein) frequently used in the literature for CLSM as an intracellular indicator for ROS in cells, for example, to detect the generation of hydrogen peroxide in the cytoplasm. This fluorochrome has also been considered a qualitative marker of oxidative stress (Yao and Greenberg, 2006; Sandalio et al., 2008).

The nonpolar compound DCFH-DA crosses the cell membranes, and the diacetate molecule is enzymatically hydrolyzed by the action of the intracellular esterases, liberating the nonfluorescent DCFH form (LeBel and Bondy, 1990). Inside the cell and in the presence of reactive oxygen species, DCFH is rapidly oxidized to highly fluorescent 2′,7′-dichlorofluorescin (DCF). However, it remains unclear which reactive oxygen species are responsible for the oxidation of DCFH. Other reactive oxygen species, such as superoxide anion ($O_2^{\cdot-}$) and hydroxyl radical (·OH), to stimulate the formation of DCF may also have some such capacity.

DCFH is unstable in intact cells, being highly susceptible to photobleaching and photo-oxidation. Photobleaching can induce an underestimation of ROS production. Moreover, the illumination of the cells by

Table 12.1 List of Fluorochromes Used for ROS and NOS Imaging

Fluorochrome	Laser excitation/ emission (nm)	Detection	References
2,7-dichlorofluorescin diacetate (DCFH-DA)	488/510–530	H_2O_2	Keston and Brandt, 1965 LeBel and Bondy, 1990 LeBel et al., 1992 Yao and Greenberg, 2006 Sandalio et al., 2008 Hernández et al., 2010
Dihydroethidium (DHE)	488/510–530	$O_2^{\bullet-}$	Bindokas et al., 1989 Carter et al., 1994 Benov et al., 1998 Tarpey et al., 2004 Camejo et al., 2011
MitoSOX Red	488/560–590	$O_2^{\bullet-}$	Miller et al., 2009 Smirnova et al., 2009 Li and Xing, 2011 Lv et al. 2011
OxyBURST Green H_2HFF BSA	488/525–560	$O_2^{\bullet-}$	Chen, 2002 Haugland, 2005 Monshausen et al., 2007 Zhang et al., 2011
Singlet Oxygen Sensor Green SOS G	488/500–600	$^1O_2$	Hideg et al., 1998 Hideg et al., 2001 Hideg et al., 2002 Flors et al., 2006 Hideg, 2008
DanePy	405/520–575	$^1O_2$	Hideg et al., 1998 Hideg et al., 2001 Hideg et al., 2002 Hideg, 2008
DAF-FM DA	488/505–525	NO	Kojima et al., 1999 Rümer et al., 2012
DAF-2D	488/505–525	NO	Kojima et al., 1998 Rümer et al., 2012

the laser can lead to ROS production and induce a light stress that can produce an ROS signal.

12.3.2.1.1 Assays for hydrogen peroxide imaging using CLSM

12.3.2.1.1.1 H_2O_2 production in root/leaf samples. Small roots/leaves or seedlings are incubated for 30 min in a fresh culture medium containing 10 μm DCFH-DA, and then washed three times with fresh medium without DCFH-DA to remove the excess fluorochrome (Hernández et al.,

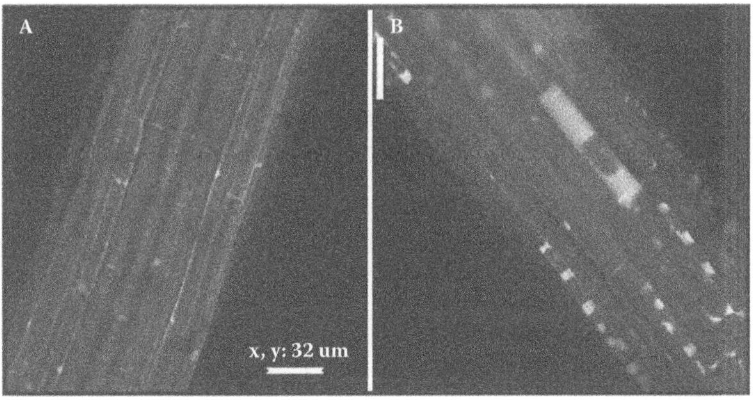

Figure 12.3 Hydrogen peroxide was located by laser confocal microscopy using DCFH-DA in control and salt stressed *Arabidopsis* roots. (a) Detail of epidermal cells from control plants. (b) Detail of salt stressed plants showing epidermal cells producing hydrogen peroxide. Samples were excited with the 488 nm line of an argon laser, and dye emission was collected at 520 ± 10 nm.

2010). Figure 12.3 Shows imaging of a root tip of *Arabidopsis thaliana*, illustrating ROS location using DCFH-DA.

12.3.2.1.1.2 H_2O_2 production in tobacco BY-2 cell cultures. Cells are incubated on an orbital shaker (120 rpm), 26°C, and in the dark for 30 min in a fresh culture medium containing 25 μM DCFH-DA, and then washed three times with a fresh medium without DCFH-DA to remove the excess fluorophore. Samples are excited with an argon laser (488 nm), and dye emission is monitored at 520 ± 10 nm. The DCF fluorescence is visualized in a single optical section of root. All images should be obtained at the same depth. A negative control should be used following incubation of the samples with ascorbate to eliminate the H_2O_2.

12.3.2.2 Superoxide imaging

12.3.2.2.1 Dihydroethidium (DHE). The presence of the intracellular superoxide radicals (O_2^-) can be visualized using the fluorescence probe dihydroethidium (DHE). The superoxide indicator DHE exhibits blue fluorescence in the cytosol until oxidized, where it intercalates within the cell's DNA, staining its nucleus a bright fluorescent red. DHE fluorescence is due to the fact that O_2^- radicals are imaged exciting with the 488 nm line of an argon laser and dye emission is monitored at 515–535 nm. The DHE is a fluorochrome that crosses the cell membranes and reacts with O_2^- to form a specific fluorescent product, oxyethidium (oxy-E) (Camejo et al., 2011). The process needs two O_2^- anions and produces ethidium radical as an intermediate (E^+). Moreover, the oxidation of DHE to E^+ has been shown to be rapid when the oxidant is O_2^-, but not when O_2, H_2o_2, HOCl, or $ONOO_2$ is used (Bindokas et al., 1989). For this reason, DHE has been applied as a detector of intracellular O_2^- (Carter et al., 1994). However, Benov et al. (1998) Note that DHE can be oxidized by cytochrome *c* and, more importantly, that DHE can catalyze the dismutation of O_2^-. The latter process interferes with the fluorimetric measurement of O_2^- with DHE. For these reasons, this technique should be recommended as a qualitative but not as a quantitative measure of O_2^- (Tarpey et al., 2004).

Finally, as with the use of any fluorochrome for tissue localization of the compound of interest, the potential for autofluorescence must be recognized and taken into account. Because of these considerations, the use of DHE for the determination of superoxide in vivo appears to have significant limitations.

12.3.2.2.1.1 Assay for superoxide using CLSM.

Superoxide production in root/leaf samples. Samples are infiltrated directly with 10 μM dihydroethidium (DHE) solution with 10 mM Tris-HCl buffer (pH 7.4) for 1 h at 37°C in the dark. Small samples are embedded in agarose (low melting point) and sectioned using a hand microtome or vibrotome. The DHE solution was removed and samples are washed three times with 10 mM Tris-HCl buffer (pH 7.4) for 30 min (Camejo et al., 2011). Samples are excited with the 488 nm line of an argon laser, and dye emission is collected at 520 ± 10 nm. The fluorescence of oxy-E is visualized in a single optical section of the sample. All images should be obtained at the same depth. Figure 12.4 shows imaging of a root tip of alfalfa, illustrating superoxide location using DHE.

12.3.2.2.2 MitoSOX red. The superoxide production by mitochondria can be monitored in fluorescence microscopy using the MitoSOX™ Red reagent, a cationic derivate of dihydroethidum (Miller et al., 2009; Smirnova et al., 2009; Li and Xing, 2011; Lv et al., 2011). This reagent has a high selective detection of superoxide where it selectively targets mitochondria. It is rapidly oxidized by superoxide, but not by other ROS and NOS. The oxidized product is highly fluorescent upon binding to nucleic acid, exhibiting red fluorescence, whereas the mitochondria of treated cells show minimal fluorescence.

The cationic triphenylphosphonium substituent of MitoSOX Red is responsible for the uptake of the probe in actively respiring mitochondria. Oxidation of MitoSOX Red indicator (or dihydroethidium) by superoxide leads to hydroxylation at the 2-position. The probe exhibits red fluorescence and an excitation peak at ~400 nm that is absent in the excitation

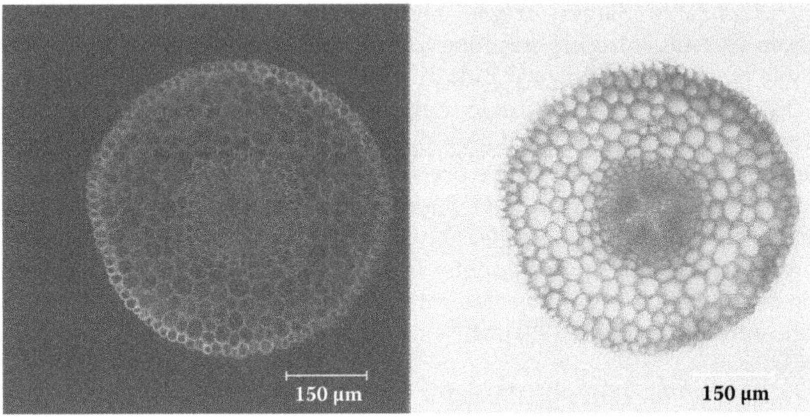

Figure 12.4 Left: Localization of superoxide by in vivo labeling using DHE and laser confocal microscopy in control bean roots. Right: The same section was also observed under bright field. Sample was excited with the 488 nm line of an argon laser, and dye emission was collected at 520 ± 10 nm.

spectrum of the ethidium oxidation product generated by an ROS other than superoxide.

MitoSox Red has been used to investigate O_2^- accumulation in plant mitochondria in response to aluminum stress, pollen grain germination, heat stress, and salt stress (Miller et al., 2009; Smirnova et al., 2009; Li and Xing, 2011; Lv et al., 2011).

12.3.2.2.3 OxyBURST green H_2HFF BSA. OxyBURST is reported to react with only certain forms of ROS, showing high sensitivity for superoxide, and is oxidized by H_2O_2 only in the presence of peroxidases (Haugland, 2005). OxyBURST Green H_2HFF Bovine Serum Albumin (BSA) is a sensitive fluorochrome for detecting extracellular release of oxidative products that can be used for fluorescence microscope. This reagent comprises BSA that has been covalently linked to dihydro-2′,4,5,6,7,7′-exafluorofluorescein (H_2HFF), making the ROS sensor cell impermeable. The intensity of OxyBURST fluorescence does not vary with environmental factors such as pH or ionic strength of the medium (Chen, 2002). This fluorochrome has been successfully used for superoxide location in the root apoplast of rice (Zhang et al., 2011) and *Arabidopsis thaliana* (Monshausen et al., 2007).

12.3.2.2.4 Singlet oxygen imaging. For singlet oxygen location, different fluorochromes have been developed (DanePy, DMAX, DPAX, Singlet Oxygen Sensor Green, MVP, and TEMPO-9-AC) (see Hideg, 2008). The most frequently used sensors are Singlet Oxygen Sensor Green (SOSG) and DanePy.

12.3.2.2.5 Singlet oxygen sensor green. Singlet Oxygen Sensor Green (SOSG) is highly selective for $^1O_2$ and does not show any appreciable response to hydroxyl radical (·OH) or superoxide (·O_2) (Flors et al., 2006; Hideg, 2008). This singlet oxygen indicator initially exhibits weak blue fluorescence, with excitation peaks at 372 and 393 nm and emission peaks at 395 and 416 nm. In the presence of singlet oxygen, it emits a green fluorescence similar to that of fluorescein with an excitation and emission at 504 and 525 nm. Singlet Oxygen Sensor Green reagent is intended for use in aqueous environments. The optimal dilution buffer and working concentration should be determined empirically; a suggested starting concentration range is 1–10 μM.

12.3.2.2.6 DanePy. DanePy (3-(*N*-diethylaminoethyl)-*N*-dansyl) aminomethyl-2,5-dihydro-2,2,5,5-tetramethyl-1*H*-pyrrole) is a double fluorescence sensor of singlet oxygen. In the presence of $^1O_2$, DanePy fluorescence is quenched, and this property has been used for in vivo imaging of $^1O_2$. DanePy was first used as a sensor for in vivo $^1O_2$ quantification using a spectrofluorimetric method in intact leaves (Hideg et al., 1998). Later, these authors have successfully used DanePy for in vivo location using laser confocal techniques (Hideg et al., 2001, 2002). In these experiments, the authors have demonstrated the production of $^1O_2$ by the chloroplast in vivo. However, DanePy is not commercially available, which supposes an important problem because it has to be obtained by synthesis in the laboratory. Moreover, DanePy reactions with $^1O_2$ are irreversible, and once quenched, the fluorescence of the fluorochrome cannot be recovered.

12.3.3 Fluorochromes used for NOS imaging

The present chapter has mainly focused on ROS imaging because this is the most studied in the literature. However, NOS location is of great importance in many different cellular systems.

12.3.3.1 Nitric oxide imaging

DAF-2 DA and DAF-FM DA are widely used for measuring NO production in plant biological studies (Delladonne et al., 1998; Durner et al., 1998; Gupta et al., 2010). Therefore, this section focuses on both compounds.

12.3.3.1.1 DAF-2 DA. 4,5-Diaminofluorescein (DAF-2) was first developed by Kojima et al. (1998). Membrane-permeable DAF-2 DA can be easily loaded into cells, and diacetate is excised by the action of esterases, liberating DAF-2 in the cytoplasm. DAF fluorescence is generally assumed to be induced by $N_2O_3^-$, which is formed by nitrosation through an NO oxidation, to form the highly fluorescing DAF triazol. The reaction of NO and DAF-2 is highly sensitive to NO quantification, whose threshold is about 5 nM. However, the fluorescence intensities of the triazole derivates of DAFs are dependent on pH and greatly decrease at pH lower than 7.

12.3.3.1.2 DAF-FM DA. In addition to the DAF-2 DA, a derivate of DAF has been also developed by Kojima et al. (1999), 3-amino-4-(N-methylamino)-2′,7′-difluorescescein (DAF-FM). This fluorochrome presents some important advantages over DAF. DAF-FM after reaction with NO, the produced triazole form, shows stable and intense fluorescence in a wide range of pH values. Also, the triazole form of DAF-FM is more photostable than that of DAF-2. Moreover, the threshold for NO detection is lower than DAF-2, about 3 nM. However, DAF-FM is not a reversible equilibrium sensor, so limiting its ability to track rapid fluctuations of the NO concentration in real time. Rümer et al. (2012) have recently analyzed whether DAF fluorescence can be produced by products other than NO. These authors have demonstrated that DAF-2 can be enzymatically converted into a variety of highly fluorescent compounds, and only a minor part of the fluorescence observed inside of the cells will be produced by NO reaction with DAF-2. In view of these results, the exclusive use of DAF fluorochromes for NO quantification and location in the cell should be carefully interpreted.

12.3.3.2 Assay for NO using CLSM

12.3.3.2.1 NO production in cell cultures. Cells are incubated on an orbital shaker (120 rpm), 26°C, and in the dark during 30 min in a fresh culture medium containing 10 μM DAF-FM DA and then washed three times with a fresh medium to remove the excess fluorophore.

12.3.3.2.2 NO production in root/leaf samples. Roots are infiltrated directly with 10 μM DAF-FM DA solution in 10 mM Tris-HCl buffer (pH 7.4) for 1 h at 25°C in the dark. Small roots/leaves are embedded in agarose (low melting point) and sectioned using a vibratome or hand microtome. The washing solution is removed and roots/leaves are washed three times with 10 mM Tris-HCl buffer (pH 7.4) for 30 min. Samples are excited with the 488 nm line of an argon laser, and dye emission is collected at 510 ± 10 nm. The DAF-FM DA fluorescence is visualized in a single optical section. All images are obtained at the same depth.

A negative control should be used for specific identification of superoxide. The most frequently used compound is a NO scavenger, 2-phenyl-4,4,5,5-tetramethylimidazoline-1-oxyl (PTIO).

12.4 Hydrogen peroxide imaging using genetically encoded sensors

We have described different fluorochromes that have been used to visualize different species of ROS in plant cells. However, many of these chemicals need to be loaded in the cell, making the measurements difficult, or they can be altered by laser intensity inducing photo-oxidation of the fluorochrome. In recent years, genetically encoded sensors have been developed that overcome some of these problems, such as loading of the fluorochromes. Genetically encoded sensors can be divided into two different groups: single FP sensors and FRET-based sensors. The most important advantages of single FP sensors are their ease of use. However, FRET-based sensors are more complicated in their technology. For ROS imaging, a single FP sensor has been developed that shows a high specificity and sensitivity to hydrogen peroxide detection; it is known as HYPER.

12.4.1 *Hyper, a specific sensor for hydrogen peroxide*

HYPER, consisting of a regulatory domain of an *Escherichia coli* transcription factor OxyR (OxyR-RD), is inserted into a circularly permutated YFP, an H_2O_2-specific genetically encoded sensor (Belousov et al., 2006; Markvicheva et al., 2011). The specificity arises as oxidation of cysteines within HYPER allows it to selectively report H_2O_2 levels. The two redox-active cysteine residues in HYPER are located in a hydrophobic pocket of the OxyR domain; hence, they are accessible only to amphiphilic molecules such as H_2O_2. The reaction OxyR-RD with H_2O_2 produces a modification of the conformational structure of the molecule, inducing a shift of the excitation maximum peak of the attached cpYFP from 420 nm to 500 nm. This permits HYPER to be used as a ratiometric sensor for H_2O_2 quantification. Moreover, these modifications are reversible, making this sensor useful for monitoring variation of hydrogen peroxide concentration in real time and spatiotemporal dynamics of hydrogen peroxide production in plant cells (Choi et al., 2012). The major problem of HYPER is its sensitivity to pH changes (Belousov et al., 2006). Therefore, it is important that the local pH should be continuously monitored using a specific dye or genetically encoded sensor for pH detection.

HYPER can be also used to monitor changes at different subcellular compartments, such as mitochondria, plastids, or peroxisomes, by driving HYPER to these organelles using signal peptides. Costa et al. (2010) have successfully used a HYPER-KSRM, which is a peroxisomal-targeted HYPER that permits the observation of H_2O_2 in peroxisomes.

References

Asada, K., Production and scavenging of reactive oxygen species in chloroplasts and their functions, *Plant Physiol.*, 141, 391–396, 2006.

Belousov, V.V., Fradkov, A.F., Lukyanov, K.A., Staroverov, D.B., Shakhbazov, K.S., Terskikh, A.V., and Lukyanov, S., Genetically encoded fluorescent indicator for intracellular hydrogen peroxide, *Nature Methods*, 3, 281–286, 2006.

Benov, L., Sztejnberg, L., and Fridovich, I., Critical evaluation of the use of hydroethidine as a measure of superoxide anion radical, *Free Radic. Biol. Med.*, 25, 826–831, 1998.

Bestwick, C.S., Brown, I.R., Bennett, M.H.R., and Mansfield J.W., Localization of hydrogen peroxide accumulation during the hypersensitive reaction of lettuce cells to *Pseudomonas syringae* pv *phaseolicola*, *Plant Cell*, 9, 209–211, 1997.

Bhattacharjee, S., Reactive oxygen species and oxidative burst: roles in stress, senescence and signal transduction in plants, *Curr. Sci.*, 89, 1113–1121, 2005.

Bindokas, V.P., Jordan, J., Lee, C.C., and Miller, R.J., Superoxide production in rat hippocampal neurons: selective imaging with hydroethidine, *J. Neurosci.*, 16, 1324–1326, 1989.

Briggs, T., Drath, D.B., Karnovsky, M.L., and Karnovsky, M.J., Localization of NADH oxidase in the surface of human polymorphonuclear leukocytes by a new cytochemical method, *J. Cell Biol.*, 67, 566–586, 1975.

Camejo, D., Martí, M.C., Jiménez, A., Cabrera, J.C., Olmos, E., and Sevilla, F., Effect of oligogalacturonides on root length, extracellular alkalinization and O_2-accumulation in alfalfa, *J. Plant Physiol.*, 168, 566–575, 2011.

Carter, W.O., Narayanan, P.K., and Robinson, J.P., Intracellular hydrogen peroxide and superoxide anion detection in endothelial cells, *J. Leukocyte Biol.*, 55, 253–258, 1994.

Chen, C.S., Phorbol ester induces elevated oxidative activity and alkalization in a subset of lysosomes, *BMC Cell Biol.*, 3, 21, 2002.

Chen, S., Olbrich, A., Langenfeld-Heyser, R., Fritz, E., and Polle, A., Quantitative x-ray microanalysis of hydrogen peroxide within plant cells, *Microsc. Res. Techniq.*, 72, 49–60, 2009.

Choi, W.G., Swanson, S.J., and Gilroy, S., High-resolution imaging of Ca^{2+}, redox status, ROS and pH using GFP biosensors, *Plant J.*, 70, 118–128, 2012.

Costa, A., Drago, I., Behera, S., Zottini, M., Pizzo, P., Schroeder, J.I., Pozzan, T., and Lo Schiavo, F., H_2O_2 in plant peroxisomes: an in vivo analysis uncovers Ca^{2+}-dependent scavenging system, *Plant J.*, 62, 760–772, 2010.

Czaninski, Y., Sachot, R.M., and Catesson, A.M., Cytochemical localization of hydrogen peroxide in lignifying cell walls, *Ann. Bot.*, 72, 547–550, 1993.

Del Rio, L.A., Sandalio, L.M., Corpas, F.J., Palma, J.M., and Barroso, J.B., Reactive oxygen species and reactive nitrogen species in peroxisomes: production, scavenging, and role in cell signaling, *Plant Physiol.*, 141, 330–335, 2006.

Delladonne, M., Xia, Y., Dixon, R.A., and Lamb, C., Nitric oxide functions as a signal in plant disease resistance, *Nature*, 394, 585–588, 1998.

Durner, J., Wendehenne, D., and Klessing, D.F., Defence gene induction in tobacco by nitric oxide, cyclic GMP, and cyclic DP-ribose, *Proc. Natl. Acad. Sci. USA*, 95, 10328–10333, 1998.

Fernandez-Garcia, N., de la Garma, J.G., and Olmos, E., ROS as biomarkers in hyperhydricity, in *Reactive oxygen species and antioxidants in higher plants*, ed. S. Dutta Gupta, 249–274, Science Publishers, Enfield, NH, 2010.

Fernandez-Garcia, N., Lopez-Perez, L., Hernandez, M., and Olmos, E., Role of phi-cells and the endodermis under salt stress in *Brassica oleracea*, *New Phytol.*, 181, 347–360, 2009.

Flors, C., Fryer, M.J., Waring, J., Reeder, B., Bechtold, U., Mullineaux, P.M., Nonell, S., Wilson, M.T., and Baker, N.R., Imaging the production of singlet oxygen in vivo using a new fluorescent sensor, Singlet Oxygen Sensor Green, *J. Exp. Bot.*, 57, 1725–1734, 2006.

Foyer, C.H., and Noctor, G., Redox sensing and signalling associated with reactive oxygen in chloroplasts, peroxisomes and mitochondria, *Physiol. Plant.*, 119, 355–364, 2003.

Fridovich, I., Biological effects of superoxide radical, *Arch. Biochem. Biophys.*, 247, 1–11, 1986.

Gupta, K.J., Fernie, A.R., Kaiser, W.M., and van Dongen, J.T., On the origins of nitric oxide, *Trends Plant Sci.*, 16, 160–168, 2010.

Halliwell, B., Reactive species and antioxidant redox biology is a fundamental theme of aerobic life, *Plant Physiol.*, 141, 312–322, 2006.

Halliwell, B., and Gutteridge, J.M.C., *Free radicals in biology and medicine*, 3rd ed., Oxford University Press, Oxford, 2001.

Haugland, R.P., *Handbook of fluorescent probes and research chemicals*, Molecular Probes, Eugene, OR, 2005.

Hernández, M., Fernández-García, N., Díaz-Vivancos, Pedro, and Olmos, E., A different role for hydrogen peroxide and the antioxidative system under short and long salt stress in *Brassica oleracea* roots, *J. Exp. Bot.*, 2, 521–535, 2010.

Hideg, E., A comparative study of fluorescent singlet oxygen probes in plant cells, *Cent. Eur. J. Biol.*, 3, 273–284, 2008.

Hideg, E., Barta, C., Kalai, T., Vass, I., Hideg, K., and Asada, K., Detection of singlet oxygen and superoxide with fluorescent sensors in leaves under stress by photoinhibition or UV radiation, *Plant Cell Physiol.*, 43, 1154–1164, 2002.

Hideg, E., Kalai, T., Hideg, K., and Vass, I., Photoinhibition of photosynthesis in vivo results in singlet oxygen production. Detection via nitroxide-induced fluorescence quenching in broad bean leaves, *Biochemistry*, 37, 11405–11411, 1998.

Hideg, E., Ogawa, K., Kalai, T., and Hideg, K., Singlet oxygen imaging in *Arabidopsis thaliana* leaves under photoinhibition by excess photosynthetically active radiation, *Physiol. Plant.*, 112, 10–14, 2001.

Imlay, J.A., and Linn, S., DNA damage and oxygen radical toxicity, *Science*, 240, 1302–1309, 1988.

Keston, A.S., and Brandt, R., The fluorometric analysis of ultramicro quantities of hydrogen peroxide, *Anal. Biochem.*, 11, 23–29, 1965.

Kim, C., Meskauskiene, R., Apel, K., and Laloi, C., No single way to understand singlet oxygen signalling in plants, *EMBO Rep.*, 9, 435–439, 2008.

Kojima, H., Sakurai, K., Kikuchi, K., Kawahara, S., Kirino., Y., Nagoshi, H., Hirata, Y., and Nagano, T., Development of a fluorescent indicator for nitric oxide based on the fluorescein chromophore, *Chem. Pharm. Bull.*, 46, 373–375, 1998.

Kojima, H., Urano, Y., Kikuchi, K., Higuchi, T., Hirata, Y., and Nagano, T., Fluorescent indicators for imaging nitric oxide production, *Angew. Chem. Int. Ed.*, 38, 3209–3212, 1999.

LeBel, C.P., and Bondy, S.C., Sensitive and rapid quantitation of oxygen reactive species in rat synaptosomes, *Neurochem. Int.*, 17, 435–440, 1990.

LeBel, C.P., Ischiropoulos, H., and Stephen, C.B., Evaluation of the probe 2′,7′-dichiorofluorescin as an indicator of reactive oxygen species formation and oxidative stress, *Chem. Res. Toxicol.*, 5, 227–231, 1992.

Li, Z., and Xing, D., Mechanistic study of mitochondria-dependent programmed cell death induced by aluminium phytotoxicity using fluorescence techniques, *J. Exp. Bot.*, 62, 331–343, 2011.

Lv, W.T., Lin, B., Zhang, M., and Hua, X.J., Proline accumulation is inhibitory to *Arabidopsis* seedlings during heat stress, *Plant Physiol.*, 156, 1921–1933, 2011.

Markvicheva, K.N., Bilan, D.S., Mishina, N.M., Gorokhovatsky, A.Y., Vinokurov, L.M., Lukyanov, S., and Belousov, V.V., A genetically encoded sensor for H_2O_2 with expanded dynamic range, *Bioorgan. Med. Chem.*, 19, 1079–1084, 2011.

Maxwell, D.P., Wang, Y., and McIntosh, L., The alternative oxidase lowers mitochondrial reactive oxygen production in plant cells, *Proc. Natl. Acad. Sci. USA*, 96, 8271–8276, 1999.

Miller, G., Honig, A., Stein, H., Suzuki, N., Mittler, R., and Zilberstein, A., Unraveling delta(1)-pyrroline-5-carboxylate-proline cycle in plants by uncoupled expression of proline oxidation enzymes, *J. Biol. Chem.*, 284, 26482–26492, 2009.

Moller, I.M., Plant mitochondria and oxidative stress: electron transport, NADPH turnover and metabolism of reactive oxygen species, *Annu. Rev. Plant Physiol. Plant Mol. Biol.*, 52, 561–591, 2001.

Moller, I.M., Jensen, P.E., and Hansson, A., Oxidative modifications to cellular components in plants, *Annu. Rev. Plant Biol.*, 58, 459–481, 2007.

Monshausen, G.B., Bibikova, T.N., Messerli, M.A., Shi, C., and Gilroy, S., Oscillation in extracellular pH and reactive oxygen species modulate tip growth of *Arabidopsis* root hairs, *Proc. Natl. Acad. Sci. USA* 104, 20996–21001, 2007.

Mullineaux, P., and Karpinski, S., Signal transduction in response to excess light: getting out of the chloroplast, *Curr. Opin. Plant Biol.*, 5, 43–48, 2002.

Neil, S.J., Desikan, R., and Hancock, J., Hydrogen peroxide signalling, *Curr. Opin. Plant Biol.*, 5, 388–395, 2002.

Olmos, E., and Hellin, E., Ultrastructural differences of hyperhydric and normal leaves from regenerated carnation plants, *Sci. Hortic.*, 75, 91–101, 1998.

Palmer, R.M., Ferrige, A.G., and Moncada, S., Nitric oxide release account for the biological activity of endothelium-derived relaxing factor, *Nature*, 327, 524–526, 1987.

Rümer, S., Krischke, M., Fekete, A., Mueller, M.J., and Kaiser, W.M., DAF-fluorescence without NO: elicitor treated tobacco cells produce fluorescing DAF-derivates not related to DAF-2 triazol, *Nitric Oxide*, 27, 123–135, 2012.

Sandalio, L.M., Rodríguez-Serrano, M., Romero Puertas, M.C., and del Río, L.A., Imaging of reactive oxygen species and nitric oxide in vivo in plant tissues, *Method Enzymol.*, 440, 397–409, 2008.

Smirnova, V., Matveyeva, N.P., Polesskaya, O.G., and Yermakov, I.P., Generation of reactive oxygen species during pollen grain germination, *Russ. J. Dev. Biol.*, 40, 345–353, 2009.

Swanson, S.J., Won-Gyu, C., Chanoca, A., and Gilroy, S., In vivo imaging of Ca^{2+}, pH, and reactive oxygen species using fluorescent probes in plants, *Annu. Rev. Plant Biol.*, 62, 273–297, 2011.

Tarpey, M.M., Wink, D.A., and Grisham, M.B., Methods for detection of reactive metabolites of oxygen and nitrogen: in vitro and in vivo considerations, *Am. J. Physiol. Regu. Integr. Comp. Physiol.*, 286, 431–444, 2004.

Thomas, J., and Trelease, R.N., Cytochemical localization of glycolate oxidase in microbodies (glyoxysomes and peroxisomes) of higher plant tissues with $CeCl_3$ technique, *Protoplasma*, 108, 39–53, 1981.

Wise, R.R., and Naylor, A.W., Chilling-enhanced photooxidation: evidence for the role of singlet oxygen and endogenous antioxidants, *Plant Physiol.*, 83, 278–282, 1987.

Xie, Y.J., and Shen, W.B., In vivo imaging of nitric oxide and reactive oxygen species using laser scanning confocal microscopy, in *Plant salt tolerance: methods and protocols, methods in molecular biology*, ed. S. Sergey and T.A. Cuin, vol. 913, Springer, New York, 2012.

Yao, N., and Greenberg, J.T., *Arabidopsis* accelerated cell death2 modulates programmed cell death, *Plant Cell*, 18, 397–411, 2006.

Zhang, C.J., Zhao, B.C., Ge, W.N., Zhang, Y.F., Song, Y., Sun, D.Y., and Guo, Y., An apoplastic H-type thioredoxin is involved in the stress response through regulation of the apoplastic reactive oxygen species in rice, *Plant Physiol.*, 157, 1884–1899, 2011.

chapter thirteen

Fluorescent ROS probes in imaging leaves

Éva Hideg and Ferhan Ayaydin

Contents

13.1 Introduction

Fluorescent probes gained immense popularity over the years in a wide range of biological applications. When combined with imaging techniques, the most important advantages of fluorescent probes are high spatial resolution and their less invasive nature. If uniform delivery of the probe can be achieved, fluorescence of relatively large areas (such as several square centimeters) can be detected by digital cameras or fluorescence scanners. However, studies at the cellular or subcellular level require high-resolution microscopy. Consequently, this review is focused on these applications, although macro-imaging has also been discussed briefly. There are various fluorescence reporter molecules for the analysis of diverse metabolites and physiological parameters, such as pH, membrane potential, ion concentration, reactive oxygen, and nitrogen species. The detection of reactive oxygen species (ROS) is of special importance due to their role in plant stress biology (Gill and Tuteja, 2010). In leaves, applications of fluorescent probes meet an extra challenge due to higher concentrations of oxygen and more extreme redox conditions than in animal cells. Moreover, the intrinsic, fluorescent plant metabolites present an additional challenge.

13.2 Pros and cons of using fluorescent probes for detecting reactive oxygen species (ROS)

13.2.1 ROS in leaves

ROS are part of plant metabolism and their major sources are (1) oxygen reduction by organellar electron transport (in chloroplasts or mitochondria) yielding superoxide anion radicals ($O_2^{-\bullet}$), (2) peroxisomes that produce hydrogen peroxide (H_2O_2) by the enzymatic activity of glycolate oxidase, and (3) plasma membrane NADPH-dependent oxidases generating $O_2^{-\bullet}$. These ROS are linked by the $O_2^{-\bullet} \rightarrow H_2O_2 \rightarrow H_2O$ series of reactions that may occur either spontaneously ($O_2^{-\bullet} \rightarrow H_2O_2$) or catalyzed by enzymes ($O_2^{-\bullet} \rightarrow H_2O_2$, by superoxide dismutases; $H_2O_2 \rightarrow H_2O$, by peroxidases). Metal ions, such as Fe^{2+} or Cu^+, catalyze the $H_2O_2 \rightarrow {}^\bullet OH$ reaction, yielding a strongly oxidizing hydroxyl radical (Apel and Hirt, 2004; Møller, 2001). In chlorophyll-containing cells an additional source is a pigment-photosensitized production of singlet oxygen (${}^1O_2$) (Triantaphylidès et al., 2008). With the exception of H_2O_2, all ROS have short, submillisecond lifetimes in leaves. Under standard metabolic conditions, ROS concentrations are strictly controlled and kept low by antioxidants in all plant tissue. Stress conditions, however, upset this balance of ROS production and neutralize it either by weakening the antioxidant system or increasing the production of pro-oxidants. Increased leaf ROS concentrations are reportedly associated with a variety of abiotic and biotic stresses (Bailey-Serres and Mittler, 2006). Excess ROS causes biological damage either directly, at their production site by oxidizing cellular components, or indirectly, by generating other oxidants capable of traveling inside cells or even passing through membranes. Such latter reactions support the idea of certain ROS or their derivatives acting as signal molecules and being part of defensive cellular responses rather than acting as damaging agents (Mittler et al., 2004; Pitzschke et al., 2006; Fischer et al., 2013).

Plant stress studies aimed at identifying either as messengers or as primary damaging agents mostly rely on indirect evidence, such as upregulation of antioxidant genes, as well as detecting higher amounts of nonenzymatic or increased activities of enzymatic antioxidants. Probes changing fluorescence properties when oxidized by ROS offer a more direct approach.

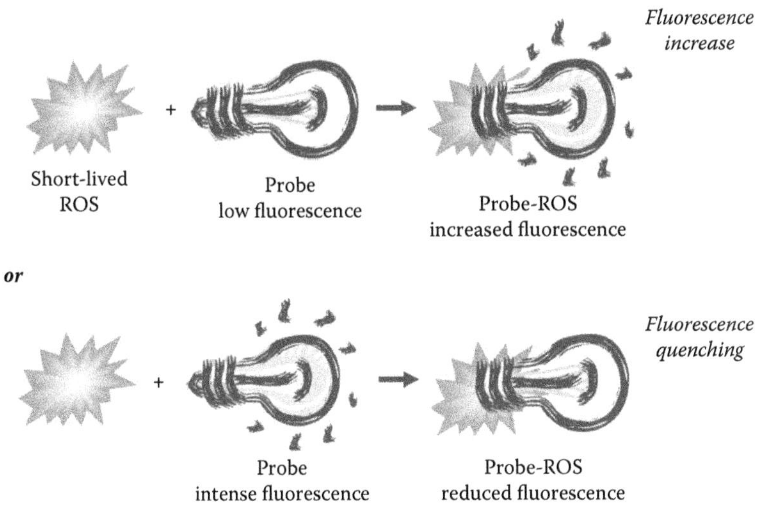

Figure 13.1 The two major types of oxidation response of ROS probes: fluorescence increase and fluorescence quenching.

13.2.2 *Fluorescent ROS probes*

Among the primary ROS products and their other radical derivatives, it is only $^1O_2$ that offers the possibility of absolute direct detection based on its unique physical properties. Detecting the infrared light emitted as $^1O_2$ phosphorescence is a well-established method in studies on mammalian cells (Ogilby, 2010), but spectral overlapping with chlorophyll phosphorescence and the presence of effective $^1O_2$ quenchers in plant cells hamper the application of this method in leaf studies (Fischer et al., 2013). Studies aimed at ROS other than $^1O_2$ can only rely on the use of special reporter molecules, which change in a well-detectable physical feature upon reacting with ROS. Fluorescent probes can react in two ways (Figure 13.1). Nonfluorescent molecules may become fluorescent as a result of oxidation by ROS, or an intrinsically fluorescent probe's emission may get quenched by ROS.

13.2.3 *Pitfalls and limitations*

As illustrated in Figure 13.2, probes need to outcompete several other reactions in order to be oxidized by ROS, and should thus be delivered as close to potential sites of ROS production as possible. Fluorescent probes freely diffusible through cell walls and membranes may simply be added to plant tissue culture medium, growth media of photosynthetic microorganisms, buffer solutions containing isolated functional preparations of organelles (e.g., chloroplasts, mitochondria), or organellar membranes

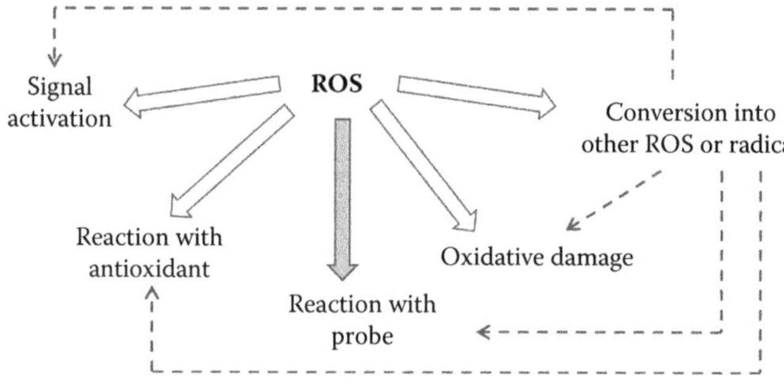

Figure 13.2 A scheme illustrating possible reactions of ROS with probes and biomolecules. White and dashed line arrows show primary and secondary reactions, respectively. It is only the reaction marked with the gray arrow that leads to changes in the probe's fluorescence and is registered as ROS detection by the probe.

(for example, thylakoids). Leaves pose difficulties due to their strong cell walls and waxy epidermal surface. Floating leaf cuttings on probe-containing solutions is usually not sufficient to deliver probes inside leaf tissues, and more active infiltration is needed. For comparisons of methods see Šnyrychová et al. (2009) and Hideg et al. (2002). Ideally, infiltrating solutions should only contain water as a solvent, but low (<1%) concentrations of alcohols or DMSO may also be present if required to achieve complete solubility of the probe. When the infiltrating medium is forced into the leaf tissue, it is mixed with intercellular fluids. In order to minimize modifications of cellular pH, infiltrating solutions should be neutral. Effects of the infiltration process itself should always be checked, by comparing values of a physiological parameter, for example, leaf photosynthesis at untreated leaf areas and at areas infiltrated with water (buffer or solvents) only. Alternatively, side effects of the infiltration process can also be monitored by co-infiltrating a cell-impermeable nucleic acid intercalating dye such as propidium iodide (PI). PI delineates cell walls in intact plant cells, and hence can be used as a convenient marker of cell borders and shapes in plant studies (Pighin et al., 2004). If plasma membrane integrity is compromised, PI enters inside the cell and labels nucleic acids. Nuclei that are rich in DNA and RNA can easily be spotted as a bright fluorescence body in these wounded cells. Figure 13.3 shows a typical example of PI-infiltrated tobacco leaf tissue where intact, cell wall-labeled cells can be clearly separated from the nuclei-labeled, membrane-compromised population.

Probes may be sensitive to a variety of factors involved in plant experiments. A typical example leading to false positive results is light sensitivity, which is usually stronger in response to higher-energy (ultraviolet) or higher-intensity light. An example is the fluorescent probe Singlet Oxygen Sensor Green (Flors et al., 2006), which, in addition to increasing

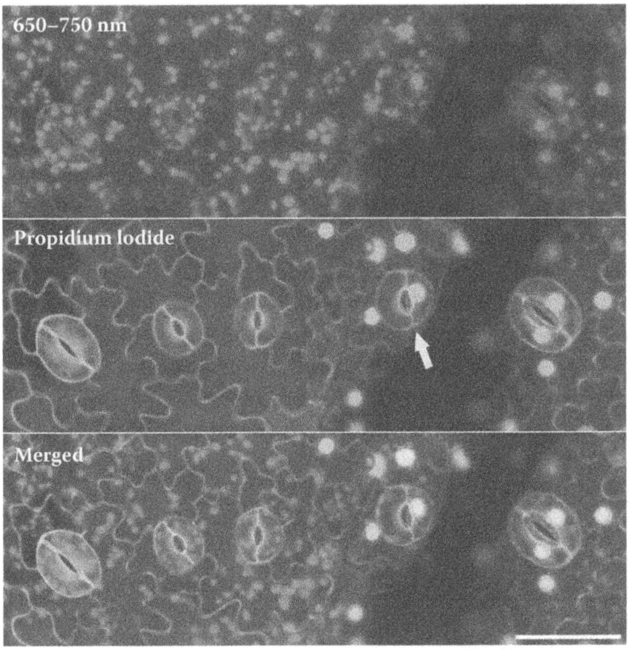

Figure 13.3 (**See color insert.**) Use of propidium iodide (PI) to monitor cell damage. PI (10 μg/ml) is infiltrated into a tobacco leaf using a syringe applicator. The region in the vicinity of the syringe injection (right half of the image) shows damaged cells with nuclear PI labeling. The left half of the middle panel shows intact leaf pavement and guard cells with PI-labeled cell walls. The arrow indicates a transition zone stoma with two guard cells, one of which is damaged, while the other one is intact. Red-colored fluorescence (mostly chlorophyll emission between 650 and 750nm) and PI (green pseudocolored, 550–650 nm emission) images are merged at the lowermost panel. Excitation wavelength = 488 nm. Scale bar = 50 μm.

its fluorescence upon illumination (Hideg, 2008), is able to produce singlet oxygen under exposure to UV or visible radiation (Ragàs et al., 2009). Other possible limitations due to interactions between probes and metabolites other than ROS are listed in Section 13.3.2.

13.3 Probing leaves in vivo

ROS detection in leaves mostly involves analyses in living tissue samples. In vivo imaging of plant leaves may pose a special challenge because of rich pigmentation, light-scattering starch granules, and multiple layers of cell walls. Wounding-induced artifacts prevent the use of thin mechanical sectioning of leaf tissue in ROS detection experiments. Due to optical sectioning ability, confocal microscopes are the preferred equipment to detect microlocalization of infiltrated fluorescent ROS probes in living tissues. There are two main types of conventional confocal microscopes: laser scanning based and spinning disk based (Pawley, 2006). While the former provides high spatial resolution, the latter excels in fast image acquisition. Both techniques require high levels of excitation that should be carefully adjusted to avoid photodamage in the leaf region under investigation. Apart from conventional confocal microscopy, two alternative methods are also available to obtain optical sections: deconvolution and multiphoton imaging. Deconvolution involves computer-based algorithms to calculate and remove out-of-focus information from images obtained by conventional wide-field fluorescence microscopes as well as confocal microscopes (Pawley, 2006). Availability of low-cost or free of charge deconvolution software and low-cost powerful computers makes this technique increasingly attractive for imaging studies. However, successful deconvolution of images requires correct simulation of the optical environment, which limits the use of this technique inside the dye-loaded, plastid-rich, and diffractive environment of parenchyma tissues of plant leaves. As an alternative to optical sectioning, multiphoton microscopy is highly advantageous, as it grants superior depth penetration. Highly pigmented green leaves efficiently absorb near-infrared (NIR) illumination; therefore, high-intensity NIR excitation used in multiphoton microscopy should be handled with particular care in experiments involving green tissues. In general, lower excitation intensities, less frequent, fast imaging, and fewer sections can greatly aid in reducing photo-oxidative damage in leaves during ROS detection experiments. In addition to causing physiological damage to leaf tissues, high excitation light intensities can also affect the performance of fluorescent ROS probes negatively. High-energy-induced alterations in probes' chemical structure may cause artifactual augmentation or reduction in fluorescence probe intensities independent of ROS concentrations inside the leaves.

In leaf tissues, high fluxes of blue excitation may render chlorophyll triplets, which are potential sources of ROS, typically singlet oxygen themselves. These phototoxic effects occur because whenever a fluorochrome (endogenous, such as chlorophyll, or exogenous, such as a fluorescent probe) is excited, there is a probability that instead of decaying to a ground state with the emission of light, the molecule will exhibit transition to a triplet state (Halliwell, 2006). Dye triplets are usually longer-lived than singlet excited states and are reactive to ground state molecular oxygen generating highly reactive singlet oxygen. The decrease of the concentration of dissolved oxygen in the medium may lessen this problem in animal cell samples (Demchenko, 2009), but it is hardly a remedy in plant samples, such as oxygen-producing leaf tissue.

13.3.1 Autofluorescence

Green tissues such as plant leaves, pollen grains, and stressed or photodamaged cells are highly fluorescent. The upper panel in Figure 13.4 shows a confocal laser scanning microscopy image of chloroplast and cell wall fluorescence in tobacco leaf epidermis. Chlorophyll fluorescence of chloroplasts occupies the red/far-red region of the spectrum (650–750 nm), while cell wall lignin fluorescence is dominant in a wide range of the visible spectrum (490–620 nm, blue to orange) (Chapman et al., 2005). During fluorescence microscopy imaging of nonetiolated leaves, bright and relatively bleach-resistant chlorophyll fluorescence offers an ideal intrinsic marker to locate the chloroplasts. Autofluorescence of cell walls can also be beneficial to locate different cell types and cell borders. On the other hand, strong chlorophyll fluorescence, with its emission maximum at 680–700 nm, limits the use of red fluorescent markers in green leaves. Using emission filters with 650 nm cutoff, the majority of strong chlorophyll fluorescence can be separated from the fluorescence of orange-red-colored probes and sensors in multicolor detection experiments. Probes emitting

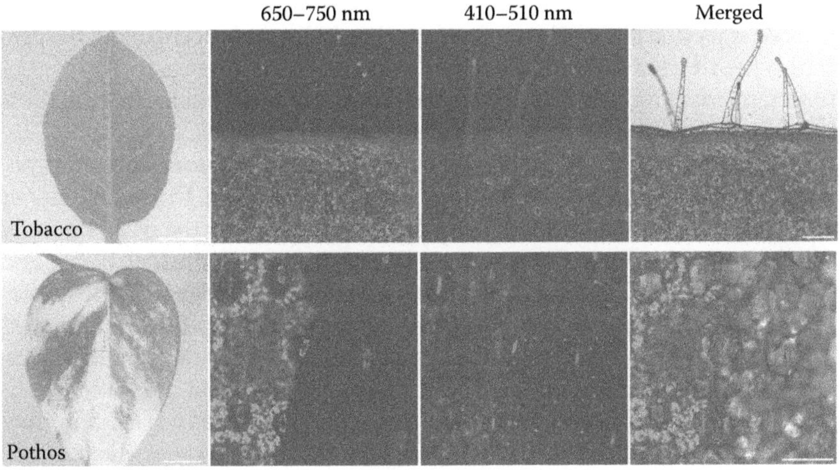

Figure 13.4 **(See color insert.)** Red (650–750 nm emission excited by 488 nm) and blue (410–510 nm emission excited by 405 nm) autofluorescence on leaf border of tobacco with trichomes (*Nicotiana tabacum*, upper panel) and variegated leaf of golden pothos (*Epipremnum aureum*, lower panel). Border region sampled for microscopy imaging is indicated with a red square on the tobacco leaf photo. The dotted red square on the pothos leaf photo indicates a variegated leaf area with green (chlorophyll-rich) and white (etiolated) sectors. Microscopy images from a variegated area are shown as close-up images of the pothos leaf. Last panels show merged images of red, blue, and bright field images. Scale bars = 1 cm (leaf macrophotos) and 100 μm (microscopy images).

red fluorescence overlapping with emission from chlorophyll can also be isolated using spectral unmixing (see below). Variegated leaves are clear examples of the metabolic heterogeneity of cells within an organ (Bartley and Scolnik, 1995), rendering it possible to compare the responses of different tissues. The lower panel in Figure 13.4 compares autofluorescence of green and white sectors of a pothos (*Epipremnum aureum*) leaf. Although strong red chlorophyll fluorescence was almost entirely missing from white leaf areas, it was present in stoma thylakoids at the surface of both types of leaf areas. Violet laser-induced (405 nm) blue autofluorescence was present on both green and white sectors, but expressed more dominantly at green sectors (Figure 13.4).

Fluorescent polyphenolic compounds such as lignin and suberin present in vascular tissues and cell walls are the dominant source of autofluorescence in roots, but they can also interfere in leaf experiments, depending on the target tissue and the degree of wounding during handling of samples. Wounding-induced accumulation of these fluorescent phenolic compounds is particularly important during ROS studies, as wounding itself stimulates transient production of ROS as a defense strategy of plants (Angelini et al., 2008). Leaf regions close to the needle-pierced injection site or the cut surface should be excluded during fluorescent sensor-based ROS measurements in leaves. As images in Figure 13.3 illustrate, mechanical cell damage itself is another reason to exclude these (usually less than 0.1 cm^2) areas during imaging studies.

Working with fluorescent sensors in a naturally fluorescent environment necessitates the use of proper control samples. Untreated leaves and water- or buffer-injected samples are essential for the assessment of the degree of intrinsic fluorescence. Depending on the target tissue and cell type, careful selection of spectral variants of fluorescent ROS sensors and suitable fluorescence filter sets can also greatly help for proper detection of externally introduced fluorescent sensors and markers in leaves. However, the number of spectrally distinct fluorescent ROS sensors is rather limited at the moment; therefore, spectral overlap with autofluorescence can be inevitable in certain situations. In such cases, a technique known as spectral imaging in combination with linear unmixing can be used to separate interfering autofluorescence from the actual signal of interest (Berg, 2004; Mylle et al., 2013). This technique makes use of the unique spectral emission properties of intrinsic fluorescence and experimentally introduced fluorescent markers. Using individual emission fingerprints of each contributing fluorochrome, mixed fluorescent signals can be unmixed by using computational methods incorporating unmixing algorithms. Figure 13.5 illustrates the details of spectral imaging and linear unmixing of propidium iodide (PI) and chlorophyll autofluorescence. PI is a convenient marker for both cell boundaries and cell viability in in vivo ROS detection experiments where dye-infiltrated leaves are

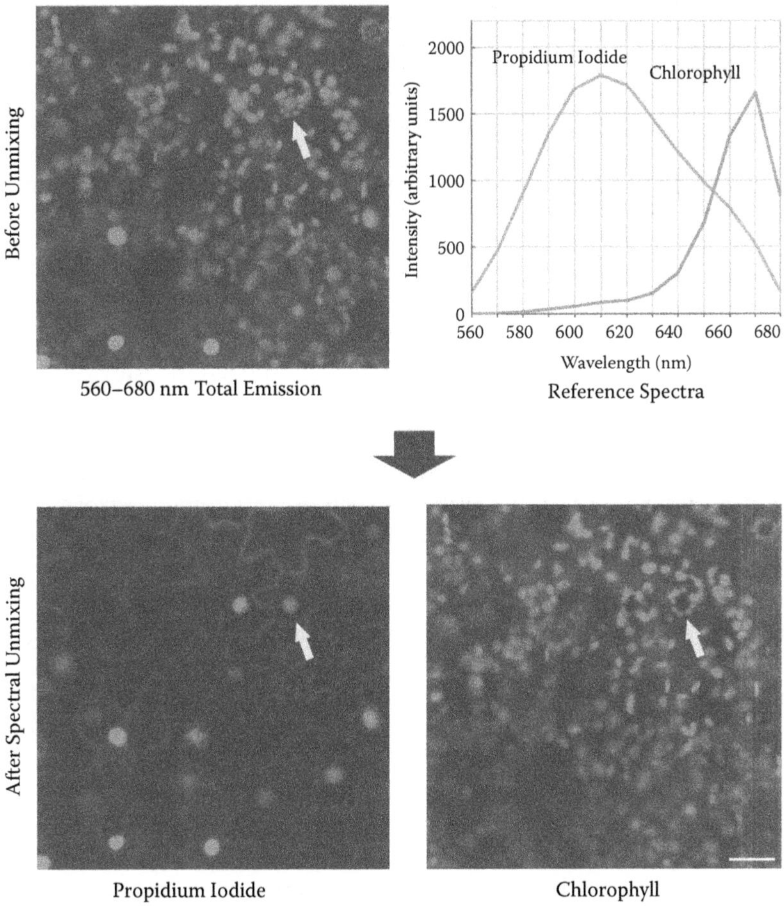

Figure 13.5 **(See color insert.)** Spectral imaging and computational unmixing of propidium iodide (PI) and chlorophyll fluorescence. Upper panel shows intensity projection over lambda axis (excitation = 488 nm; emission = 560–680 nm, 10 nm/step, 20 nm resolution) of a PI-infiltrated tobacco leaf with chlorophyll autofluorescence. A labeled nucleus and a well-isolated chloroplast are used to capture in vivo reference spectra for PI (orange) and chlorophyll (red), respectively (graph, upper panel). A mixed image is unmixed into PI and chlorophyll channels using the reference spectra and the spectral unmixing software of Olympus Fluoview (lower panel). Arrows indicate clear separation of a PI-labeled nucleus surrounded by chloroplasts. All images are pseudocolored. Scale bar = 20 μm.

used (Figure 13.3; see also Section 13.2.3). On the other hand, its broad orange-red-colored emission overlaps with chlorophyll fluorescence at $\lambda >$ 650 nm. Note the nuclear propidium iodide signal bleed-through into the chlorophyll channel in the top panel of Figure 13.3. To record the emission fingerprint of autofluorescence or the dye of interest, unlabeled control samples are needed. Alternatively, well-isolated reference regions within the image to be unmixed can also be used to gather reference spectra of the fluorochromes to be unmixed (Figure 13.5). Using these reference spectra and unmixing procedure, dye fluorescence (e.g., PI, ROS probes) and autofluorescence (chlorophyll, polyphenols) signals can be efficiently separated in leaves (Figure 13.5).

13.3.2 Probe and sample interactions

Spectral overlapping with autofluorescence is not the only way leaf samples interact with fluorescent probes. They can also interact physically by intercepting the excitation light targeted to the fluorescent probe. For example, chlorophyll-containing tissues have strong absorption in both blue and red; therefore, fluorescent probes utilizing such excitation may perform poorly simply because they are spectrally shielded. Similarly, fluorescence emission from the probe may be absorbed by the sample, limiting its use. ROS probes that function via fluorescence quenching (Figure 13.1) are easy to test for these interferences, as they are already fluorescent before any ROS generating perturbation (Hideg, 2008). Unoxidized probes that have low fluorescence or none at all cannot be traced this way, and a positive control should be established by infiltrating a preoxidized, highly fluorescent form of the probe. This strategy was successfully incorporated to show plastid localization of the H_2O_2-sensitive probe Amplex Red (Šnyrychová et al., 2009).

In addition to physical interactions, fluorescent ROS probes can also interact with leaf tissues chemically. This can happen two ways: leaf metabolites may modify probes, or the presence of probes may alter leaf metabolism. An example of the first possibility typically manifests as ROS-unrelated quenching of the fluorescence of the probe. The second possibility is illustrated in Figure 13.6. In a study aimed at discovering novel fluorescent $^1O_2$ probes (Kálai et al., 2013), we found that a 1,8-naphthalimide derivative (compound-4384; T. Kálai, personal communication) proved chaotropic when infiltrated in leaf tissue. Figure 13.6 illustrates that this chemical, which was eventually unused as an ROS probe, destroyed cellular structures. The intracellular membrane network and the central vacuole are both disintegrated to form several highly dynamic micro- and macrovesicles distributed all over the cell after a few minutes of infiltration. This was most likely due to the membrane-destabilizing, detergent-like structure of compound-4384, illustrating a rather extreme example of undesired probe-sample interaction.

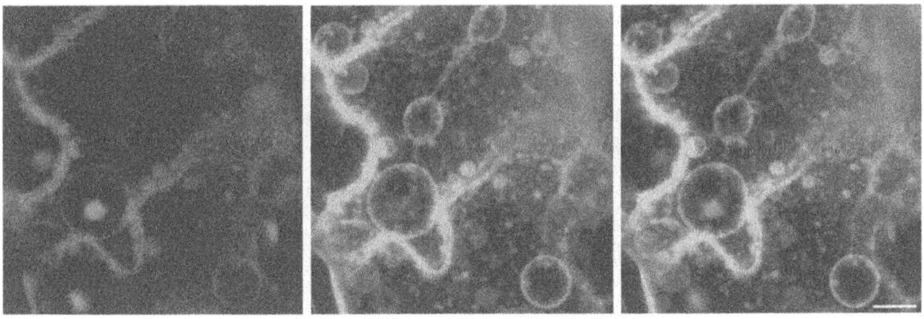

Figure 13.6 (**See color insert.**) Tobacco leaves infiltrated with a 1,8-naphthalimide derivative (compound-4384). Last panel shows the merged images of red emission (excitation = 488 nm, emission = 650–750 nm, mostly chlorophyll autofluorescence) and green pseudocolored chemical fluorescence (excitation = 405 nm, emission = 500–590 nm). Scale bar = 10 μm.

13.3.3 Interpreting results

Microscopic images of ROS probe fluorescence provide an excellent tool to identify sites of ROS production in the leaf. Qualitative conclusions on actual ROS production at these sites, however, require extreme caution and should not be made based solely on microscopy images. This is because probes report on trapped amounts of ROS, which are only a fraction of the amount actually produced (Figure 13.2). Moreover, probe fluorescence may be lessened via both physical and chemical interactions detailed above (Section 13.3.2.). Based on these, one would conclude that detected amounts of oxidized fluorescent ROS probes can always give lower estimates of ROS concentrations in vivo. This is, in fact, a more complicated issue, and fluorescence intensities detected in leaves cannot be quantified on the basis of fluorescence measured from probe solutions in vitro, for example, in a spectrofluorometer.

Fluorescence intensity of the same molecule may vary, as it is influenced by a number of factors, such as hydrophilic/hydrophobic nature or the pH of its environment. These interactions are complex and can either increase or decrease fluorescence (Lakowiz, 2006). Furthermore, microscopic close-up images represent data from a relatively small part of the whole leaf sample; hence, they should be accompanied by low-magnification macroimages that offer a better overview and characterization at the whole leaf level. For example, UV-induced bright yellow-green fluorescence from the $^1O_2$ probe DanePy (Kálai et al., 1998) is most pronounced at the whole leaf level (data not shown), while parallel confocal experiments with optical sectioning showed that the probe was present in chloroplasts of mesophyll cells (Hideg, 2008).

13.4 Concluding remarks

Fluorescent ROS probes are powerful tools to address various aspects of leaf physiology. On the other hand, detailed understanding of their in vivo behavior, their physical and chemical interactions with plant tissues, is essential to minimize misinterpretation of the results. In addition to careful analyses of their reactivity, stability, and subcellular localization, appropriate control experiments accompanied by multifaceted in vivo and in vitro analyses are also essential to make use of their full potential.

Acknowledgments

The authors thank Prof. Tamás Kálai (Department of Organic and Medicinal Chemistry, Faculty of Medicine, University of Pécs) for kindly providing compound-4384 and related information regarding its chemical structure.

References

Angelini, R., Tisi, A., Rea, G., Chen, M. M., Botta, M., Federico, R., and Cona, A., Involvement of polyamine oxidase in wound healing, *Plant Physiol.*, 146, 162–177, 2008.

Apel, K., and Hirt. H., Reactive oxygen species: metabolism, oxidative stress, and signal transduction, *Annu. Rev. Plant Physiol.*, 55, 373–399, 2004.

Bailey-Serres, J., and Mittler. R., The roles of reactive oxygen species in plant cells, *Plant Physiol.*, 141, 311, 2006.

Bartley, G. E., and Scolnik, P. A., Plant carotenoids: pigments for photoprotection, visual attraction, and human health, *Plant Cell*, 7, 1027–1038, 1995.

Berg, R. H., Evaluation of spectral imaging for plant cell analysis, *J. Microsci.*, 214, 174–181, 2004.

Chapman, S., Oparka, K. J., and Roberts, A. G., New tools for in vivo fluorescence tagging, *Curr. Opin. Plant Biol.*, 8, 565–573, 2005.

Demchenko, A. P., Introduction to fluorescence sensing, in *Sensing inside living cells and tissues*, 455–506, Springer Science + Business Media B. V., the Netherlands, 2009.

Fischer, B., Hideg, É., and Krieger-Liszkay, A., Production, detection and signaling of singlet oxygen in photosynthetic organisms, *Antioxid. Redox Signaling*, 18, 2145–2162, 2013.

Flors, Ch., Fryer, M. J., Waring J., Reeder B., Bechtold U., Mullineaux P. M., Nonell, S., Wilson, M. T., and Baker, N. R., Imaging the production of singlet oxygen in vivo using a new fluorescent sensor, Singlet Oxygen Sensor Green®, *J. Exp. Bot.*, 57, 1725–1734, 2006.

Gill, S. S., and Tuteja, N., Reactive oxygen species and antioxidant machinery in abiotic stress tolerance in crop plants, *Plant Physiol. Biochem.*, 48, 909–930, 2010.

Halliwell, B., Reactive species and antioxidants: redox biology is a fundamental theme of life aerobic life, *Plant Physiol.*, 141, 312–322, 2006.

Hideg, É. A., Comparative study of fluorescent singlet oxygen probes in plant leaves, *Cent. Eur. J. Biol.*, 3, 273–284, 2008.

Hideg, É., Barta, C. S., Kálai, T., Vass, I., Hideg, K., and Asada, K., Detection of singlet oxygen and superoxide with fluorescent sensors in leaves under stress by photoinhibition or UV radiation, *Plant Cell Physiol.*, 43, 1154–1164, 2002.

Kálai, T., Hideg, É., Ayaydin, F., and Hideg, K., Synthesis and potential use of 1,8-naphthalimide type $^1O_2$ sensor molecules, *Photochem Photobiol. Sci.*, 12, 432–438, 2013.

Kálai, T., Hideg, É., Vass, I., and Hideg, K., Double (fluorescent and spin) sensors for detection of reactive oxygen species in the thylakoid membrane, *Free Radical Biol. Med.*, 24, 649–652, 1998.

Lakowicz, J. R., *Principles of fluorescence spectroscopy*, 3rd ed., Springer, Berlin, 2006.

Mittler, R., Vanderauwera, S., Gollery, M., and Van Breusegem, F., Reactive oxygen gene network of plants, *Trends Plant Sci.*, 9, 490–498, 2004.

Møller, I. M., Plant mitochondria and oxidative stress: electron transport, NADPH turnover, and metabolism of reactive oxygen species, *Annu. Rev. Plant Physiol.*, 52, 561–591, 2001.

Mylle, E., Codreanu, M.-C., Boruc, J., and Russinova, E., Emission spectra profiling of fluorescent proteins in living plant cells, *Plant Methods*, 9, 10, 2013.

Ogilby, P. R., Singlet oxygen: there is indeed something new under the sun, *Chem. Soc. Rev.*, 39, 3181–3209, 2010.

Pawley, J. B., ed., *Handbook of biological confocal microscopy*, Springer Verlag, Berlin, 2006.

Pighin, J. A., Zheng, H. Q., Balakshin, L. J., Goodman, I. P., Western, T. L., Jetter, R., Kunst, L., and Samuels, A. L., Plant cuticular lipid export requires an ABC transporter, *Science*, 306, 702–704, 2004.

Pitzschke, A., Forzani, C., and Hirt, H., Reactive oxygen species signaling in plants, *Antioxid. Redox Signaling*, 8, 1757–1764, 2006.

Ragàs, X., Jimènez-Banzo, A., Sánchez-Garcia, D., Batllori X., and Nonell, S., Singlet oxygen photosensitisation by the fluorescent probe Singlet Oxygen Sensor Green®, *Chem. Commun.*, 2009, 2920–2922, 2009.

Šnyrychová, I., Ayaydin, F., and Hideg, É., Detecting hydrogen peroxide in leaves in vivo—a comparison of methods, *Physiol. Plant.*, 135, 1–18, 2009.

Triantaphylidès, C., Krischke, M., Hoeberichts, F. A., Ksas, B., Gresser, G., Havaux, M., Van Breusegem, F., and Mueller, M. J., Singlet oxygen is the major reactive oxygen species involved in photooxidative damage to plants, *Plant Physiol.*, 148, 960–968, 2008.

chapter fourteen

Analysis of root growth using image analysis

Andrew P. French and Michael P. Pound

Contents

14.1 Introduction

Root growth is a crucial aspect to study when trying to understand plant growth. Today, with an increased emphasis on automated phenotyping approaches to plant research, great headway is being made in developing new automated approaches to measuring root growth parameters from images. New technologies are being developed to capture the images themselves, and of course, new algorithms designed to analyze these new image data—which are often generated in vast quantities—and produce useable numerical data for biologists and mathematical modelers. In this chapter, we aim to orient the reader with the most commonly used technologies of image data capture for roots, and present some example image analysis techniques for analyzing images captured from different modalities. We will consider these analysis techniques by looking at examples of the processing necessary to extract data from different scales of root images, from subcellular images through to full architecture extraction.

Our aim is to provide a shallow yet broad introduction to the burgeoning research area of root-based image analysis.

Broadly speaking there are two aspects to consider when researching roots using imaging approaches. On the one hand, we can study the growth mechanisms of individual roots, looking in detail at the cellular scale, at cell division and elongation processes, or indeed, at subcellular details such as colocalized proteins or nuclei location. We might also be interested in measuring morphological features on the root, root hairs, for example, or the appearance of the root cap, or the emergence of lateral roots. On the other hand, we can look at a much larger scale at the architecture of the root system, or how the roots are arranged in space. From this we can infer how the root is exploring the soil, and therefore the plant's ability to capture nutrients. The shape of this root architecture is important to successful plant development. The relative scarcity of water on the surface of the earth, and local variations in nutrient availability mean the physical arrangement of the roots greatly affects a plant's ability to capture such resources, and thus can determine whether a plant will flourish or perish (Lynch, 1995). Dependent on the kind of investigation, it may be that the fine geometric details of the architecture are required, such as angles of emergence, or number of root branches. Alternatively, a cruder distribution gradient may be sufficient, which captures the amount of roots within certain depth bands, for example.

We may wish to examine combinations of multiple image scales in the same experiment; this is likely to be much more commonplace as technology evolves. Figure 14.1 shows some example images at extreme scales. From an image analysis perspective, clearly different scales produce data that must be processed differently. There is no one single approach that can be used for analysis across the scales, producing a consistent result. There are, however, techniques that can be used to segment different features across multiple scales. An example of this might be active contour segmentation methods (e.g., Kass et al., 1988), which could be used to segment the root tip in an organ-scale image, or a cell wall in a subcellular image (Sethuraman et al., 2012).

Because of their usually hidden nature, it can be more challenging to image root systems than it is to image their aboveground counterparts. Traditional methods to circumvent this problem and visualize roots for manual measurement, such as growing them on gel plates, can be adapted for automated imaging. New technology is also introducing new ways of acquiring root image data, which do not have a manual equivalent, or at least certainly not a nondestructive manual alternative. Examples of such technologies are micro-CT x-ray imaging (Mairhofer et al., 2012; Mooney et al., 2012) and MRI scanning, which can visualize roots in situ (Poorter et al., 2012). This allows roots to be seen growing in their natural habitat, soil, as opposed to the simplified, 2D world of the agar plate. Even with plate-based

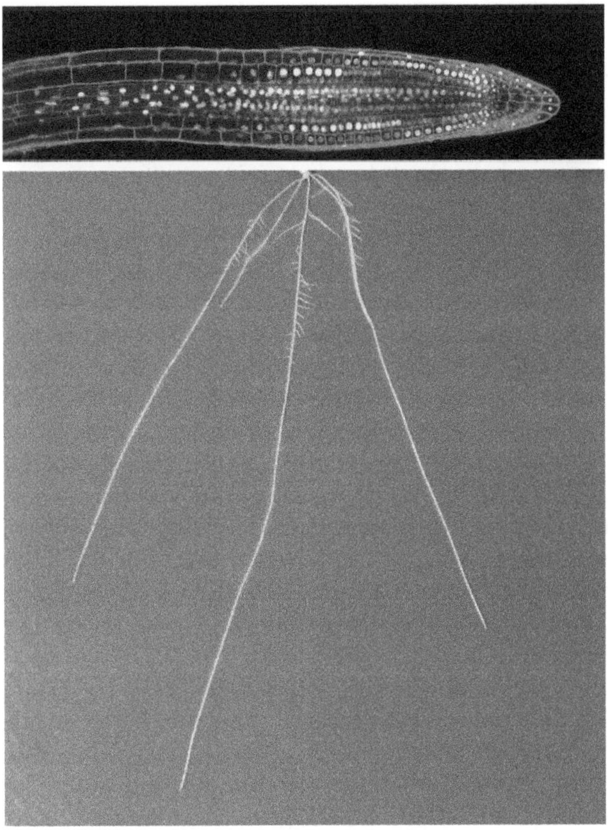

Figure 14.1 (**See color insert.**) Example images of roots at the cellular (top) and whole architecture (bottom) scales. The top image is a confocal image of an *Arabidopsis thaliana* root tip with fluorescent markers on cell walls and nuclei. The bottom image is a wheat root system growing on a filter paper system, imaged with a digital camera. (Images courtesy of, respectively, Dr. Antoine Larrieu and Mr. Jonathan Atkinson, University of Nottingham.)

imaging, though, techniques can be employed with digital imaging that are not possible with manual observation. Near-infrared lighting, cameras, and filters, for example, can be used to image plated roots in darkness during the night phase of day-night cycles (Wells et al., 2012). Some more details on these methods will be presented in the following section, which considers some approaches to image capture for studying roots.

14.2 Capturing root images

Analysis of root growth using image analysis techniques clearly starts with the image capture itself. Very different approaches can be used to capture the roots, and each approach produces a particular kind of image data. For example, traditional phenotyping of small seedlings such as *Arabidopsis* roots typically takes place by growing the seedlings on agar gel on small plates between 10 and 30 cm across. This takes place in growth rooms or chambers, which may have constant light or day-night cycles. The manual approach to quantifying root growth may have consisted of perhaps daily measurements, where each seedling on each plate was manually measured, or marks were made on the plate where the root tip had reached, with a view to later measurement of the marks. From this, daily growth rates could be inferred, as well as total root length. Automating this process is relatively straightforward. A digital camera can be set up in the growth room pointing at a plate, and set to automatically take photos at whatever frequency is required. This might be daily, as with the manual measures, or perhaps every hour or more dependent on the time resolution of the process under study. This approach can clearly generate a great number of images for a time sequence—one plate on a 1-week time series taking images every 30 minutes produces 336 images (~1 Gb), for example.

If more than one plate at a time is required in the imaging—perhaps due to the large number of plants that need to be phenotyped—then a robotic system may be appropriate. This typically involves either moving the camera, or the plates, such that images of multiple samples can be taken at each time point. There are advantages and disadvantages to both approaches. Moving the samples (e.g., Nagel et al., 2012) means imaging conditions (e.g., lighting) remain the same for each sample, as the camera is not moving, and allows for imaging of higher numbers of samples, but the movement itself may cause problems for growth studies from, for example, the associated vibrations. The alternative, moving the camera, requires either replicated lighting on a per-sample basis or moving the lighting source with the camera (e.g., Wang et al., 2009). A possible third alternative is to have one camera per sample, though this has obvious financial and sample number limitations.

Whatever the approach, a way of automating the image capture and storing the images on a computer is required in practice. This collection of images then represents a series of per-sample time-lapse sequences that can be stored ready for image analysis processing. One important technical aspect to note here is the quality at which the images are stored. As discussed in Chapter 1, different file types can degrade data quality in ways not immediately clear to the observer. For example, JPEG compression should not be used (or at least, not on a low-quality setting), as it discards much spatial color information that we as humans do not perceive well. For automated processing, such data loss can be detrimental to the success of the output, and so image format should be considered carefully. While it is possible to later compress data to reduce storage space, it is impossible to later increase the quality of a poorly captured dataset. Therefore, it is worth ensuring before image capture has taken place that a suitable data storage protocol is adhered to, as this cannot be corrected in retrospect.

Whether automated or not, lighting the roots on the plates to produce good images can be a challenge. Locating the light source is the first concern. Placing the light near the camera itself (for example, located in the position of a typical flash on a camera) can produce large reflections on the plate lids. Rear lighting (locating the light source behind the plates) needs to be diffused evenly across the back of the plate in order to not produce bright spots. Side lighting can highlight the 3D nature of the roots, highlighting texture and structure, but can also cause distracting shadows. The particular choice of lighting lies with the kind of processing that is going to be required.

A recurrent problem with imaging root system time-series sequences when growing on agar plates is the formation of condensation (Finer and Finer, 2007). Clearly this presents a problem for any image analysis that is required, as the roots are obscured by the condensation itself. One practical solution is to image the clear plates from behind, through the gel—though this does depend on how clear the particular growth medium is that is being used. We have found that altering the position of the plates in the growth room, effectively changing the airflow around them, can help reduce condensation. A cooling-warming cycle of air from the temperature regulator in the growth rooms can cause this; thus, repositioning in this case can help.

When working with gel media, more than just simple 2D images can be acquired by growing inside the gel rather than on its surface. It is possible to image roots in 3D by growing them inside agar in a clear cylinder. By rotating the cylinder, the root system can be easily imaged from multiple viewpoints—using either a camera-based system (Clark et al., 2011) or other technologies, such as a laser scanner (Fang et al., 2009). Some traits of growth, such as circumnutation—an effect whereby the root tip generates a spiral path as it grows—are more comprehensively captured using a 3D approach to data capture.

As an alternative to using a digital image to photograph plated roots, some approaches use a document scanner for the same effect. The advantage with this technique is that the very controlled lighting conditions and high resolution produce very high-quality, low-noise images. The downsides are that the approach is hard to automate and involves moving the sample to the scanner bed each time an image is required.

As the fundamental purpose behind all these approaches is to understand the process of plant growth, clearly it is sometimes necessary to accurately mimic day/night cycles by varying the lighting in the growth chambers. If this lighting is being used for image capture, there is an impact of this cycle for the image acquisition system. In addition, when imaging roots, it may be beneficial to keep the root system in darkness, with the upper section of the plant still in light, to mimic the natural environment more closely. Both of these problematic situations can be mitigated using near-infrared (IR) lighting, cameras, and filters. The sensors in most digital cameras are sensitive to a wide range of wavelengths, commonly limited to what we consider visible light, by having a filter installed. IR cameras do not contain this IR blocking filter. Therefore, they are sensitive to IR light. A visible light blocking filter (i.e., IR pass filter) can then be installed on the camera, meaning it is able to image IR and only IR light. In combination with a near-IR light source (approximately 800–900 nm), the camera will be largely blind to visible light but able to image the IR light that we cannot see. Such an IR illuminated setup in the growth room is insensitive to the ambient lighting in the room (or lack of it at night). In addition, visible light blocking/IR pass filters can be installed on plates at the root location, meaning most ambient visible light is excluded (as is the case underground in real life), yet IR light can pass through, and hence image the roots beyond (Wells et al., 2012). The resulting images are grayscale, and can essentially be treated as any other grayscale digital image for the purposes of processing. Note that the near-IR imaging described here is not the same as thermal IR imaging (which can be used to measure temperature); for a description of this technique, see Chapter 8.

Another common capture technique is using microscopy. Two common types of microscopy used for the capture of root images are brightfield microscopy and confocal laser scanning microscopy. Brightfield microscopy is one of the simplest forms of microscopy, where the sample is illuminated from behind on the stage. An example can be seen in the main panel in Figure 14.2. While it is hard to resolve cellular details, it clearly shows morphological features, and is useful to identify stained areas, for example. Images can be taken by attaching a digital camera. Much higher-resolution and higher-contrast details of select structures can be seen using confocal laser scanning microscopy (CLSM). Modern CLSM techniques allow us to image growing roots. The imaging itself requires using a fluorescent protein within the plant, or fluorescent dyes to stain structures, which are then made to fluoresce via the laser in the confocal microscope. The confocal term refers to the fact that it is possible to selectively image thin planes within a sample (known as optical sectioning); from this, a 3D dataset can be generated if required.

Acquiring time-series confocal microscope images of growing roots presents a number of challenges. First, due to the gravitropic nature of root growth, and depending on the particular growth mechanism that is under study and the length of sequence required, a vertical stage on the microscope may be required (Band et al., 2012). Additionally, for prolonged sequences on the microscope, a light source may be required

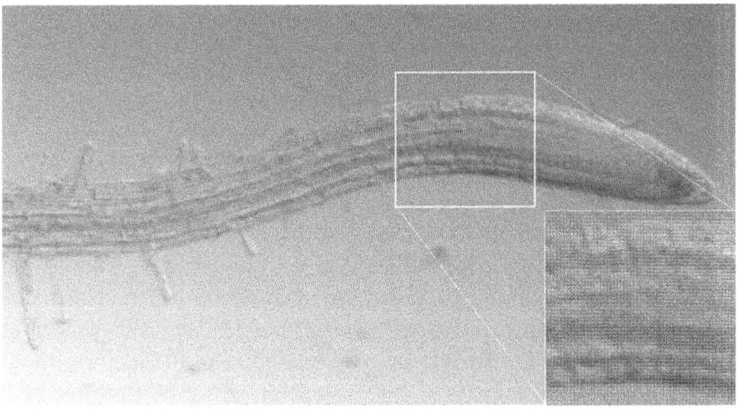

Figure 14.2 An example of an optic flow field generated from two time steps of a growth sequence captured on a light microscope. The image shows an *Arabidopsis thaliana* root. The inset panel shows the raw optic flow output calculated from the image data. (Images courtesy of Dr. Antoine Larrieu, University of Nottingham.)

nearby to allow the plant to continue growing. A further complication is the bleaching effect that overexposure to the confocal laser can have on the fluorescent protein (Pawley, 2006).

14.3 *Image analysis across modalities and scales*

As we have seen, there are several possible imaging modalities available when phenotyping root systems. Of course, different capture mechanisms mean different analysis approaches, because the types of images, and the underlying data, can vary widely. Clearly, a microscope-scale image of a root reveals qualitatively different features than a scanner-produced image, for example.

When it comes to this variation in scale, the images fall mainly into two categories: those in which complete root systems as a whole are visible, and those in which cellular details are visible. Perhaps a third category is appropriate for the kind of light microscopy techniques used to examine morphology, such as lateral root emergence, or image growth in roots over a matter of hours. In this third, organ-scale case, a visible texture on the surface of the roots is important, and some cellular features may be visible, such as some cell walls, but these individual features might not be considered as reliably present in all cases. These three categories of images, which fall across different scales, will be briefly discussed in the following three sections, and we point the reader to some existing examples in the literature for each case. This is then followed by two case studies working through cell-scale and whole-organism-scale image analysis.

14.3.1 Cellular-scale images

Typically, cell-scale images are captured with a CLSM microscope. Once acquired, there are a number of measures we have found to be commonly required from the images. One of these is growth rate, for which it may make sense to track the root tip—for the tip alone, this can be done using a light microscope as well as CLSM (Wells et al., 2012; Garcia et al., 2004). If more detail about the growth rate is required, we can track more than just the tip, considering individual cells and their growth as a network, for example (Sethuraman et al., 2012; Roberts et al., 2006), or by examining the motion of the nuclei within the cells (Sena et al., 2011).

It may be that the identification of *subcellular* details is necessary. This may include nuclei location (e.g., Esteves et al., 2012) or the more general colocalization of fluorescent markers representing different structures (for literature, see French et al., 2008). It is also possible to detect qualitative cell-scale growth events such as cell division (Marcuzzo et al., 2009). This is of course a rapidly developing area; as the quality of imaging technology improves year by year, so new image analysis techniques become available. For an example of the cell-scale approach to image analysis, see the CellSeT case study later in this chapter.

14.3.2 Surface texture- or organ-scale images

In some circumstances, while it may not be convenient or necessary to image individual cells, we may still wish to see which areas of the root are producing a growth response. This may be the case if it is necessary to investigate a gravitropic response, for example. Here, light microscope images can be used, but instead of resolving cellular details, it is the surface detail that we wish to image. This detail, a combination of surface texture and elements of the underlying cellular structure, can produce enough information to see how parts of the root are moving, in this case as a result of growth. These fine-scale details effectively produce a pattern across the surface of the root that can be tracked over local regions, by which we mean local areas of the pattern are unique enough in appearance that they can be followed over small steps in time as the root grows. This approach of following the movement of local regions of the root across the whole surface is referred to as *optic flow* (see Figure 14.2). The technique assumes lighting remains constant, and so any change in the intensity of pixels can be attributed to motion. It has formed the basis of several previously described tools to study root growth (van der Weele et al., 2003; Blanchfield and Chen, 2010). One form of optic flow measure, particle image velocimetry (PIV), has been used to measure both cell-scale root growth and displacement of the surrounding growth medium (Bengough et al., 2010). This is achieved by tracking the displacement of small patches of pixels across the image plane between consecutive frames. Care has to be taken when examining fine motion details in confocal images, due to the more complex image noise present, and the lack of spatial texture between fluorescent marker locations (Roberts et al., 2009). However, making and accounting for appropriate assumptions, it is certainly possible to perform this kind of growth analysis on confocal images of growing *Arabidopsis* roots (Wuyts et al., 2011). A more thorough examination of optic flow, and ways of using it to measure plant growth, can be seen in Barron et al. (1994) and Barron and Liptay (1994).

During a root growth image sequence, then, optic flow techniques can produce local growth rate information. In some root biology literature, this is referred to as relative elemental growth or elongation rates, and can be viewed as a color heat map depicting the magnitude and direction of growth over the surface of the root (e.g., Walter et al., 2002). This can be interpreted as a map of where expansion is occurring in the tissue.

14.3.3 Root architecture-scale images

Part of the understanding of root growth development processes requires a holistic view of the root system. We cannot understand the big picture of how roots organize themselves for nutrient collection, for example, by considering only the cellular scale. This means there is a need to image what the root is doing at the whole organism scale. This might mean focusing on an individual root or root tip, in the context of the rest of the root system, for imaging gravitropic or hydrotropic responses, for example, or studying the arrangement of the entire root system as a whole. This second perspective is often referred to as the root architecture. Here again, there are analysis approaches appropriate to both techniques, and both will be discussed here.

In the first instance, it may be necessary to image primary root growth, i.e., imaging the growth of one root per plant. There may be many seedlings on an agar plate, but the processing required can simply be repeated for each plant (as long as they are not overlapping each other); see Figure 14.3 for an example. What is required is to somehow segment the root from the background, usually so that an approximation of the midline of this root can then be found, and hence a length and curvature measures can be produced.

Several examples exist of software designed to look at root growth at this scale. RootTrace (French et al., 2009; Naeem et al., 2011) is designed to work with multiple roots on a plate over a time series and will measure growth in response to simple stimuli such as that due to gravity. A look at the underlying image analysis approaches of RootTrace is presented in more detail in the case study below. EZ-Rhizo (Armengaud, 2009) works with similar images and takes a semiautomated approach to the problem,

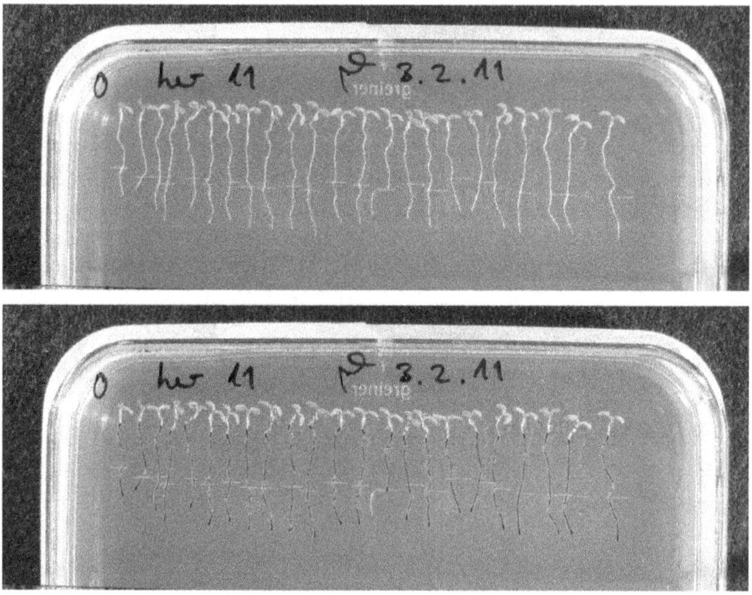

Figure 14.3 Example of plated *Arabidopsis thaliana* roots. Top: Images using a digital camera. Bottom: The centerlines have been found using the RootTrace software. Note the success of the trace despite various distractions, including reflections, and white lines marked manually on the plate by the experimentalist. Note Root 13's trace fails due to a violation of the linearity assumption. Curvature of the root is depicted as a heat map displayed along the centerline. (Images courtesy of Dr. Daniela Dietrich, University of Nottingham.)

as it guides the user through a series of stages for thresholding and skeletonizing root-like features on the plate. The segmentation is tidied up using noise reduction (such as a Gaussian blur), morphological operations (such as dilation), and finally giving the user the ability to retouch the image manually.

When considering how to capture the entire architecture from a root image (Figure 14.1), it is necessary to not only segment individual roots, but also identify and record how they are related to each other. In other words, we must capture the branching structure of the system. Here again, we need to segment the root from the background, and then trace not only the centerline of one root, but also the complete system, and maintain relationships between the individual root "branches." Because the entire architecture must be captured in detail, scanning is a common imaging technology for this scale. Dedicated analysis software such as that described in Arsenault et al. (1995) measures parameters such as total area and root length, as well as detecting branching points. An alternative growth method to using plates is to grow on filter paper. An example of this is the wheat plant in Figure 14.1. The root system geometry in these images can be well captured by considering the system as a branching graph structure. One way to use such a system is using the RootNav software tool (Pound et al., 2013), which provides a semiautomated approach to tracing the roots from scanned or digitally imaged roots, allowing the user to set start points on the seed, and end points on root tips. This a priori information is then used to fully trace the roots by first segmenting them using expectation-maximization (Dempster et al., 1977) and then tracing the paths using a shortest-path algorithm (Dijkstra, 1959). An interactive approach is then adopted, and the user can easily reclassify incorrect paths by dragging and dropping key points. The resulting structure is saved in a database, and can be queried to produce a variety of metrics (total root length, average branching angle, number of roots within depth bands, etc.).

Another image analysis tool that can extract architecture measures is SmartRoot (Lobet et al., 2011). Working with aeroponically grown roots imaged with a digital camera or a scanner, root tracing is triggered by a user click inside of a root in an area of interest (the whole system is not traced in this case, but rather, samples are taken from the whole system). The roots are described using a vector-based representation, which can include width of the roots.

However, the images are analyzed, when working with architecture descriptions there are many possible measurements that can be reported (de Dorlodot et al., 2007). Fine-detail descriptors can be reported, such as number of branches, angles of branches, number of lateral roots, root diameter, growth rate of individual roots, etc. Measures that capture the overall shape and distribution include the convex hull (created by drawing straight lines around the root system to fully enclose it), root type counts and root area counts at different soil depths, fractal-based descriptors of parts of the system, etc. What exactly is required will depend on the final application of the data, and especially whether the data are for phenotyping or mathematical model development.

In order to present some grounded examples of the underlying image analysis approaches, we now present two case studies of root image analysis spanning different scales, and describe the steps used to achieve the final results.

14.3.4 Whole-organ-scale case study: Using RootTrace to measure growing primary roots

Here, we consider one approach to the problem of extracting growth information from images of plated roots taken as part of a time series. Each plate can contain around 40 seedlings, and there could be several hundred images in the complete time series. A typical image from one time point can be seen in Figure 14.3.

The first approach to tackling this problem might be to try thresholding (see Chapter 1). That is, we try to differentiate root and background on each frame simply by choosing the roots that are brighter than a fixed value. Ideally, this will separate or segment each root from the background. However, this technique has severe limitations, for example, that if there is nonuniform illumination, the background may have a large gradient in intensity values, which may mean we cannot differentiate roots and background on parts of the image. Note that it may look like the image is uniformly illuminated, when in fact there can be a large intensity gradient across the image; using image processing software to measure an intensity profile across a region will reveal this type of artifact.

Instead of thresholding, in RootTrace (French et al., 2009) the approach taken is derived from tracking-based literature (Isard and Blake, 1998), but instead of following an object through a time series of images, we are following our root object through the image space. Tracking as a methodology within image analysis involves the location of a target in an image, and a prediction about where the target will be in the next frame. A variety of algorithms are available (and are still being developed) to make this predictive step. Here, a *particle filtering* (Isard and Blake, 1998) approach is taken, where a number of estimated locations (particles) are made from each frame, each one producing a new velocity for the target. At the next frame, a search is made for the target at each predicted location using a model of expected appearance. The best particle is selected for the location, and velocity and position updated accordingly. Here, as mentioned previously, instead of tracking an object through a sequence of frames, we are using a tracking approach to control a process of "walking" down the root as a sequence of steps.

In order to make the problem tractable, it is necessary to make some prior assumptions about the images (Pridmore et al., 2012). These assumptions represent the necessary properties of the data that mean the underlying algorithms will work correctly. For example, a simple thresholding technique may require the assumption that the roots are brighter than the background. It is hard to think of any image analysis algorithm not making assumptions about the image data, out of computational necessity. Here, we make the following general assumptions:

1. The roots are locally linear, and brighter than the background.
2. The position of the top of the root does not change between time points.

Point 2 leads to a secondary assumption that it is possible to define a point on the root from which we can start tracing the root. If a simplistic thresholding approach were being used, we might have to make more assumptions; e.g., the background pixels all fall below an intensity value of 50, and everything above 50 is a root (i.e., no clutter, reflections, etc., which might confuse a simpler, bottom-up approach). The hope here is that by applying a top-down model of what we are looking for, we can make the segmentation of the roots more robust.

In order to produce traces such as Figure 14.3, a combination of image analysis methods are used together. First, the image can be filtered to suppress noise; using a Gaussian filter is common, but median filtering sometimes performs better, dependent on the type of noise present in the photographs. The user is required to set a start point to begin the tracing of each root. Tracing continues down the image from this point, so the point must be inside the root. Assumption 2 means we can use this position to initialize tracking from frame to frame. The user is also asked to identify an area of background that is used for a background appearance model, implemented to help identify the root tip.

Tracking continues by stepping down the root using the described particle filtering approach (Isard and Blake, 1998), and an appearance model that differentiates the ideal root from the background. The tracking continues until the tip of the root is reached, identified using a statistical approach that identifies a change in the appearance at the location of the tracker. A hysteresis approach is implemented to allow the appearance to change over small periods (if the gray level of the root drops due to noise for a small period, for example), but if consistent background pixels are traversed, the algorithm will backtrack to the last identified root model-derived pixel, and halt tracing there. Once the tip is found, the complete particle set that has been generated is then used to produce an estimate of the centerline of the root. Low-probability particles are removed, and the resulting network traversed using Dijkstra's shortest-path algorithm (Dijkstra, 1959). This approach is repeated for each time frame, each centerline extracted, and measures calculated across the resulting dataset to produce growth rates, tip angles and curvatures, etc. More details can be found in related publications (French et al., 2009; Naeem et al., 2011).

The purpose of this case study is to show that even to solve a relatively simple root image analysis problem—separating a bright white primary root from a fairly uniform and controlled background in a digital image—a diverse collection of algorithms can be used together to good effect. In situations when simple thresholding would not work, these types of approaches can succeed. The take-home message here, then, is one not just limited to root analysis, but image analysis as a whole—if a simple approach does not work, do not discount image analysis as an approach. It may just be that an alternative method, or a combination of approaches used together, will yield the desired result.

14.3.5 Cell-scale case study: Using CellSeT to analyze cells in CLSM images of roots

Here we consider the problem of quantifying the properties of a complex network of plant cells in a confocal microscope image. Each image contains only a single region of plant, often a single root tip (see Figure 14.1 (top) for an example). However, the number of cells in each image can vary greatly, and the quality of the image is also highly variable, dependent on the fluorescent markers or dyes used, the particular tissue being imaged, and the depth of the tissue. Confocal images are capable of focusing on a thin "slice" through a plant root; however, as the root itself is seldom completely parallel to the imaging plane, some areas of the image will inevitably be seen in better clarity than others, and "ghosting" of cell walls partially visibly from upper and lower cell files causes distractions in the image. Therefore, the confocal image is challenging to analyze and there is no global approach that will guarantee segmenting all cell walls. We make the basic assumption that cell walls will appear brighter than the surrounding pixels, and that they are interconnected. We also assume that neighboring cells are extremely close together and, as such, treat walls between cells as a single entity for the purposes of detection and quantification. The tool we describe here to segment these cellular networks and allow further processing of them (to identify nuclei inside each cell, for example) is called CellSeT, and is described more fully in previous literature (Pound et al., 2012); we provide a summary of the underlying analysis steps below.

The first processing step is to apply image filters to reduce the image noise. Confocal laser scanning microscopes produce image noise characterized by a number of statistical distributions (Roberts et al., 2007). Some of that noise is similar to Gaussian noise in distribution. There are often some additional speckles of light where fluorescence has been unusually high, or low, similar to salt-and-pepper noise. Therefore, a sequence of noise reduction filters is used. A median filter can be used to remove this salt-and-pepper-like noise, followed by applying a Gaussian filter with a small standard deviation. The result is an image that is substantially less noisy, and a better candidate for segmentation.

As with RootTrace, an initial approach might be to threshold the image, to attempt to separate the cell wall sections from the background. However, given the varied intensities and thicknesses of each wall, combined with salt-and-pepper-like noise, this would not be a successful solution, and a more complex segmentation method is required. Watershed segmentation will efficiently extract closed regions, in this case cells, that are distinguished by region boundaries, in this case cell walls (Vincent and Soille, 1991). "Two-level" watershed proves a more robust solution, due to its ability to overcome any noise that remains after filtering. The watershed approach considers an image to be a landscape, with height equal to intensity. Water is seeded in basins, and lakes form as the landscape "floods." Where lakes meet (the watershed points), boundaries are formed. Two-level watersheds simply always flood over boundaries within a threshold of the current water level: this helps prevent oversegmentation, where too many regions are created, which is a common problem with the watershed algorithm.

Following watershed segmentation, there remain regions that are either under- or oversegmented. CellSeT allows the user to intervene at this point, drawing walls manually where segmentation has failed. It was decided that an intuitive and fast manual step was preferable to some cells being omitted by a fully automatic segmentation. However, manual segmentation introduces subjectivity into the analysis; it might be that two users may segment the same cells differently. To counteract this subjectivity, the output of the segmentation and additional user information is then processed using an automatic tracking approach, network snakes (Sethuraman et al., 2012). Active contours, often called snakes (Kass et al., 1988), are energy-minimizing splines that move over an image to "best fit" the features, in this case the high-intensity, generally linear sections defining the cell walls. In CellSeT, a snake is initialized over each wall, and each snake is joined to nearby neighbors by junction nodes. Each snake bends and moves over its local area, objectively refining its position to match the position of the cell wall in the image, while avoiding unnecessary deformations in its shape. The result is a similar, but refined network of cell walls, from which the cell geometry can be extracted. This process and the resulting network of snakes can be seen in Figure 14.4.

After wall refinement, each cell is detected by examining nearby walls for loop structures. Each cell is associated with a number of walls, and the area, circumference, and other properties of the cell can be extracted. Cells can be semantically tagged by the user, and the entire geometry exported for use in modeling packages. Further processing can then carried out on the subcellular features of the cell, such as looking at distribution of fluorescent proteins on an individual wall, or searching for nuclei inside each cell, and measuring their fluorescence.

The approach taken in CellSeT was one of semiautomation. As alluded to earlier, there is a trade-off in image analysis between truly high-throughput, automated methods that are able to process incredible numbers of images with practically no user input, and much slower, more manual approaches. However, in all but the simplest and best-controlled cases, the main trade-off is the accuracy of the results. How do we know the measures produced are actually the result of the correct image analysis output? Is the software segmenting all the cells, and doing so correctly?

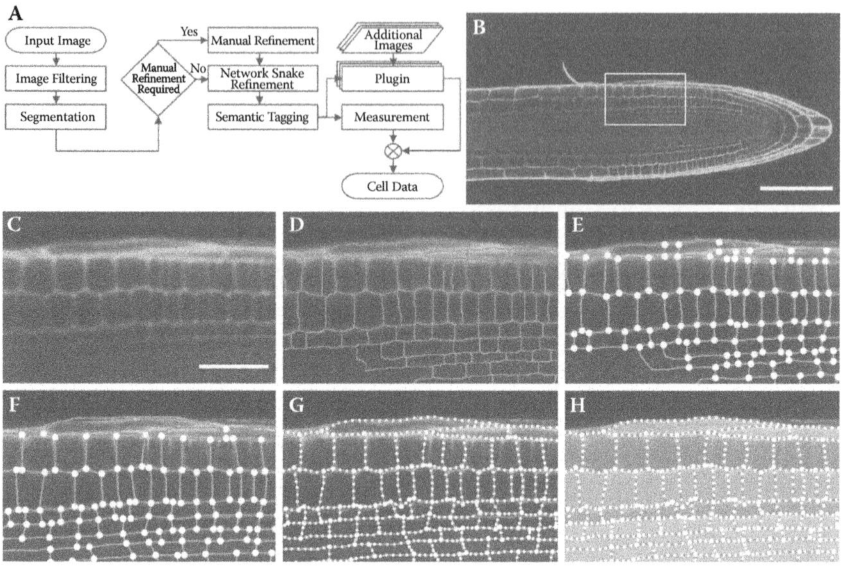

Figure 14.4 **(See color insert.)** The CellSeT processing pipeline. (a) The steps involved in confocal image analysis using CellSeT. (b) A typical input CLSM image. Cell walls have been stained with propidium iodide (here displayed green) to reveal their cellular organization. (c) A zoomed portion of the input image after filtering. (d) Two-level watershed segmentation is used to initially distinguish cells. In cases where cell walls are poorly defined, this may lead to under- or oversegmentation. (e) The refinement graph structure generated using the segmented image. (f) Manual refinement of the graph structure; oversegmentation in the epidermal cells has been removed, and undersegmentation in the stele cells has been corrected. (g) CellSeT uses a network snake algorithm to refine the results of manual segmentation. (h) Cells can be semantically tagged by the user to provide additional information during the output stage. Bar in (b) = 100 µm, bar in (c) = 30 µm. (From Pound et al., *Plant Cell*, 24, 1353–1361, 2012. Copyright © American Society of Plant Biologists.)

Without any user interaction, extreme confidence must be placed on the software, or alternatively, a statistical approach adopted whereby it is accepted that an amount of the results are known to be erroneous, and thus handled appropriately. Here we feel something similar to the 80/20 rule applies. In practice, we find that removing 80% of the work the biologist has to do to analyze the images still frees up much of his or her time, and typically removes the most laborious aspects of the analysis, and leaves him or her more than willing to do the 20% of work still required to produce meaningful, medium-throughput data. The more we can automate, the better, as it also removes subjectivity, but often it is valuable to keep the human expert in the loop.

14.4 Conclusion

In this chapter, we have presented to the reader an introduction to the analysis of root growth using image analysis. Compared to other plant organs, roots by their nature are normally hidden, but a number of ways of imaging them in lab conditions have been discussed. After introducing some of these capture methods, we then presented some typical image analysis methods appropriate to images produced by different imaging modalities, and across image scales. Examples were presented from the cellular scale, organ scale, and whole organism scale. We ended the chapter with two practical case studies, which presented the steps of processing required for the analysis of data in a cellular-scale example and a whole-organ example, drawn from the authors' own software tools for the image analysis of roots.

References

Armengaud, P., EZ-Rhizo software, *Plant Signal. Behav.*, 4, 139–141, 2009.

Arsenault, J.-L., Poulcur, S., Messier, C., and Guay., R., WinRHIZOTM™, a root-measuring system with a unique overlap correction method, *HortScience*, 30, 906–906, 1995.

Band, L.R., Wells, D.M., Larrieu, A., et al., Root gravitropism is regulated by a transient lateral auxin gradient controlled by a tipping-point mechanism, *Proc. Natl. Acad. Sci.*, 109, 4668–4673, 2012.

Barron, J.L., Fleet, D.J., and Beauchemin, S.S., Performance of optical flow techniques, *Int. J. Comput. Vis.*, 12, 43–77, 1994.

Barron, J.L., and Liptay, A., Optical flow to measure minute increments in plant growth, *Bioimaging*, 2, 57–61, 1994.

Bengough, A.G., Hans, J., Bransby, M.F., and Valentine, T.A., PIV as a method for quantifying root cell growth and particle displacement in confocal images, *Microsc. Res. Tech.*, 73, 27–36, 2010.

Blanchfield, P., and Chen, X.L., Improving the automatic measurement of plant root growth, in *MELECON 2010—15th IEEE Mediterranean Electrotechnical Conference*, 2010, pp. 363–367.

Clark, R.T., MacCurdy, R.B., Jung, J.K., et al., Three-dimensional root phenotyping with a novel imaging and software platform, *Plant Physiol.*, 156, 455–465, 2011.

de Dorlodot, S., Forster, B., Pagès, L., Price, A., Tuberosa, R., and Draye, X., Root system architecture: opportunities and constraints for genetic improvement of crops, *Trends Plant Sci.*, 12, 474–481, 2007.

Dempster, A.P., Laird, N.M., and Rubin, D.B., Maximum likelihood from incomplete data via the EM algorithm, *J. R. Stat. Soc. Ser. B Stat. Methodol.*, 39, 1–38, 1977.

Dijkstra, E.W., A note on two problems in connection with graphs, *Numer. Math.*, 1, 269–271, 1959.

Esteves, T., Quelhas, P., Mendonça, A.M., and Campilho, A., Gradient convergence filters and a phase congruency approach for in vivo cell nuclei detection, *Mach. Vis. Appl.*, 23, 623–638, 2012.

Fang, S., Yan, X., and Liao, H., 3D reconstruction and dynamic modeling of root architecture in situ and its application to crop phosphorus research, *Plant J.*, 60, 1096–1108, 2009.

Finer, J., and, Finer, J., A simple method for reducing moisture condensation on Petri dish lids, *Plant Cell Tiss. Org. Cult.*, 91, 299–304, 2007.

French, A.P., Mills, S., Swarup, R., Bennett, M.J., and Pridmore, T.P., Colocalization of fluorescent markers in confocal microscope images of plant cells, *Nat. Protocols*, 3, 619–628, 2008.

French, A., Ubeda-Tomás, S., Holman, T.J., Bennett, M.J., and Pridmore, T., High-throughput quantification of root growth using a novel image-analysis tool, *Plant Physiol.*, 150, 1784–1795, 2009.

Garcia, B., Campilho, A., Scheres, B., and Campilho, A., Automatic tracking of *Arabidopsis thaliana* root meristem in confocal microscopy, in *Image analysis and recognition*, ed. A. Campilho and M. Kamel, 166–174, Springer, Berlin, 2004.

Isard, M., and Blake, A., Condensation—conditional density propagation for visual tracking, *Int. J. Comput. Vis.*, 29, 5–28, 1998.

Kass, M., Witkin, A., and Terzopoulos, D., Snakes: active contour models, *Int. J. Comput. Vis.*, 1, 321–331, 1988.

Lobet, G., Pagès, L., and Draye, X., A novel image-analysis toolbox enabling quantitative analysis of root system architecture, *Plant Physiol.*, 157, 29–39, 2011.

Lynch, J., Root architecture and plant productivity, *Plant Physiol.*, 109, 7–13, 1995.

Mairhofer, S., Zappala, S., Tracy, S.R., Sturrock, C., Bennett, M., Mooney, S.J., and Pridmore, T., RooTrak: automated recovery of three-dimensional plant root architecture in soil from x-ray microcomputed tomography images using visual tracking, *Plant Physiol.*, 158, 561–569, 2012.

Marcuzzo, M., Guichard, T., Quelhas, P., Mendonca, A.M., and Campilho, A.C., Cell division detection on the *Arabidopsis thaliana* root, in *Pattern Recognition and Image Analysis—Lecture Notes in Computer Science 5524*, 2009, pp. 168–175.

Mooney, S.J., Pridmore, T.P., Helliwell, J., and Bennett, M.J., Developing x-ray computed tomography to non-invasively image 3-D root systems architecture in soil, *Plant Soil*, 352, 1–22, 2012.

Naeem, A., French, A.P., Wells, D.M., and Pridmore, T.P., High-throughput feature counting and measurement of roots, *Bioinformatics*, 27, 1337 –1338, 2011.

Nagel, K.A., Putz, A., Gilmer, F., et al., GROWSCREEN-Rhizo is a novel phenotyping robot enabling simultaneous measurements of root and shoot growth for plants grown in soil-filled rhizotrons, *Funct. Plant Biol.*, 39, 891–904, 2012.

Pawley, J., *Handbook of biological confocal microscopy*, 3rd ed., Springer Science + Business Media, 2006.

Poorter, H., Bühler, J., van Dusschoten, D., Climent, J., and Postma, J.A., Pot size matters: a meta-analysis of the effects of rooting volume on plant growth, *Funct. Plant Biol.*, 39, 839–850, 2012.

Pound, M.P., French, A.P, Atkinson, J., Wells, D.M., Bennett, M.J., and Pridmore, T.P., RootNav: navigating images of complex root architectures, *Plant Physiol.*, 162, 1802–1814, 2013.

Pound, M.P., French, A.P., Wells, D.M., Bennett, M.J., and Pridmore, T.P., CellSeT: novel software to extract and analyze structured networks of plant cells from confocal images, *Plant Cell*, 24, 1353–1361, 2012.

Pridmore, T.P., French, A.P., and Pound, M.P., What lies beneath: underlying assumptions in bioimage analysis, *Trends Plant Sci.*, 17, 688–692, 2012.

Roberts, T.J., McKenna, S.J., Du, C., Wuyts, N., Valentine, T.A., and Bengough, A.G., Estimating the motion of plant root cells from in vivo confocal laser scanning microscopy images, *Mach. Vis. Appl.*, 21, 921–939, 2009.

Roberts, T.J., McKenna, S.J., Hans, J., Valentine, T.A., and Bengough, A.G., Part-based multi-frame registration for estimation of the growth of cellular networks in plant roots, in *18th International Conference on Pattern Recognition, ICPR 2006*, 2006, vol. 2, pp. 378–381.

Roberts, T., McKenna, S., Wuyts, N., Valentine, T., and Bengough, A., Performance of low-level motion estimation methods for confocal microscopy of plant cells in vivo, in *IEEE Workshop on Motion and Video Computing*, WMVC '07, 2007, pp. 13–13.

Sena, G., Frentz, Z., Birnbaum, K.D., and Leibler, S., Quantitation of cellular dynamics in growing *Arabidopsis* roots with light sheet microscopy, *PLoS ONE*, 6, e21303, 2011.

Sethuraman, V., French, A., Wells, D., Kenobi, K., and Pridmore, T., Tissue-level segmentation and tracking of cells in growing plant roots, *Mach. Vis. Appl.*, 23, 639–658, 2012.

van der Weele, C.M., Jiang, H.S., Palaniappan, K.K., Ivanov, V.B., Palaniappan, K., and Baskin, T.I., A new algorithm for computational image analysis of deformable motion at high spatial and temporal resolution applied to root growth: roughly uniform elongation in the meristem and also, after an abrupt acceleration, in the elongation zone, *Plant Physiol.*, 132, 1138–1148, 2003.

Vincent, L., and Soille, P., Watersheds in digital spaces: an efficient algorithm based on immersion simulations, *IEEE Trans. Pattern Anal. Mach. Intell.*, 13, 583–598, 1991.

Walter, A., Spies, H., Terjung, S., Küsters, R., Kirchgessner, N., and Schurr, U., Spatio-temporal dynamics of expansion growth in roots: automatic quantification of diurnal course and temperature response by digital image sequence processing, *J. Exp. Bot.*, 53, 689–698, 2002.

Wang, L., Uilecan, I.V., Assadi, A.H., Kozmik, C.A., and Spalding, E.P., HYPOTrace: image analysis software for measuring hypocotyl growth and shape demonstrated on *Arabidopsis* seedlings undergoing photomorphogenesis, *Plant Physiol.*, 149, 1632–1637, 2009.

Wells, D.M., French, A.P., Naeem, A., Ishaq, O., Traini, R., Hijazi, H., Bennett, M.J., and Pridmore, T.P., Recovering the dynamics of root growth and development using novel image acquisition and analysis methods, *Philos. Trans. R. Soc. B Biol. Sci.*, 367, 1517–1524, 2012.

Wuyts, N., Bengough, A.G., Roberts, T.J., Chengin, D., Bransby, M.F., McKenna, S.J., and Valentéal, T.A., Automated motion estimation of root responses to sucrose in two *Arabidopsis thaliana* genotypes using confocal microscopy, *Planta*, 234, 769–784, 2011.

chapter fifteen

Advances in imaging methods on plant chromosomes

Toshiyuki Wako, Seiji Kato, Nobuko Ohmido, and Kiichi Fukui

Contents

15.1 Introduction

Chromosomes are formed only at the M-stage of a cell cycle, and even in the M-stage, they continuously and dynamically change morphology. Thus, the analysis of chromosome morphology has always been challenging. One method that is used to overcome this difficulty is to condense the chromosomes using a pretreatment method, such as cold water, or colchicine treatment, etc. After pretreatment, the chromosomes are condensed enough and the morphological changes are minimized as the shortest possible state. To date, most chromosome analyses of chromosome numbers within the complements, their morphology, and karyotyping used chromosomes in this state. Effective but simple parameters of chromosome length and ratio of the length of the long and short arms have been easily obtained without using imaging methods for chromosomes in this state. Nomenclature of the chromosomes, such as metacentric and submetacentric, and definition of the relative length (percent of the length of an individual chromosome in the total chromosomal length) and arm ratio (length of long arm/length of short arm) have been defined (Levan, 1964).

Fukui (1985, 1986) described an imaging method for the analysis of plant chromosomes. Fukui and Mukai (1988) identified nine pairs of *Atriplex rosea* chromosomes ($2n$ = 18) by focusing on the uneven chromosomal condensation that appears uniquely on individual chromosomes at the prometaphase stage. Previously, no method, other than the digital imaging method, could quantify the density parameter, making this the first report of the digital identification of plant chromosomes by uneven condensation at the plant prometaphase chromosome. Using this method, Fukui and Iijima (1991) objectively identified all 12 pairs of rice chromosomes based on the condensation pattern (CP) of each rice (*Oryza sativa* ssp. *japonca*) chromosome. The efficacy of a combination of digital imaging and CP (density profile at the axis of each chromatid) analysis in identifying individual chromosomes and in the development of quantitative chromosome maps has been demonstrated unequivocally in rapeseed (Kamisugi et al., 1998), spinach (Ito et al., 2000a), *Arabidopsis thaliana* (Ito et al., 2000b), *Lotus japonicus* (Ito et al., 2000c), and red clover (Kataoka et al., 2012).

Because of the development of imaging methods, fluorescence in situ hybridization (FISH) (Schwarzacher and Heslop-Harrison, 1991; Fukui et al., 1994; Ohmido et al., 2010) and three-dimensional (3D) analysis of chromosomes and nuclei (Cremer and Cremer 2001; Pecinka et al., 2004) are now commonly used techniques in the analyses of the detailed structure and function of chromosomes and nuclei. Imaging methods are indispensable tools when morphology, density, and color play important roles in the biological systems.

In this chapter, we review the history of chromosome imaging, focusing mainly on the systems and the software developed by our group.

Then, the applications of imaging methods for the analysis of banding patterns, CPs, and pachytene chromosomes are described. Recent applications of imaging methods in multicolor FISH and immune staining are also described.

15.2 History of chromosome imaging

15.2.1 Brief history of the chromosome image analyzing system (CHIAS)

Identification of individual chromosomes within a complement using morphological characteristics is often difficult, and usually only a specialist can accurately identify each chromosome using its visual characteristics. Image analysis, conversely, is expected to have the capability to analyze chromosomes and develop their maps quantitatively. However, the imaging methods and systems have been developed mainly for the analysis of human chromosomes from the earliest time of chromosome imaging because of the exigent requirements of the patients in the medical field. In plant chromosome studies, there are difficulties that arise because of the variations in chromosome numbers and sizes.

The project to develop a chromosome image analyzing system for plant chromosomes started in 1982 (Fukui, 1999). Although chromosome maps are indispensable for plant genetics and breeding because they provide the basic information, chromosome maps based on the complete identification of individual chromosomes were not available at that time, not even for the major crops and vegetables, such as barley, rice, and Brassicas. Plants possess different chromosome numbers with varying sizes, and ploidy levels. Such variation made it difficult to develop a versatile imaging system required for the development of plant chromosome maps. Thus, in developing the Chromosome Image Analyzing System (CHIAS), the following three prerequisites were considered:

1. The main targets will be plant chromosomes with the numbers from 4 to 120.
2. Routine works of image processing should be automated as much as possible.
3. Interactivity should be maintained to use the experience and skills of the researchers.

With the continued improvements in computer performance, the CHIAS, which was originally implemented on a mainframe computer, can now be run on an ordinary personal computer. As a result, the cost of image analysis decreased with each new version of CHIAS, and now it is within the budget of every researcher.

The first version of CHIAS was developed in 1985 (Fukui, 1986, Table 15.1), and in 1990, a personal computer version of CHIAS-mini was released, making chromosome imaging both affordable and accessible to researchers (Kamisugi and Fukui, 1990). The cost was still problematic because a specific image processing unit was required. In 1993, the CHIAS project focused on developing low-cost high-performance image processing, and CHIAS II was released (Nakayama and Fukui, 1997). However, with the rapid improvements in computing performance, image analysis required nothing more than a personal computer, and so immediately after the release of CHIAS II, CHIAS III was released (Kato et al., 1997; Kato and Fukui, 1998). CHIAS III was similar to CHIAS-mini and used a Macintosh personal computer and the publicly available NIH Image software (http://rsb.info.nih.gov/nih-image/). Apart from the computer, all that was required was a device for capturing images. The NIH Image software can construct quantitative idiograms that accurately depict chromosome morphology with degree of condensation along the chromosome. The early versions of CHIAS were dependent on the hardware, while CHIAS III is less reliant on the hardware, thus making it applicable for wider applications. As a result, CHIAS III is now recognized as one of the best standard systems to construct idiograms for many plant species, from diploid to tetraploid, with small or large chromosomes (Table 15.1), including *Saccharum spontaneum* (Ha et al., 1999), *Oryza sativa* ssp. *indica* (Apisitwanich et al., 2000), *Spinacia oleracea* (Ito et al., 2000a), *Arabidopsis thaliana* (Ito et al., 2000b), and *Pennisetum glaucum* (Goel et al., 2003). Furthermore, CHIAS III is a flexible system in which the procedures for analyses can be modified according to the situation and the materials with only a little programming knowledge. The system can be used not only to construct idiograms, but also for quantitative analysis of signal locations in FISH experiments (Ohmido et al., 1998). Measurement of the DNA content at a specific chromosome region is also possible (Akiyama et al., 2004). CHIAS III with an easy-to-use online manual is freely available (http://www2.kobe-u.ac.jp/~ohmido/cl/chiasIII/) and currently is being further improved and extended (Kato et al., 2003).

15.2.2 Current version of CHIAS (CHIAS IV)

Imaging methods are continuously being advanced, and as a result, the tools required to analyze them need to be updated regularly. First, to analyze multichanneled images, for example, RGB color images after multicolor FISH and multicolor fluorochrome staining, the image analysis system must be able to deal directly with multichanneled images. Second, because most digital images are now de facto standard, now it is usually necessary to present quantitative data using adequate image

Table 15.1 History of the Development of the Chromosome Image Analyzing System (CHIAS)

	CHIAS	CHIAS-mini	CHIAS II	CHIAS III			CHIAS IV
Released year	1985	1990	1997	1998	1998	2003	2009
CPU	8 bit	16 bit	16 bit	32 bit			64 bit or 32 bit
OS	CP/M	MS-DOS	MS-DOS	MacOs Classic			Windows, Mac OS X, LINUS (JAVA)
Image processor	○	○	○				
Analysis cases	CP (rice, *Atriplex*) C-band (*Crepis*) N-band (Barley)	CP	CP (*Brassica* spp.)	CP (sugarcane)	FISH (FP; *Arabidosis*)	Pachytene (rice, *Lotus japonicus*)	FISH (FP; red clover) Pachytene (rice)

analysis methods. Thus, hardware and software for image analysis systems that are lower in cost and easier to use than the earlier versions are required.

The newly released CHIAS IV runs as a plug-in of the ImageJ software (http://rsbweb.nih.gov/ij/) developed by the NIH (Abramhoff et al., 2004), which has become a powerful imaging tool for medical and biological researchers (Kato et al., 2009). ImageJ can handle directly 8-, 16-, and 32-bit gray images and also 24-, 48-, and 96-bit RGB color images. This feature means that almost all microscopic images, including multicolored FISH/immune stain images, can be analyzed. A new image parameter, the fluorescence profile (FP), was defined for fluorescence images that reflect uneven condensation along a chromosome. The FP is equivalent to the gray values of the CP after Giemsa staining of the fluorescence image. Because ImageJ is a JAVA application program, ImageJ and CHIAS IV can now be run on Windows and Linux and not on Mac OS X computers only, as was previously the case. As a result, almost all the hardware for image analysis can be used to run the system. CHIAS III requires two softwares to run, the public domain NIH image software and the commercial Microsoft Excel software. CHIAS IV needs only the public domain ImageJ software. Thus, the cost of image software is now negligible.

Compared to CHIAS III, image processing using CHIAS IV is significantly more automated. The new CHIAS IV system was developed to reduce the difficulties associated with the analyses of small plant chromosomes. As an example of how CHIAS IV works, a three-step analysis of the chromosome images of red clover obtained at the prometaphase is listed in Table 15.2. Step 1 consists of several subprocesses, such as extraction of chromosome regions using the counterstained image, automatic addition of chromosomal ID by area order, automatic sorting of chromosomes by ID, detection of the centromeric position, and measurement of CP or FP (Figure 15.1a). The simultaneous measurement of the FPs for each channel of the RGB image is performed. Fluorescence intensities of counterstaining and multicolor signals are measured, and the profile image of each chromatid is developed automatically. The profile image is referred to as a colorgram and represents each FP of the RGB chromosomal image.

Step 2 in CHIAS IV is significantly different from the method that was used in CHIAS III. In CHIAS IV, the data of the colorgram are saved in a TIFF format (Figure 15.1), and the operations that were performed in Microsoft Excel in CHIAS III have been integrated within ImageJ in CHIAS IV. In addition, each channel of RGB and the index label can be processed simultaneously and has a function similar to that of the pachytene chromosome analysis in CHIAS III.

In Step 3, condensed regions are digitally identified, and the construction of the idiogram or quantitative chromosome map is completed

Table 15.2 Three-Step Analysis Using Chromosome Image Analyzing System (CHIAS) IV of the chromosome images of red clover obtained at the Prometaphase

	Name of operation	Explanation of operation
1.1	Open image	Image for analysis is opened as an RGB stack.
1.2	Extract chromosomal regions	Chromosomal regions are extracted from the image after regional adjustment of the threshold.
1.3	Separate close or overlapping chromosomes	Chromosomes that are close to each other and for overlapping are separated manually.
1.4	Extract signal regions	FISH signals of the BAC clones are extracted. If the signal is scattered, this step of the process may be omitted.
1.5	Add ID	Chromosome numbers are automatically added to all the chromosomes by order of area.
1.6	Sort (and normalize)	Chromosomes are sorted into a region of interest (ROI) according to their name order. Then, the chromosomal orientations are corrected to a vertical alignment. The image is enlarged by a factor of 2 to prevent image quality deterioration during the rotation process. Simultaneously, the range of the gray level of the chromosome region is normalized to 1–255.
1.7	Rotate chromosome	The top (short arm) and bottom (long arm) of the chromosome are turned upside down.
1.8	Detect centromere	The chromosomal arms are separated at the centromeric region.
1.9	Index labeling	An index label layer is added to the 3 RGB layers. Labels are inserted for segments of the chromosome regions. This operation should be executed only when regional segmentation is necessary.
1.10	The measurement of CP or FP 2—create a standard profile	A line is drawn along the mid-rib of the chromatid, and the CP or the FP on the line is measured.
2.1	Average the profile (AVG tool)	The profile data of homologous chromosomes are averaged.
2.2	Align all colorgrams of the chromosomes (Align tool) 3—create an idiogram	All measured chromosome colorgrams are aligned to one image by using their centromere regions.

Continued

Table 15.2 (Continued) Three-Step Analysis Using Chromosome Image Analyzing System (CHIAS) IV of the Chromosome Images of Red Clover Obtained at the Prometaphase

	Name of operation	Explanation of operation
3.1	Create graygram	The averaged graygram of the counterstain layer obtained in step 2 is created by using the Otsu method for automatic thresholding.
3.2	Fix condensed regions	The first threshold in the gray level is determined. The gray areas (i.e., condensed regions within a chromosome) of the chromosomes are identified.
3.3	Fix highly condensed region	The most condensed area is determined. The black areas of the chromosomes are identified.

(Figure 15.1c). The condensed regions, slightly condensed regions, and noncondensed or dispersed regions are displayed as black, gray, and white bars, respectively. CHIAS IV employs automatic thresholding according to the Otsu method (Otsu, 1979), which is different from the method used in CHIAS III (Figure 15.1c). Idiogram is then created using the counterstain layer. Comparison of the FISH signals is easy, because the different FISH signals are saved in the individual layers of the colorgram.

15.3 *Quantification of large mitotic chromosomes with bands*

15.3.1 *N-banded barley chromosomes and their maps*

Some plant species show clear and distinct chromosome bands, such as the C- and N-bands that are used to identify individual chromosomes. In wheat, 21 pairs of chromosomes were identified completely by their C-bands without resorting to image analysis (Endo, 1986).

The early versions of CHIAS helped researchers to identify chromosomes and construct objective chromosome maps of the C-banded rye chromosomes ($2n = 14$) (Fukui, 1986) and N-banded barley chromosomes ($2n = 14$) (Fukui and Kakeda, 1990). Both these plants have large chromosomes about 10–15 μm in length. Digitally scanned photographs of the rye chromosomes were analyzed, whereas digitally captured images of the barley chromosomes were used. The digital camera systems enable researchers not to work in darkrooms, which was very time-consuming, thus making it possible to analyze 200 metaphase image spreads of the barley chromosomes within a short period of time (Fukui and Kakeda, 1990). CHIAS helped researchers to identify banded regions by implementing the following three criteria: (1) uneven lighting both by camera work and

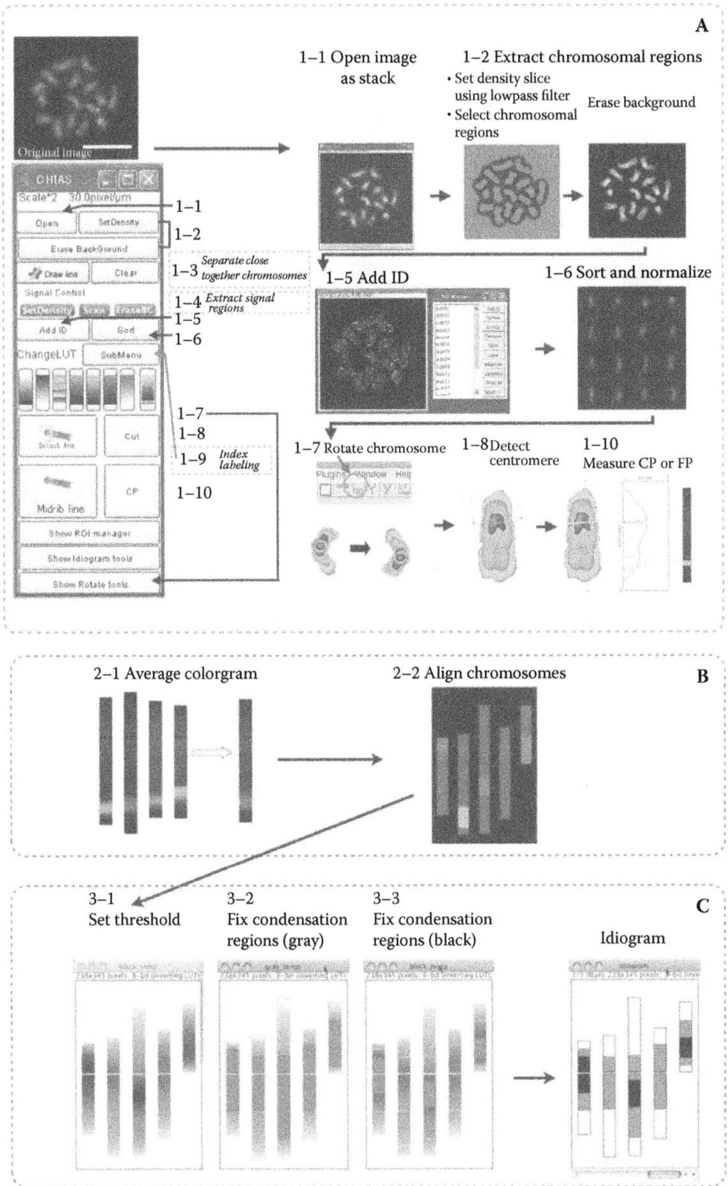

Figure 15.1 **(See color insert.)** Operation diagram of Chromosome Image Analyzing System (CHIAS) IV showing the three-step analysis of the chromosome images of the red clover obtained at the prometaphase. (a) Step 1: Measurement of condensation pattern (CP) or fluorescent pattern (FP) of the chromosomes. (b) Step 2: Create the standard profile from the CP or FP of each chromosome. (c) Step 3: Create the ideogram based on the standard profile.

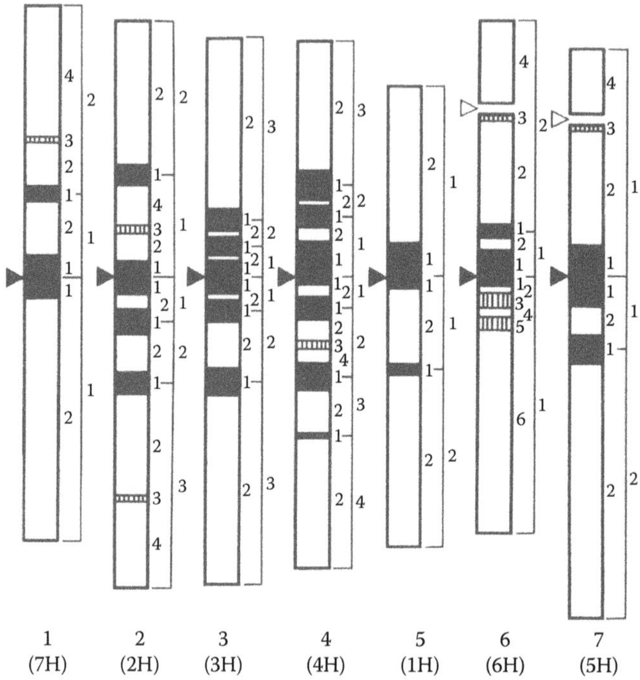

Figure 15.2 Quantitative idiogram of barley chromosomes based on N-banded chromosomes. ▷, nucleolar organizing region; ▶, centromere. Solid and hatched bands indicate stable and unstable bands, respectively. The values in and out of the parentheses under the idiogram indicate the barley chromosome numbers and the homoeologous groups, respectively. (Modified from Fukui and Kakeda, *Genome*, 33, 450–458, 1990. Copyright © 2008, Canadian Science Publishing or its licensors. Reproduced with permission.)

microscopy was compensated; (2) image contrast was enhanced by a normalization process; and (3) the light and shade regions were enhanced clearly using a pseudocolored image. The objective map of the barley chromosomes was developed using these tools (Figure 15.2) (Fukui and Kakeda, 1990).

The digital analysis of chromosomes allowed researchers to obtain objective chromosome maps within a shorter period of time and with less effort than previously. Image analysis methods make a strict distinction between the short and long arms, even when both arms are almost the same length, as is the case for barley chromosome 7H (chromosome 1 according to the previous nomenclature). This karyotyping method was later used by other researchers to construct chromosome maps for other plant species with different banding patterns, for example, the C-banded chromosomes of *Liliaceae* (Miyamoto et al., 1991).

15.3.2 Simulation of human vision in identifying and quantifying C-bands

The banded regions on chromosomes are sometimes difficult to analyze because some bands do not always appear in the same manner, and variations between individual researchers are unavoidable. Thus, a standardized method for identifying the banded regions was developed using the C-banded *Crepis* chromosomes ($2n = 6$, 4–7 μm in length) as a model (Fukui and Kamisugi, 1995).

A simple method that used a single threshold to discriminate the banded regions usually failed. When a bright threshold was used to identify the faint bands effectively, it failed to separate the dark bands (Figure 15.3a and b, line *a*), and when a dark threshold was used to separate the dark bands adequately, it failed to identify the faint bands (Figure 15.3b, lines *b* and *c*). The point is that the human visual sense can easily distinguish the dark and faint bands. Applying an unsharp mask filter, which consists of a low-pass filter with an adequate matrix size (Figure 15.3c and d), and image subtraction improved the image brightness and darkness (Figure 15.3e and f). In the processed image of a C-banded *Crepis* chromosome, a single threshold could successfully identify the dark and faint bands simultaneously from the background; the bands accurately corresponded to the bands that were identified by the human vision (Figure 15.3e and f). By using the processed images, an objective map of the banded *Crepis* chromosomes was obtained (Fukui and Kamisugi, 1995). The unsharp mask filter has also been applied for the identification of chromocenters of various sizes in the interphase nucleus (Wako and Fukui, 2003).

15.4 Identification of small mitotic chromosomes and development of their maps

15.4.1 Condensation pattern (CP) on rice prometaphase chromosomes

An alternative method to objectively karyotype plant species with small size chromosomes, for example, *Atriplex rosea* and *Oryza sativa* (Fukui and Mukai, 1988; Fukui and Iijima, 1991), was also developed. Rice chromosomes were objectively identified 80 years after the accurate determination of the chromosome number in 1910 (Kuwada, 1910; Fukui and Iijima, 1991; Iijima et al., 1991). The method included two major improvements: (1) prometaphase chromosomes were prepared instead of ordinal chromosome spreads of the metaphase, and (2) averaged density profiles along the two axes of the chromatids axes, referred to as CP, were measured digitally.

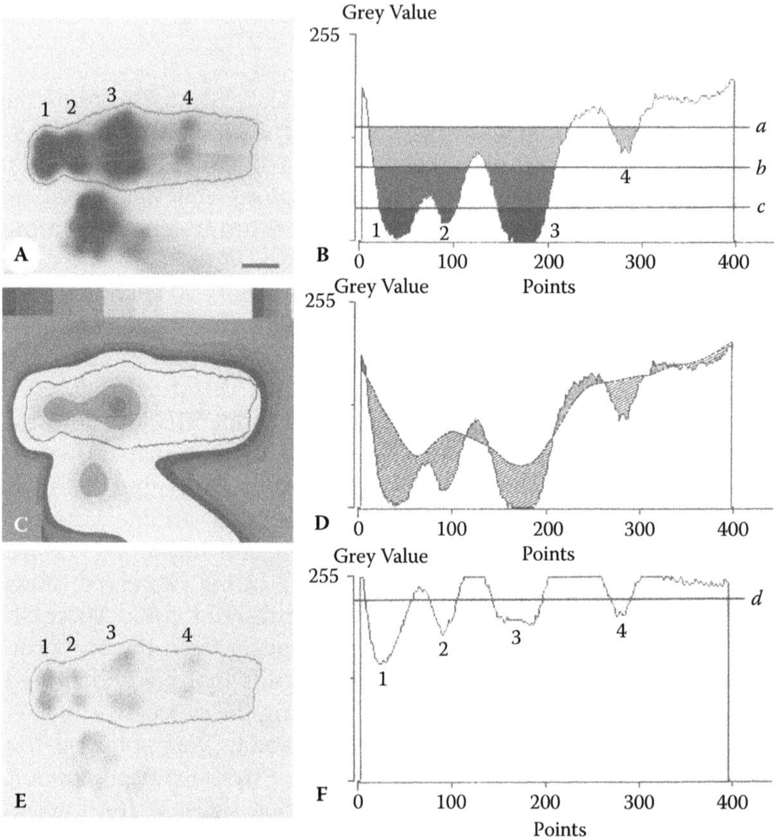

Figure 15.3 **(See color insert.)** Identifying C-banded *Crepis* chromosomes using simulated human vision. (a) Digital image of a C-banded *Crepis* chromosome (2C). The chromosome contour line and the chromatid mid-rib line were superimposed with an overlay line. The human eye can easily recognize four bands. Bar = 1 μm. (b) Density profile of the gray values along the mid-rib of the chromatid. The *x*- and *y*-axes of the graph represent the number of pixels and the gray values, respectively. High values correspond to light gray values. Lines *a*, *b*, and *c* indicate gray values of 140, 90, and 40, respectively. Along lines *a* and *b*, bands 1 and 2 are combined as a single large band, and along lines *b* and *c*, band 4 is not recognized as a band. (c) Pseudocolor representation of the low-pass filtered image. (d) Difference between the gray values of the original image and the values for the low-pass filtered image. Two representative density profiles at the mid-rib of the chromatid were used in the calculation. (e) Standardized gray image generated by subtracting the low-pass filtered image from the original image. (f) Density profile at the mid-rib of the standardized image. In this processed image, all four bands were successfully picked up and separated along line *d*. (Modified from Fukui and Kamisugi, *Chromosome Res.*, 3, 79–86, 1995. With kind permission from Springer Science and Business Media.)

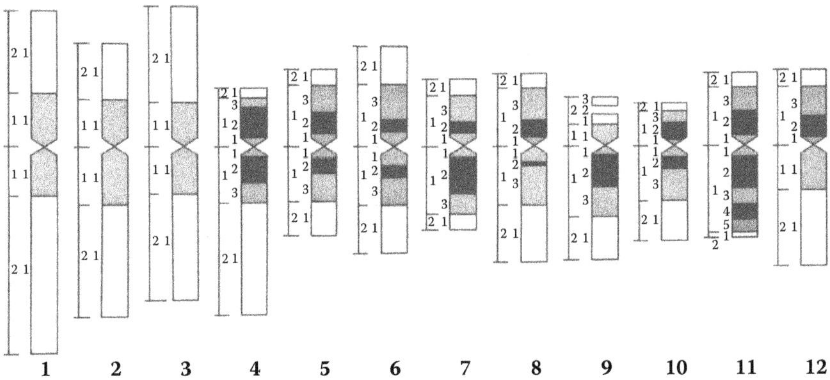

Figure 15.4 Quantitative ideogram of somatic rice chromosomes. (Modified from Fukui and Iijima, *Theor. Appl. Genet.*, 81, 589–596, 1991. With kind permission from Springer Science and Business Media.)

Prometaphase chromosomes are often longer than metaphase chromosomes, and each individual chromosome has a unique CP, which is effective for chromosome identification, especially for plant species that have small metaphase chromosomes. Preparation methods were improved to obtain good prometaphase spreads. The pretreatment step that was developed by Blackslee and Avery in 1937 and used as the standard procedure for chromosome preparation over the years was omitted to obtain intact prometaphase chromosomes. An enzymatic cocktail of cellulase and pectolyase was used to prepare the prometaphase chromosome spreads, and an air-drying method, instead of the more common flame-drying method, was adopted to minimize damage to the chromosomes. As a result, the enzymatic maceration/air-drying (EMA) method was developed (Iijima and Fukui, 1991).

Prometaphase rice chromosome spreads without any overlapping were photographed under two different conditions: normal exposure to obtain the exact chromosome shapes, especially for the more weakly stained chromosome ends, and underexposure to obtain detailed information about the staining of the chromatids. Then, the CP on each chromatid was measured for the complete identification of the rice chromosomes (Figure 15.4) (Fukui and Iijima, 1991; Fukui 1996). Digital data revealed 11 critical morphology characteristics for the systematic identification of rice chromosomes.

15.4.2 *Efficacy evaluation of chromosome identification based on CP*

The efficacy of the automatic identification of small plant chromosomes was evaluated using rice. Various image parameters obtained from the CPs and numerical parameters such as chromosome lengths and arm ratios were used for the fully automatic identification of 332 rice chromosomes using three different methods (Kamisugi et al., 1993). Linear discrimination, which is the most common method for identification, scored only 84.4% of the chromosomes correctly, even though 31 parameters were used. The discrimination chart and the minimum distance classifier based on standardized Euclidean distance correctly scored 91.1 and 92.2%, respectively. Scores were calculated based on the chromosome identification performed by experienced researchers as the control or as the correct answer. The minimum distance method produced the fewest errors and also provided the second and third candidates so that any errors could be corrected manually. Thus, the method was estimated to be the most effective identification method for fully automated identification. These results indicated that CP is a highly reliable image parameter for chromosome identification.

15.5 *Simultaneous analysis using CPs and fluorescence in situ hybridization (FISH) signals*

15.5.1 *Simultaneous analysis of CPs and location of FISH signals in legumes*

Fabaceae is the second most important family of crop plants after Poaceae for humans, because they provide sources of food, feed for livestock, and raw materials such as oil and protein for industries (Graham and Vance, 2003). Among the legumes, the subfamily Papilionoideae contains the majority of pulse crops such as pea (*Pisum sativum*, $2n = 14$, 5000 Mb), alfalfa (*Medicago sativa*, $2n = 16$, 1600 Mb), and soybean (*Glycine max*, $2n = 40$, 1100 Mb), and *Lotus japonicus* and *Medicago truncatula* are used as model legume species (Young et al., 2005).

The density profiles at the centerline of both chromatids (mid-rib line) of the prometaphase chromosomes of the two model legumes helped to establish the first chromosome maps of several plant species with small chromosomes (Yanagisawa et al., 1991; Ito et al., 2000, 200a, 200b, 200c; Sato et al., 2005; Ohmido et al., 2007; Kataoka et al., 2012). In this section, cytogenetic and molecular chromosome studies of *Lotus* and red clover are described. *L. japonicus* is characterized by a small genome ($2n = 2x = 12$; genome size per haploid, 472 Mb) and a relatively short life cycle (2–3 months), and because it is an autogamous diploid plant, it is easy to manipulate genetically (for example, transformation) (Jiang and Gresshoff, 1997; Udvardi et al., 2005). When the identification of individual mitotic prometaphase chromosomes of *L. japonicus* based on CPs became feasible, the chromosome maps were developed (Ito et al., 2000c; Hayashi et al., 2001; Pedrosa et al., 2002).

Prometaphase chromosomes and the corresponding idiograms for two *Lotus* accessions, Miyakojima and Gifu, are shown in Figure 15.5. The idiograms are based on imaging analysis by CHIAS III of three Miyakojima and seven Gifu prometaphase plates selected for their similar total chromosome lengths, which suggested that they were at a similar stage in the cell cycle. Both accessions have six chromosome pairs

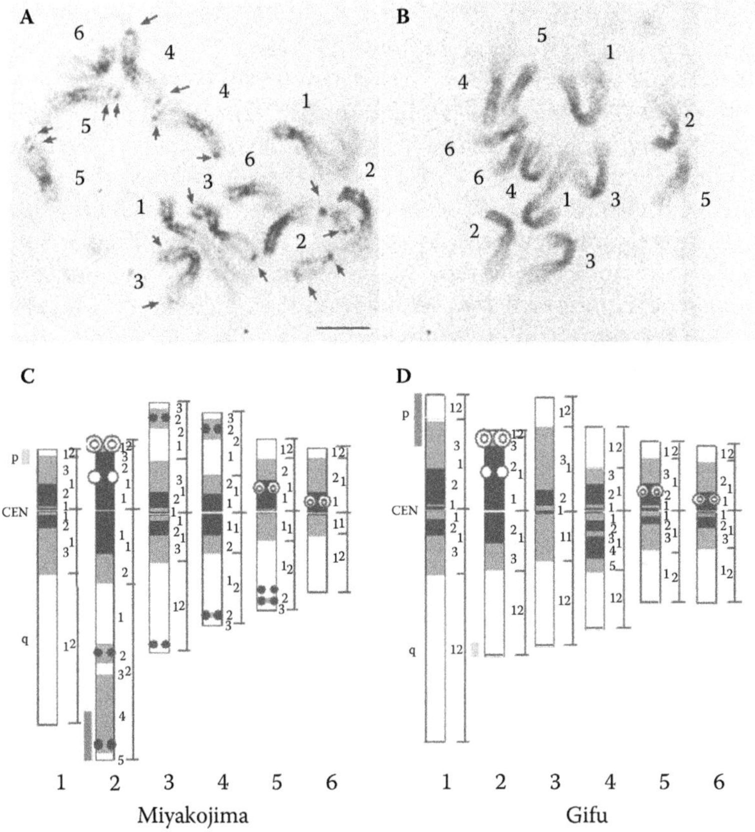

Figure 15.5 Cytological analysis of *Lotus japonicus* (Regel) Larsen, accessions Miyakojima and Gifu, according to Ito et al. (2000c). (a) Mitotic prometaphase chromosomes of accession Miyakojima stained with Giemsa solution. Arrows indicate constitutive heterochromatin. Bar = 5 μm. (b) Mitotic prometaphase chromosomes of accession Gifu stained with Giemsa solution. (c, d) Chromosome maps of accessions Miyakojima (c) and Gifu (d). The positions of the 5S rDNA (single circles), 45S rDNA (double circles), and faint, unstable, and small condensed (FUSC) regions (solid circles) are presented as small black circles. Light and dark blue bars indicate the translocation regions between Miyakojima and Gifu. (Reproduced from Ito et al., *J. Plant Res.*, 113, 435–442, 2000c. With kind permission from Springer Science and Business Media.)

with distinct CPs after Giemsa staining. Faint, unstable, and small condensed (FUSC) regions were observed at some chromosome termini of Miyakojima (Figure 15.5a, arrows) but not of Gifu (Figure 15.5b). The nuclear DNA content of both accessions measured by flow cytometry was estimated to be 472.1 ± 3.98 Mb for Miyakojima and 442.8 ± 2.32 Mb for Gifu. The FUSC regions on the chromosomes of Miyakojima might have contributed to its larger genome size. The 45S rDNA loci were on chromosomes 2, 5, and 6 (Ito et al., 2000c), and the 5S rDNA locus was mapped on the short arm of chromosome 2 (Figure 15.5c and d).

The two accessions differ particularly with respect to the size and morphology of chromosomes 1 and 2. In Gifu, chromosome 1 is submetacentric and the largest chromosome, while in Miyakojima, the homologous chromosome 1 is acrocentric and the second largest chromosome. In Miyakojima, a terminal nucleolus organizing region (NOR) is located on the largest, submetacentric chromosome 2, while in Gifu, the homologous chromosome 2 is smaller. TAC (transformation-competent artificial chromosome) clones, which were assigned to linkage group 1 on the distal part of the short arm of chromosome 1 of Gifu, were localized at the distal region of the long arm of chromosome 2 of Miyakojima by FISH (Hayashi et al., 2001). Thus, a translocation between the short arm of chromosome 1 of Miyakojima and the long arm of chromosome 2 in Gifu was determined.

15.5.2 Development of a quantitative cytological map for red clover

Red clover has a small genome ($2n = 2x = 14$, 440 Mb) and is an allodiploid. Taylor and Chen (1988) reported that the length of metaphase chromosomes ranged from 1.9 to 2.9 μm; however, seven chromosomes could not be identified by the simple length parameter. Later, karyotyping was carried out based on microscopic observation of the prometaphase chromosomes stained with DAPI (4′,6-diamidino-2-phenylindole), and the lengths of the prometaphase chromosomes were found to range from 5.1 to 7.4 μm. CPs that have also proven to be useful in chromosome identification were clearly observed. The 26S rDNA loci were detected as the most intense signals on chromosome 1 and as weak signals on the short arm of the interstitial regions of chromosome 7 (Figure 15.6) (Kataoka et al., 2012). A 5S rDNA locus was detected in the proximal region on the short arm of chromosome 1 adjacent to a 26S rDNA locus, and two minor 5S rDNA loci were detected on the short arm of chromosome 2.

Fourteen chromosome-specific BAC (bacterial artificial chromosome) clones were detected close to both ends of each linkage group (LG) by FISH. They were mapped as LG1–7, on the cytological map of red clover

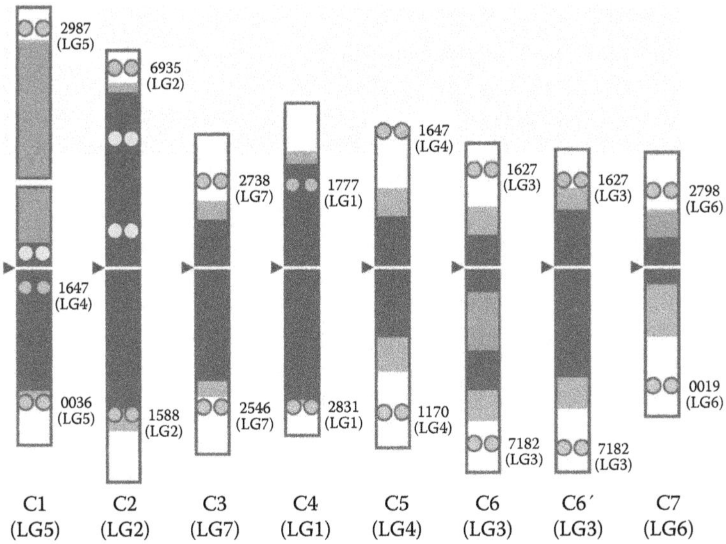

Figure 15.6 **(See color insert.)** Chromosome map of red clover. The loci of then BAC clones that harbor linkage group-specific microsatellite markers are indicated by green circles; 26S rDNA loci are indicted by red boxes; and 5S rDNA loci are indicated by orange circles. Arrowheads indicate the positions of the centromeres. Reproduced from Kataoka et al., *Cytogen. Genome Res.*, 137, 60–69, 2012; with kind permission from Karger Publishers.

HR (Figure 15.6) (Kataoka et al., 2012). It was found that the BAC clones for LG1, LG2, LG3, LG4, LG5, LG6, and LG7 hybridized exclusively to chromosomes 4, 2, 6, 5, 1, 7, and 3, respectively, and all 14 signals, one for each of the BAC clones, were detected on each of the seven chromosomes. The chromosomes were identified based on their CPs and the presence of the specific signals for the BAC clones. Moreover, the positions of the respective BAC markers were consistent with the positions of the corresponding linkage groups, which ensured a one-to-one relationship between each of the seven linkage groups and the seven chromosomes. This study is the first report of a red clover chromosome map constructed by a combined cytogenetic and molecular approach. The integration of physical, genetic, and quantitative cytological chromosome maps of red clover provides a valuable platform for the different genetic data.

15.6 Quantification of plant meiotic chromosomes

Pachytene chromosomes have a much higher resolution than somatic chromosomes because of their elongated structure in meiosis. Moreover, linearly arranged chromomeres, which serve as landmarks for chromosome maps, are observed. Pachytene chromosomes are especially useful in S-type chromosome analyses and have been studied for many years in plants like rice, tomato, and *Arabidopsis* that have small chromosomes.

Development of trisomic series in tomato (Rick and Barton, 1954) and rice (Iwata et al., 1970; Khush et al., 1984), detection of chromosome aberrations in tomato (Havekes et al., 1994), and a detailed characterization of *A. thaliana* (Fransz et al., 2000) were all achieved using pachytene chromosomes.

Construction of quantitative pachytene chromosomes has, however, been difficult because of their tangled nature, which is caused by their elongated structures. Thus, an imaging method for the construction of quantitative pachytene chromosome maps was developed using rice chromosome 9 because it has a NOR that stains weakly with DAPI but strongly with propidium iodide (PI). Chromosome 9 can thus be identified easily from among the 12 rice chromosomes, making it a good model chromosome for the development of a pachytene chromosome map after DAPI and PI double staining (Figure 15.7a). The averaged FPs along the mid-rib of pachytene chromosome 9 after DAPI and PI double staining are shown in Figure 15.7b. The FPs after PI (FP_{PI}) and DAPI (FP_{DAPI}) staining were similar through all the chromosomes except in the NOR. The highest FP_{PI} value was detected in the NOR, while the FP_{DAPI} had a lower value than FP_{PI} in the same region. To measure the lengths and FPs of the chromosomes, step 1 of the improved CHIAS III macro-program (Kato et al., 1997) was used with Object-Image (version 2.06, http://simon.bio.uva.nl/Object-Image/object-image.html), which is an extended version of NIH Image for the Mac OS.

The image analysis protocol in CHAIS III is almost the same as in CHIAS IV. A characteristic feature of the pachytene chromosome map is that it was developed based on the relative gray value for each chromomere. Figure 15.8 depicts the three major steps that were used to develop the pachytene chromosome map for rice (Kato et al., 2003). By using filtered intermediate images, as shown in Figure 15.8, the most condensed region within each chromomere was determined. The same method that was developed to determine the C-band-positive regions in *Crepis* (Fukui and Kamisugi, 1995) was applied. By integrating the chromomeric and condensed regions within the chromomeric regions, a quantitative pachytene chromosome map was developed, as shown in Figure 15.9 (Kato et al., 2003).

It is worth noting that the three maps of rice chromosome 9, namely, the somatic prometaphase chromosome map based on the CP, the pachytene chromosome map, and the linkage map based on the BAC molecular markers, are all different from each other, as shown in Figure 15.9. Interestingly, the short arm and the satellite regions are both negligibly depicted in the linkage map, indicating that recombination is almost completely suppressed at this region, and the relative lengths of the short arms differ between the somatic and pachytene maps.

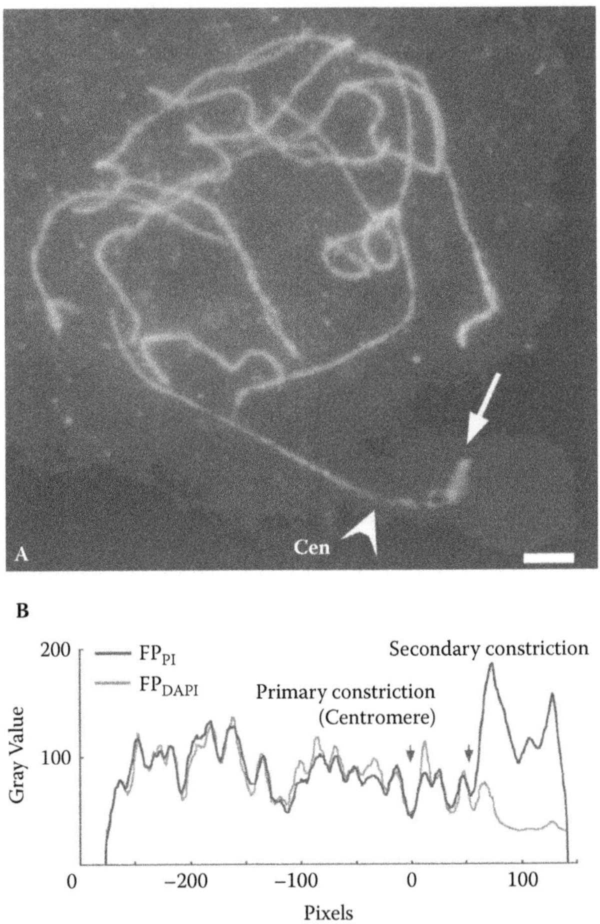

Figure 15.7 Rice chromosome 9 visualized by double staining with PI and DAPI. (a) Pachytene chromosome complement with double staining. Arrow indicates the NOR where DAPI stained weakly and PI strongly. "Cen" indicates the centromeric region. Bar = 5 μm. (b) Averaged fluorescence profiles (FPs) along the mid-rib of pachytene chromosome 9. FP$_{PI}$, the FPs after PI staining (light gray curve); FP$_{DAPI}$, the FPs after DAPI staining (dark gray curve). (Modified from Kato et al., *Genes Genet. Syst.*, 78, 155–161, 2003. With kind permission from the Genetics Society of Japan.)

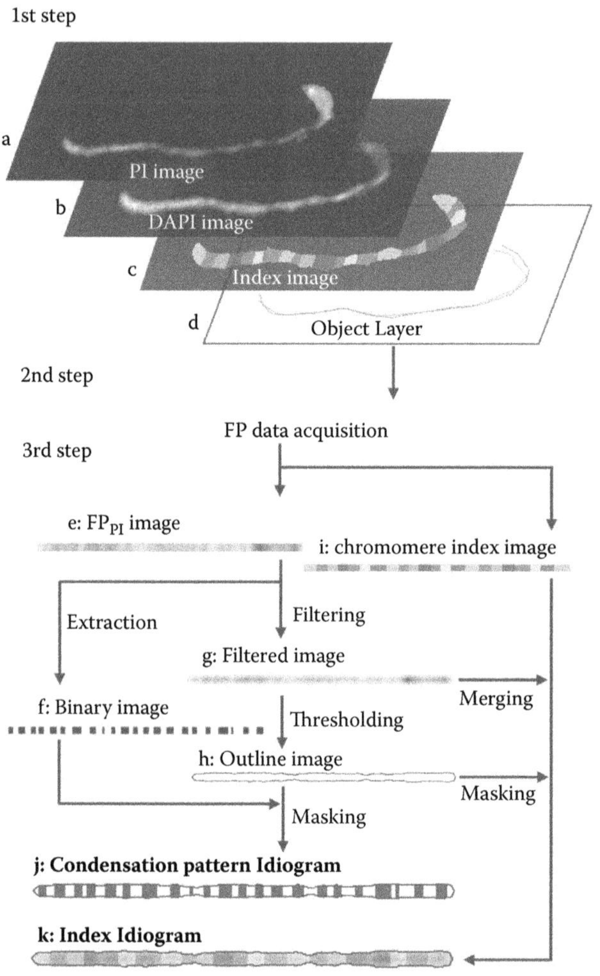

Figure 15.8 Representative steps for the development of a pachytene chromosome map. Starting with two images of chromosome 9 stained with PI and DAPI, two idiograms showing heavily condensed regions within each chromomere (CP ideogram) and each chromomere (index ideogram) were generated through three major steps. (Reproduced from Kato et al., *Genes Genet. Syst.*, 78, 155–161, 2003. With kind permission from the Genetics Society of Japan.)

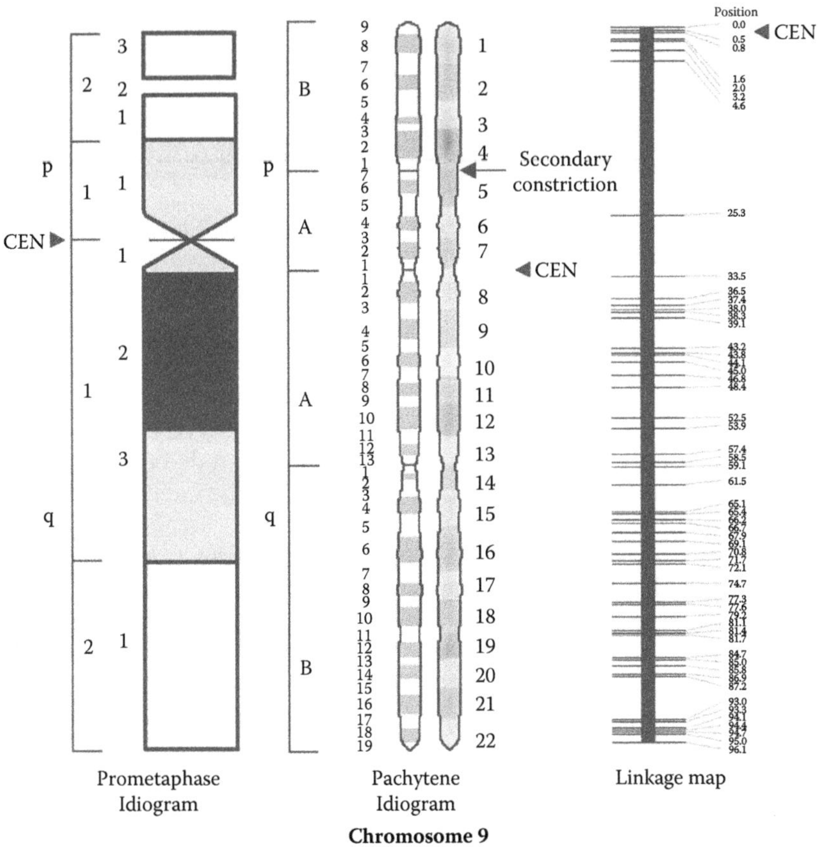

Figure 15.9 Comparison of three maps of rice chromosome 9. From left to right, ideogram for the somatic prometaphase stage, idiogram for the pachytene chromosomes (CP ideogram, index ideogram), and linkage map. (Reproduced from Kato et al., *Genes Genet. Syst.*, 78, 155–161, 2003. With kind permission from the Genetics Society of Japan.)

15.7 Analysis of three-dimensional image information

Chromosomes change their positions and morphology three-dimensionally. Thus, 3D observation of chromosome dynamics as they interact with target DNAs or proteins is required. Three-dimensional observation was achieved using a confocal microscopy with a motor-controlled stage and confocal optics. Then, either image processing software that is installed in the microscope system or ImageJ is used to reconstruct a 3D image view from a sequential two-dimensional (2D) image set. The reconstructed 3D image provides rotated and sliced images that can be

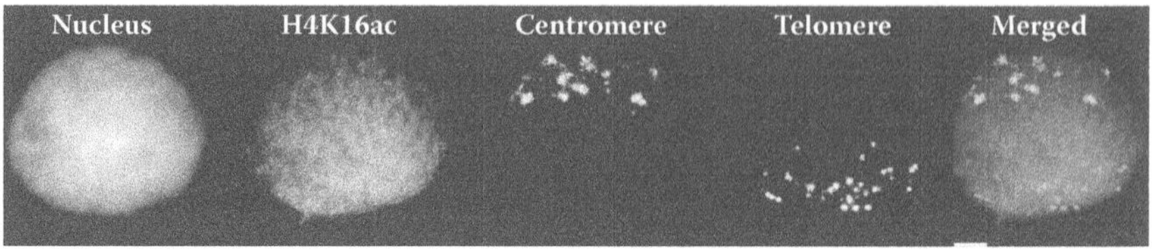

Figure 15.10 **(See color insert.)** Immuno-FISH images of barley nucleus. Each panel is a stacked image from optical slices obtained with a confocal microscope. The image was reconstructed using the 3D viewer plug-in of ImageJ. Each 2D image was processed using a median filter, normalization, and thresholding. Reproduced from Wako et al., *Plant Mol. Biol.*, 49, 645–653, 2002; with kind permission from Springer Science and Business Media.

viewed from directions that are impossible to observe directly. Thus, localization and morphology of the fluorescent signals of DNA probes and targeted proteins on the chromosomes and in the nuclei can be visualized clearly.

A reconstructed immuno-FISH image of the barley nucleus with the simultaneous detection of histone H4 acetylated at Lys16, the centromere, and the telomere is shown in Figure 15.10 (Wako et al., 2002). The image has four channels, including the nucleus counterstained with DAPI. The 3D image clearly shows the typical Rabl orientation, in which the centromeres are clustered in a narrow area near the nuclear surface, and that the acetylated region includes telomeres but not centromeres. The H4 acetylation levels in the mitotic cells of barley were measured based on fluorescent intensity through M-phase, and the dynamics of acetylation through the cell cycle was revealed. Image analysis of the 3D images revealed that the different acetylation sites showed stage-specific patterns during the cell cycle (Wako et al., 2002, 2003, 2005).

Chromosomes display dynamic behavior during mitosis by condensation and decondensation, and rapid movement within the cellular space for separation of sister chromatids. Three-dimensional live cell imaging is a powerful tool for investigating chromosome dynamics because changes of chromosomal and related protein localizations in a single living cell can be visualized. Chromosomes are often visualized in living cells using either live DNA staining with fluorochromes (e.g., Hoechst 33258) or histones with fluorescent protein tags (e.g., GFP (green fluorescent protein)-tagged H3). Targeted proteins are visualized by transforming the fluorescent protein-tagged target protein.

Live chromosome images are often acquired as 2D images with high temporal resolution or as a 3D image set with low temporal resolution using a confocal microscope and a deconvolution system. The deconvolution system is an alternative 3D observation method that uses mathematical

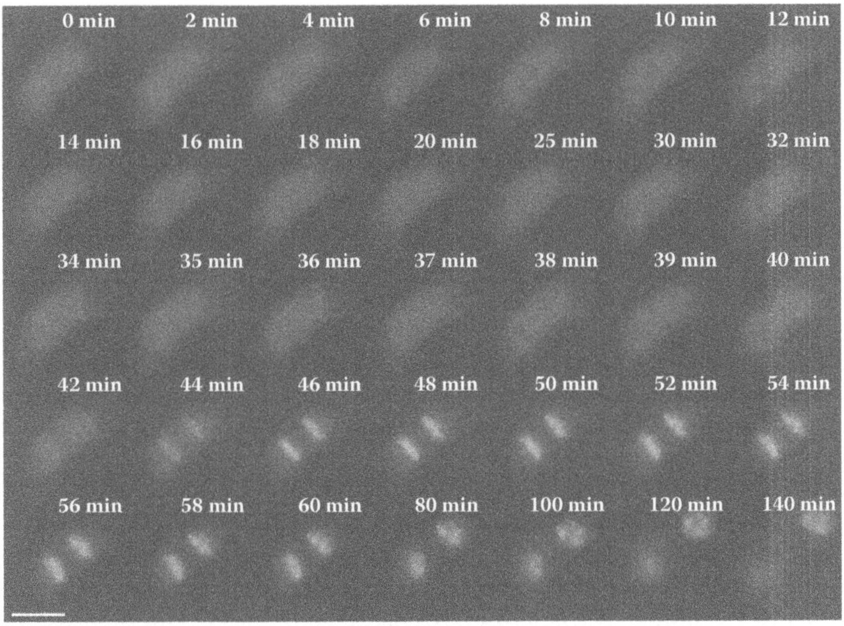

Figure 15.11 **(See color insert.)** Time-lapse image of tobacco chromosomes (blue) during mitosis transformed with GFP-fused AtLHP1 (green). Bar = 30 μm.

image processing and the point spread function of the microscope. The system reconstructs clear 3D images without blur from the outside of the focal planes generated by the optical system. The confocal microscope, conversely, eliminates the blur optically by setting a confocal pinhole.

The chromosomal protein *Arabidopsis* Like Heterochromatin Protein 1 (AtLHP1) binds to histone H3 methylated at Lys 27 and is involved in epigenetic gene suppression (Turck et al., 2007; Zhang et al., 2007). AtLHP1 is a homolog of the HP1 family proteins that characterize heterochromatins (Eissenberg and Elgin, 2000). A time-lapse image set of GFP-tagged AtLHP1 in tobacco mitotic cells revealed the dynamics of AtLHP1 in the chromosomes (Figure 15.11) (Matsunaga et al., 2006). Live imaging and 3D analysis of GFP-LHP1 in the tobacco mitotic cells were performed by the deconvolution method using the DeltaVision version 2.00 software (Applied Precision, Issaquah, WA). The fluorescent foci of AtLHP1 overlapped at the DAPI foci on the chromosomes in the mitotic interphase, disappeared at the prophase to anaphase (Figure 15.11, 0–42 min) (Matsunaga et al., 2006), and recovered again at the telophase (Figure 15.11, 44 min). These results indicate that localization of AtLHP1 on chromosomes depends on the mitotic stage, and are similar to the results obtained for human and mouse HP1 (Sugimoto et al., 2001).

15.8 Conclusion

Imaging methods enable quantitative chromosome analysis. The plant imaging system CHIAS has developed objective chromosome maps in crops, vegetables, and other plant species. The key technologies for plant image analysis are as follows: (1) omitting the pretreatment step for sample preparation, (2) enzymatic maceration and air-drying (EMA) method, and (3) digitally analyzed CPs or FPs that are newly introduced parameters by the CHIAS imaging method. Image analysis of FISH signals obtained with rDNA or other probes is useful for complete chromosome identification when the CP/FP method is insufficient. Image processing with an unsharp mask filter that simulates human vision helps in the identification of densely stained regions such as C- and N-bands of metaphase chromosome, chromocenters in interphase nuclei, and chromomeres of pachytene chromosomes.

Three-dimensional microscopy with higher resolution is required for chromosome analysis because chromosomes are dynamic and 3D in configuration and structure. The confocal microscopy and the deconvolution system have often been used for 3D observation. Newly developed super-resolution fluorescence microscopy such as stimulated emission depletion (STED) and structured illumination microscopy (SIM) is capable of producing better resolutions than those obtained in confocal microscopy (Schermelleh et al., 2010). These new microscopic techniques will provide more detailed information on chromosome structure and localization within the cell.

A combination of FISH with chromosome-specific probes and 3D microscopy revealed positions of individual chromosomes in a nucleus in yeast, mammals, and *Arabidopsis* (Cremer and Cremer, 2010; Pecinka et al., 2004). Each chromosome occupied discrete territories in the cell nucleus that have been referred to as chromosome territories (Cremer and Cremer, 2001; Tanabe et al., 2002). The development of chromosome arm-specific probes and the chromosomal region-specific probes will make it possible to detect the chromosome positions in the interphase nucleus with higher resolution.

Because chromosomes are composed of various proteins such as histone, condensin, and topoisomerase, visualization of the chromosomal proteins is indispensable for the analysis of chromosome structure and function. Indirect immunofluorescence methods using specific antibodies could detect chromosomal proteins or the posttranscriptional modification of proteins. Further, using fluorescence proteins (GFP, mCherry) or tag proteins (Flag) fused to specific gene products will allow protein dynamics in the living cell. Recently, specific chromosomal domains can be visualized in the living cells of *Arabidopsis* (Kato and Lam, 2001; Esch et al., 2003; Pecinka et al., 2005) by introducing fluorescence-tagged repressor protein and its binding DNA sequences to the genome (Straight et al., 1996; Robinett et al., 1996). This technology could be applied to visualize specific chromosome domains in living cells of various species to reveal the chromosome dynamics.

Image analysis has undergone remarkable progress in the field of chromosome science. Future chromosome image analysis will include the integration of sequence data, linkage maps, and quantitative chromosome maps, as has been tried in rice and some legumes.

References

Abramhoff, M. D., Magelhaes, P. J., and Ram, S. J., Image processing with ImageJ, *Biophotonics Int.*, 11, 36–42, 2004.

Akiyama, Y., Conner, J. A., Goel, S., Morishige, D. T., Mullet, J. E., Hanna, W. W., and Ozias-Akins, P., High-resolution physical mapping in *Pennisetum squamulatum* reveals extensive chromosomal heteromorphism of the genomic region associated with apomixes, *Plant Physiol.*, 134, 1733–1741, 2004.

Apisitwanich, S., Shishido, R., Akiyama, Y., and Fukui, K., Quantitative chromosome map of a representative indica rice, *Euphytica*, 116, 161–166, 2000.

Cremer, T., and Cremer, C., Chromosome territories, nuclear architecture and gene regulation in mammalian cells, *Nat. Rev. Genet.*, 2, 292–301, 2001.

Cremer, T., and Cremer, M., Chromosome territories. *Cold Spring Harb. Perspect. Biol.*, 2, a003889, 2010.

Eissenberg, J. C., and Elgin, S. C., The HP1 protein family: getting a grip on chromatin, *Curr. Opin. Genet. Dev.*, 10, 204–210, 2000.

Endo, T. R., Complete identification of common wheat chromosomes by means of the C-banding technique, *Jpn. J. Genet.*, 61, 89–93, 1986.

Esch, J. J., Chen, M., Sanders, M., Hillestad, M., Ndkium, S., Idelkope, B., Neizer, J., and Marks, M. D., A contradictory GLABRA3 allele helps define gene interactions controlling trichome development in *Arabidopsis*, *Development*, 130, 5885–5894, 2003.

Fransz, P., Armstrong, S., de Jong, J. H., Parnell, L. D., Van Drunen, C., Dean, C., Zabel, P., Bisseling, T., and Jones, G. H., Integrated cytogenetic map of the short arm of chromosome 4 of *Arabidopsis thaliana*: structural organization of a heterochromatic knob and pericentromeric heterochromatin, *Cell*, 100, 367–376, 2000.

Fukui, K., Identification of plant chromosomes by image analysis method, *Cell* (Tokyo), 17, 145–149, 1985 (in Japanese).

Fukui, K., Standardization of karyotyping plant chromosomes by a newly developed chromosome image analyzing system (CHIAS), *Theor. Appl. Genet.*, 72, 27–32, 1986.

Fukui, K., Recent advances in rice chromosome research, in *Rice Genetics III*, ed. G. S. Khush, International Rice Research Institute, Manila, Philippines, 1996, pp. 117–130.

Fukui, K., Advances in image analysis of plant genomes and chromosomes, *Breeding Res.*, 1, 157–164, 1999 (in Japanese with English summary).

Fukui, K., and Iijima, K., Somatic chromosome map of rice by imaging methods, *Theor. Appl. Genet.*, 81, 589–596, 1991.

Fukui, K., and Kakeda, K., Quantitative karyotyping of barley chromosomes by image analysis methods, *Genome*, 33, 450–458, 1990.

Fukui, K., and Kamisugi, Y., Mapping of C-banded *Crepis* chromosomes by imaging methods, *Chromosome Res.*, 3, 79–86, 1995.

Fukui, K., and Mukai, Y., Condensation pattern as a new image parameter for identification of small chromosomes in plants, *Jpn. J. Genet.*, 63, 359–366, 1988.

Fukui, K., Ohmido, N., and Khush, G. S., Variability in rDNA loci in the genus *Oryza* detected through fluorescence in-situ hybridization, *Theor. Appl. Genet.*, 87, 893–899, 1994.

Goel, S., Chen, Z., Conner, J. A., Akiyama, Y., Hanna, W. W., and Ozias-Akins, P., Delineation by fluorescence in situ hybridization of a single hemizygous chromosomal region associated with aposporous embryo sac formation in *Pennisetum squamulatum* and *Cenchrus ciliaris*, *Genetics*, 163, 1069–1082, 2003.

Graham, P. H., and Vance, C. P., Legumes: importance and constraints to greater use, *Plant Physiol.*, 131, 872–877, 2003.

Ha, S., Moore, P. H., Heinz, D., Kato, S., Ohmido, N., and Fukui, K., Quantitative chromosome map of the polyploid *Saccharum spontaneum* by multicolor fluorescence in situ hybridization and imaging methods, *Plant Mol. Biol.*, 39, 1165–1173, 1999.

Havekes, F. W., de Jong, J. H., Heyting, C., and Ramanna, M. S., Synapsis and chiasma formation in four meiotic mutants of tomato (*Lycopersicon esculentum*), *Chromosome Res.*, 2, 315–325, 1994.

Hayashi, M., Miyahara, A., Sato, S., Kato, T., Yoshikawa, M., Taketa, M., Hayashi, M., Pedrosa, A., Onda, R., Imaizumi-Anraku, H., Bachmair, A., Sandal, N., Stougaard, J., Murooka, Y., Tabata, S., Kawasaki, S., Kawaguchi, M., and Harada, K., Construction of a genetic linkage map of the model legume *Lotus japonicus* using an intraspecific F_2 population, *DNA Res.*, 8, 301–310, 2001.

Iijima K., and Fukui, K., Investigation of the conditions for the image analysis of plant chromosomes, *Bull. Natl. Inst. Agrobiol. Resour.*, 6, 1–58, 1991.

Iijima, K., Kakeda, K., and Fukui, K., Identification and characterization of somatic rice chromosomes by imaging methods, *Theor. Appl. Genet.*, 81, 597–605, 1991.

Iwata, N., Omura, T., and Nakahara, M., Studies on the trisomics in rice plants (*Oryza sativa* L.). I. Morphological classification of trisomics, *Jpn. J. Breed.*, 20, 230–236, 1970.

Ito, M., Ohmido, N., Akiyama, Y., Fukui, K., and Koba, T., Characterization of spinach chromosomes by condensation patterns and physical mapping of 5S and 45S rDNAs by FISH, *J. Am. Soc. Hort. Sci.*, 125, 59–62, 2000a.

Ito, M., Ohmido, N., Akiyama, Y., and Fukui, K., Quantitative chromosome map of *Arabidosis thaliana* L. by imaging methods, *Cytologia*, 65, 325–331, 2000b.

Ito, M., Miyamoto, J., Mori, Y., Fujimoto, S., Uchiumi, T., Abe, M., Suzuki, A., Tabata, S., and Fukui, K., Genome and chromosome dimensions of *Lotus japonicas*, *J. Plant Res.*, 113, 435–442, 2000c.

Jiang, Q., and Gresshoff, P. M., Classical and molecular genetics of the model legume *Lotus japonicus*, *Mol. Plant Microbe Interact.*, 10, 59–68, 1997.

Kamisugi, Y., and Fukui, K., Automatic karyotyping of plant chromosomes by imaging techniques, *Biotechniques*, 8, 290–295, 1990.

Kamisugi, Y., Furuya, N., Iijima, K., and Fukui, K., Computer-aided automatic identification of rice chromosomes by image parameters, *Chromosome Res.*, 1, 189–196, 1993.

Kamisugi, Y., Nakayama, S., O'Neil, C. M., Mathias, R. J., Trick, M., and Fukui, K., Visualization of the *Brassica* self-incompatibility S-locus on identified oilseed rape chromosomes, *Plant Mol. Biol.*, 38, 1081–1087, 1998.

Kataoka, R., Hara, M., Kato, S., Isobe, S., Sato, S., Tabata, S., and Ohmido, N., Integration of linkage and chromosome maps of red clover (*Trifolium pratense* L.), *Cytogen. Genome Res.*, 137, 60–69, 2012.

Kato, N., and Lam, E., Detection of chromosomes tagged with green fluorescent protein in live *Arabidopsis thaliana* plants, *Genome Biol.*, 2, research 0045, 2001.

Kato, S., and Fukui, K., Condensation pattern (CP) analysis of plant chromosomes by an improved chromosome image analysing system, CHIAS III, *Chromosome Res.*, 6, 473–479, 1998.

Kato, S., Hirose, T., Akiyama, Y., O'Neill, C. M., and Fukui, K., Manual on the Chromosome Image Analyzing System III, CHIAS III, *Res. Rep. Hokuriku Natl. Agr. Exp. Stn.*, 36, 1–76, 1997.

Kato, S., Ohmido, N., and Fukui, K., Development of a quantitative pachytene chromosome map in *Oryza sativa* by imaging methods, *Genes Genet. Syst.*, 78, 155–161, 2003.

Kato, S., Ohmido, N., Hara, M., Kataoka, R., and Fukui, K., Image analysis of small plant chromosomes by using an improved system, CHIAS IV, *Chromosome Sci.*, 12, 43–50, 2009.

Khush, G. S., Singh, R. J., Sur, S. C., and Librojo, A. L., Primary trisomics of rice: origin, morphology, cytology and use in linkage mapping, *Genetics*, 107, 141–163, 1984.

Kuwada, Y., A cytological study of *Oryza sativa* L., *Bot. Mag. (Toyko)*, 24, 267–281, 1910.

Levan, A., Nomenclature for centrometric position on chromosomes, *Hereditas*, 52, 201–220, 1964.

Matsunaga, S., Ohmido, N., and Fukui, K., Chromosome dynamics in tobacco BY-2 cultured cells, in *Biotechnology in agriculture and forestry, tobacco BY-2 cells: from cellular dynamics to omics 58*, ed. T. Nagata, K. Matsuoka, and D. Inze, 51–63, Springer-Verlag, Berlin, 2006.

Miyamoto, J., Kurita, S., and Fukui, K., Image analysis of C-banding patterns in two herbs: *Paris tetraphylla* A. Gray and *Paris verticillata* M. v Bieb. (*Liliaceae*), *Jpn. J. Genet.*, 66, 335–345, 1991.

Nakayama, S., and Fukui, K., Quantitative chromosome mapping of small plant chromosomes by improved imaging on CHIAS II, *Genes Genet. Syst.*, 72, 35–40, 1997.

Ohmido, N., Akiyama, Y., and Fukui, K., Physical mapping of unique nucleotide sequences on identified rice chromosomes, *Plant Mol. Biol.*, 38, 1043–1052, 1998.

Ohmido, N., Fukui K., and Kinoshita, T., Recent advances in rice genome and chromosome structure research by fluorescence in situ hybridization (FISH), *Proc. Jpn. Acad. Ser. B Phys. Biol. Sci.*, 86, 103–116, 2010.

Ohmido, N., Sato, S., Tabata, S., and Fukui, K., Chromosome maps of legumes, *Chromosome Res.*, 15, 97–103, 2007.

Otsu, N., Threshold selection method from gray-level histograms, *IEEE Trans. Syst. Man. Cybern.*, 9, 62–66, 1979.

Pecinka, A., Kato, N., Meister, A., Probst, A. V., Schubert, I., and Lam, E., Tandem repetitive transgenes and fluorescent chromatin tags alter local interphase chromosome arrangement in *Arabidopsis thaliana*, *J. Cell Sci.*, 118, 3751–3758, 2005.

Pecinka, A., Schubert, V., Meister, A., Kreth, G., Klatte, M., Lysak, M. A., Fuchs, J., and Schubert, I., Chromosome territory arrangement and homologous pairing in nuclei of *Arabidopsis thaliana* are predominantly random except for NOR-bearing chromosomes, *Chromosoma*, 113, 258–269, 2004.

Pedrosa, A., Sandal, N., Stougaard, J., Schweizer, D., and Bachmair, A., Chromosomal map of the model legume *Lotus japonicus*, *Genetics*, 161, 1661–1672, 2002.

Rick, C. M., and Barton, D. W., Cytological and genetical identification of the primary trisomics of the tomato, *Genetics*, 39, 640–666, 1954.

Robinett, C. C., Straight, A., Li, G., Willhelm, C., Sudlow, G., Murray, A., and Belmont, A. S., In vivo localization of DNA sequences and visualization of large-scale chromatin organization using lac operator/repressor recognition, *J. Cell Biol.*, 135, 1685–1700, 1996.

Sato, S., Isobe, S., Asamizu, E., Ohmido, N., Kataoka, R., Nakamura, Y., Kaneko, T., Sakurai, N., Okumura, K., Klimenko, I., Sasamoto, S., Wada, T., Watanabe, A., Kohara, M., Fujishiro, T., and Tabata, S., Comprehensive structural analysis of the genome of red clover (*Trifolium pratense* L.), *DNA Res.*, 12, 301–364, 2005.

Schermelleh, L., Heintzmann, R., and Leonhardt, H., A guide to super-resolution fluorescence microscopy, *J. Cell Biol.*, 190, 165–175, 2010.

Schwarzacher, T., and Heslop-Harrison, J. S., In situ hybridization to plant telomeres using synthetic oligomers, *Genome*, 34, 317–323, 1991.

Straight, A. F., Belmont, A. S., Robinett, C. C., and Murray, A. W., GFP tagging of budding yeast chromosomes reveals that protein-protein interactions can mediate sister chromatid cohesion, *Curr. Biol.*, 6, 1599–1608, 1996.

Sugimoto, K., Tasaka, H., and Dotsu, M., Molecular behavior in living mitotic cells of human centromere heterochromatin protein HP1 alpha ectopically expressed as a fusion to red fluorescent protein, *Cell Struct. Funct.*, 26, 705–718, 2001.

Tanabe, H., Müller, S., Neusser, M., von Hase, J., Calcagno, E., Cremer, M., Solovei, I., Cremer, C., and Cremer, T., Evolutionary conservation of chromosome territory arrangements in cell nuclei from higher primates, *Proc. Natl. Acad. Sci. USA*, 99, 4424–4429, 2002.

Taylor, N. L., and Chen, K., Isolation of trisomics from crosses of diploid, triploid, and tetraploid red clover, *Crop Sci.*, 28, 209–213, 1988.

Turck, F., Roudier, F., Farrona, S., Martin-Magniette, M. L., Guillaume, E., Buisine, N., Gagnot, S., Martienssen, R. A., Coupland, G., and Colot, V., *Arabidopsis* TFL2/LHP1 specifically associates with genes marked by trimethylation of histone H3 lysine 27, *PLoS Genet.*, 3, e86, 2007.

Udvardi, M. K., Tabata, S., Parniske, M., and Stougaard, J., *Lotus japonicus*: legume research in the fast lane, *Trends Plant Sci.*, 10, 222–228, 2005.

Wako, T., Fukuda, M., Furushima-Shimogawara, R., Belyaev, N. D., and Fukui, K., Cell cycle dependent and lysine residue specific dynamic changes of histone H4 acetylation in barley, *Plant Mol. Biol.*, 49, 645–653, 2002.

Wako, T., and Fukui, K., Quantitative analysis of nuclear chromocenter in *Spiranthes sinensis* (Pers.) Ames, *Bioimages*, 11, 97–103, 2003.

Wako, T., Houben, A., Furushima-Shimogawara, R., Belyaev, N. D., and Fukui, K., Centromere-specific acetylation of histone H4 in barley detected through three-dimensional microscopy, *Plant Mol. Biol.*, 51, 533–541, 2003.

Wako, T., Murakami, Y., and Fukui, K., Comprehensive analysis of dynamics of histone H4 acetylation in mitotic barley cells, *Genes Genet. Syst.*, 80, 269–276, 2005.

Yanagisawa, T., Tano, S., Fukui, K., and Harada, H., Marker chromosomes commonly observed in the genus *Glycine*, *Theor. Appl. Genet.*, 81, 606–612, 1991.

Young, N. D., Cannon, S. B., Sato, S., Kim, D., Cook, D. R., Town, C. D., Roe, B. A., and Tabata, S., Sequencing the genespaces of *Medicago truncatula* and *Lotus japonicus*, *Plant Physiol.*, 137, 1174–1181, 2005.

Zhang, X., Germann, S., Blus, B. J., Khorasanizadeh, S., Gaudin, V., and Jacobsen, S. E., The *Arabidopsis* LHP1 protein colocalizes with histone H3 Lys27 trimethylation, *Nat. Struct. Mol. Biol.*, 14, 869–871, 2007.

chapter sixteen

Machine vision in estimation of fruit crop yield

A. Payne and K. Walsh

Contents

16.1 Introduction

As a "poor cousin," technology application in agriculture often follows development in other fields, such as defense or health, as complexity of operation and cost decreases. The use of machine vision is one such technology. The last decade has seen a revolution in the adoption of digital still and video cameras, with the effective demise of preceding technologies. Developments in hardware for near-infrared and infrared imaging, and hyperspectral imaging, have been spectacular. Machine vision capability has advanced to the point of face recognition capacity built in to mass market camera applications, and into social networking websites, as well as to the more prosaic application of toll applications on roadways. Likely areas of application development include networked cameras, e.g., in security applications. In horticulture, the driver for adoption of this technology lies in the increasing cost of labor, the decreasing proportion of income consumers are willing to pay for food, and the demand for consistent lots of produce. These pressures result in the need to reduce cost of production while improving quality control.

Machine vision technology began to be widely implemented in fruit packhouses in the 1970s, replacing grading machines based on mechanical principles (diverging belt for size grading, counterbalance weight tipping buckets for weight grading, and the human eye for color grading). By the 1990s, image classification functions (e.g., using a neural network-based classifier) were being employed for surface defect detection on fruit moving along the pack line, distinguishing between acceptable features (e.g., a stalk) and blemishes. A major challenge in this application was speed of object movement. With pack line speeds of 1 m/s common, a given fruit passes through the camera image area at 1 mm/ms. Mechanisms to rotate the fruit during passage through the image area, and use of angled side mirrors to reveal the sides of the fruit to the camera, were innovations to improve machine vision performance.

While demanding in terms of speed, the pack line environment offers advantages to the field environment in terms of machine vision. On the pack line, fruits are singulated (i.e., separated fruits are moving on the conveyor) and in a structure. In contrast, assessment of fruit on trees in the field environment involves an unstructured environment.

A wide range of machine vision applications have been proposed in the field of horticulture, including weed identification to guide targeted spraying, fruit load estimation, and autonomous harvesting. However, these goals are yet to be realized, given the complexities of the field environment (variable lighting and background).

Estimation of fruit crop yield prior to harvest is an application that, at face value, is well suited to the application of machine vision. Growers seek to estimate the fruit load within an orchard in order to guide agronomic treatments. An early identification of yield variation can be used for management interventions, such as controlled flower and fruitlet thinning, fertilization, and irrigation regimes (Aggelopoulou et al., 2011). Further, growers seek to estimate fruit load to assist in the estimation of resource requirements, e.g., number of pickers, boxes, and transport means, and also to forward plan marketing. Indeed, such information is required by an increasingly sophisticated supply chain.

A machine vision count of fruit based on images of trees of the entire orchard should allow better estimation than the current method, which is based on visual estimation of load on a small subset of trees in the orchard. Further, this application can tolerate a higher error in fruit location than a number of other applications, such as harvesting. An error of 10% on fruit count would be acceptable for crop load estimation.

The present review describes the state of the art in machine vision application to fruit crop yield estimation.

16.2 The extent of the literature

Interest in machine vision of fruit has increased markedly between 1990 and 2012, particularly since 2004, as gauged by the number of publications listed by a Scopus search on keywords *machine vision* and *fruit* (Figure 16.1, bottom panel), although many of these papers relate to sorting of fruit on a pack line, rather than on trees. This topic area represents approximately 5% of all papers on *machine vision* (Figure 16.1, top panel).

An excellent review on the use of computer vision to locate fruit on trees was provided by Jiménez et al. (2000), although this resource is over a decade old now, i.e., it was created before the explosion in publications in this category. Jiménez et al. (2000) categorized image classification routines into local or shape approaches, and also noted that range imaging

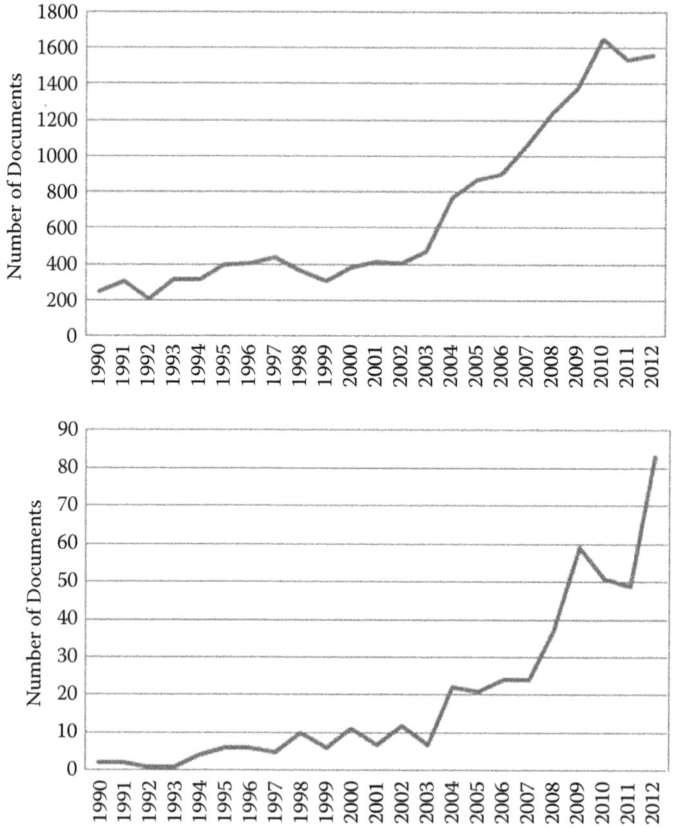

Figure 16.1 Scopus search on 1990–2012 publications using keywords *machine vision* (top panel) and *machine vision* and *fruit* (bottom panel).

was a promising technique (specifically for watermelons). It was noted that occlusion of fruit (by leaves, other fruit, etc.) was a significant issue with any approach, and that new techniques were required to address the issue. They concluded that it was possible to identify more than 85% of fruit within images of tree canopies, with low error rates (<5%). We accept this estimate, but only for applications in which fruit is of a significantly different color than the image background. Fruit that is similar in color to the background, or variable in color, offers a greater challenge.

Subsequent to this seminal review, Li et al. (2011) published a short review of work relevant to fruit harvesting robots, including a section in the area of image analysis. They have noted image segmentation was usually achieved using color, shape, machine learning, and statistical or neural network analyses. Significantly, however, while Jiménez et al. (2000) were generally reporting on these approaches in isolation, Li et al. (2011)

were reporting combination approaches involving techniques from multiple disciplines as the norm.

Given the rapid advances in the machine vision field in other application areas, it is timely to provide an updated review of the state of play in this field, focusing on publications since 2000. In this chapter, reports involving machine vision detection of fruit on trees, in field conditions, have been reviewed, excluding those papers associated with grading line sorting.

16.3 Imaging sensors

There are a number of sensing technologies that can be used to detect fruit load on trees. These include CCD or CMOS still, video, or stereo imagery, thermal imagery, hyperspectral imagery, multispectral imagery, ultrasound and LIDAR (light detection and ranging) imagery. All imaging techniques require recourse to image analysis techniques to interpret the image.

16.3.1 CCD/CMOS still imaging

As the task of fruit crop load estimation is currently undertaken by human visual assessment, it is amenable to imaging detectors that operate within the visible wavelengths. Silicon-based charged-coupled device (CCD) or complementary metal oxide semiconductor (CMOS)-based imaging technology is now widely and cheaply available. The sensors are based on silicon, with detection of wavelengths through and beyond the visible range (400–700 nm, and up to about 1000 nm). Many digital cameras include a filter before the detector to block the longer, nonvisible wavelengths. To achieve color vision, filters are used on alternate pixels, with image data stored in a so-called RGB (red, green, blue) format. CCD or CMOS imaging is the dominant technology reported for the application of machine vision estimation of fruit load.

16.3.2 CCD/CMOS video imaging

CCD- or CMOS-based video cameras are also widely available, storing a rapid sequence of individual RGB images. Video imaging allows the rapid acquisition of multiple views of a tree (and therefore of the fruit), with processing done on an image basis, or of multiple assembled images following a three-dimensional model. Although such video imaging is commonly employed in fruit grading applications, only the report of Tabb et al. (2006) has noted its use for an in-field application. They approximated background colors using a global mixture of Gaussians algorithm, and allocated each pixel to either apple fruit or background. A blob count then determined the number of fruit, with 85–96% of fruit in video frames identified.

16.3.3 CD/CMOS three-dimensional imaging

Three-dimensional (3D) imaging involves the inclusion of depth in the consideration of output. Such consideration is important for an autonomous fruit picking system—such an application requires depth information given the 3D spread of fruit within a tree canopy. However, a fruit load estimation system would also benefit from distance information, e.g., to localize the count to fruit on one side of the canopy, ignoring fruit belonging to the other side of the canopy. Three-dimensional information can be extracted from a stereo image, acquired using two cameras placed at a particular distance apart. Such systems are increasingly available "off the shelf," as manifested in 3D options in cinema and television.

Plebe and Grasso (2001) utilized stereo images in a processing algorithm that achieved detection of up to 85% of fruit (cloudy day). However, other studies that have collected stereo images have used only a single best image per tree in processing (MacArthur et al., 2006; Chinchuluun and Lee, 2006; Wenhua et al., 2009).

16.3.4 Infrared (thermal) imaging equipment

The general principle of detection using thermal imaging lies in the thermal lag in fruit temperature relative to leaf temperature under conditions of changing ambient temperature, due to the greater thermal mass of the fruit. However, sunlight penetration through a canopy is not uniform, such that fruit will vary in temperature. In consequence, a simple analysis involving segregation of an image based on a single temperature value is not a viable means of separating fruit from foliage, with an overlap of fruit and foliage temperatures (Figure 16.2). Thermal imaging is probably best used in combination with a second approach, such as RGB, in order to give additional points of variation for analysis between the fruit and background.

Bulanon et al. (2008, 2009b), Stajnko et al. (2004), and Wachs et al. (2010) reported the use of thermal imaging to localize fruit on trees. All achieved adequate results, but none recommend the technique as "the solution" to this application.

16.3.5 Multi- and hyperspectral imaging

Multi- and hyperspectral imaging has found great use in remote sensing applications. In this approach, a set of images is produced for each field of view, with images acquired over a set of typically three to five wavelength bands (multispectral) or over many wavelengths (hyperspectral).

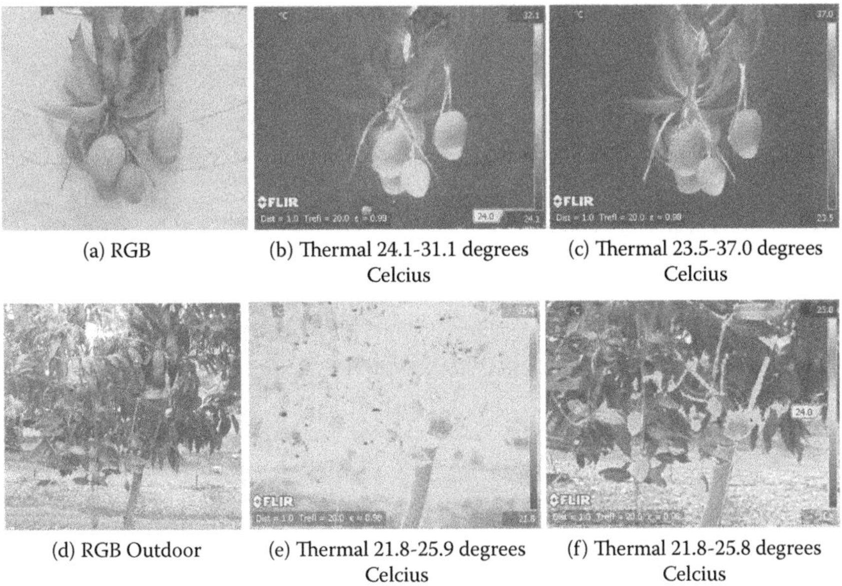

(a) RGB (b) Thermal 24.1-31.1 degrees Celcius (c) Thermal 23.5-37.0 degrees Celcius

(d) RGB Outdoor (e) Thermal 21.8-25.9 degrees Celcius (f) Thermal 21.8-25.8 degrees Celcius

Figure 16.2 (**See color insert.**) (b, c) Thermal images of fruit and foliage brought from ambient (hot) conditions into a cool (20°C) environment (a) with RGB image for reference. (e, f) Thermal images of fruit on tree collected in late afternoon (ambient temperatures decreasing) (d) with RGB image for reference.

A multispectral scanner may involve several detectors, each with a different wavelength filter, or a single detector with a rotating filter wheel. Hyperspectral scanners are typically line scanners, i.e., a 1D (line) input is accepted, with this light dispersed in the second dimension across a detector array, giving a complete spectrum for each point in the line. The line scan is moved forward (along the canopy row in this case) to create a hyper cube of data.

16.3.5.1 Case studies on hyperspectral imaging

The reflectance spectra of mango fruit and leaves, for example, have differences, both in the visible spectrum and around 960 nm (less reflectance, more absorbance—presumably associated with the water content of fruit) (Figure 16.3). Using such point spectra as input, a principal component analysis can be undertaken, and the principal components used to classify the image, revealing pixels likely to be associated with fruit (Figure 16.4). However, while such a process shows technical promise, the current cost of the hyperspectral imaging equipment is sufficiently high and, in general, beyond the limit of the horticultural industry.

Kane and Lee (2006) and Okamoto and Lee (2009) have reported on the use of hyperspectral imaging to identify fruit on trees. Kane and Lee

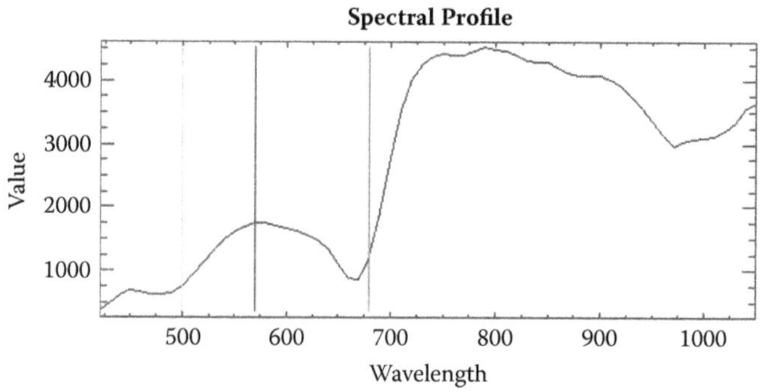

(a) Mango Spectra

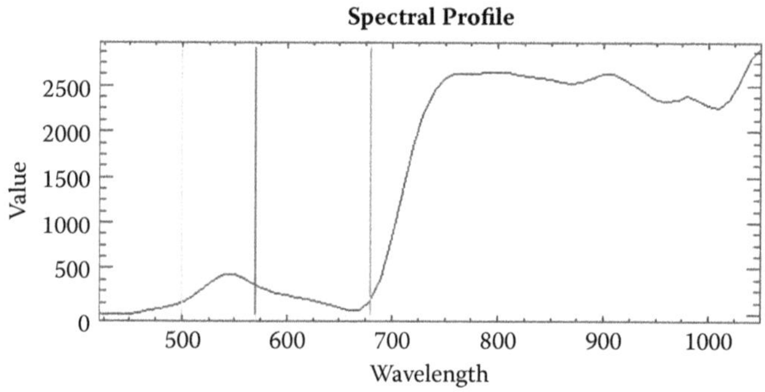

(b) Leaf Spectra

Figure 16.3 Spectra of (a) mango fruit and (b) leaves (hyperspectral imager from Integrated Spectronics, HyVista, Sydney).

(2006) processed only several images, but correctly identified all fruit in the images. Okamoto and Lee (2009) achieved correct identification of 70–85% of pixels, and correct identification of 80–89% of fruit.

16.3.5.2 Multispectral imaging example

It may also be possible to distinguish between the fruit and background using specific regions of the visible or shortwave near-infrared spectrum, or the ratio or differences between these regions. For example, in images of mango fruit on tree, the fruits are very poorly distinguished in the blue end of the visible spectrum, and in the shortwave NIR region to 900 nm (Figure 16.5).

Of the papers reviewed, four reported use of multispectral imaging. Alchanatis et al. (2007) and Safren et al. (2007) carried out small-scale trials with apples, achieving 87% correct detection of fruit, with 15% errors. Bulanon et al. (2010) and Kane and Lee (2007) achieved similar results with citrus.

(a) RGB Image (b) Band 4 Filtering

Figure 16.4 Mango fruit and foliage, and a white tile reference: (a) RGB image and (b) grayscale image based on values of principal component 1 from a PCA of the hyperspectral images (example spectra shown in Figure 16.3).

16.3.6 Ultrasound, LIDAR, and ToF imaging

Other technologies with potential for the application of fruit load estimation are ultrasound, LIDAR, and time of flight (ToF) imaging. These techniques produce images that can be processed to present three-dimensional models of the viewed area. Ultrasound sensors use oscillating sound pressure, at a frequency greater than the human ear can hear. The sound is used to detect objects. For example, Jiménez et al. (2000) identified fruit within ultrasound images of a watermelon crop, while Zaman et al. (2006) correlated orange tree canopy volume estimated using ultrasound with later harvest yield.

LIDAR sensors rapidly measure the distance between the sensor and the objects in the surrounding environment, enabling the construction of three-dimensional point clouds. Using computationally intensive algorithms, the information in these point clouds is used to reconstruct a three-dimensional rendering of the environment.

A ToF or range imaging camera measures the time for light to pass between the camera and each point of the image, and from this information distance is calculated based on the known speed of light. Thus, a ToF camera is a type of scannerless LIDAR. However, in ToF imaging the entire scene is captured with laser or LED pulse (typically 100 MHz

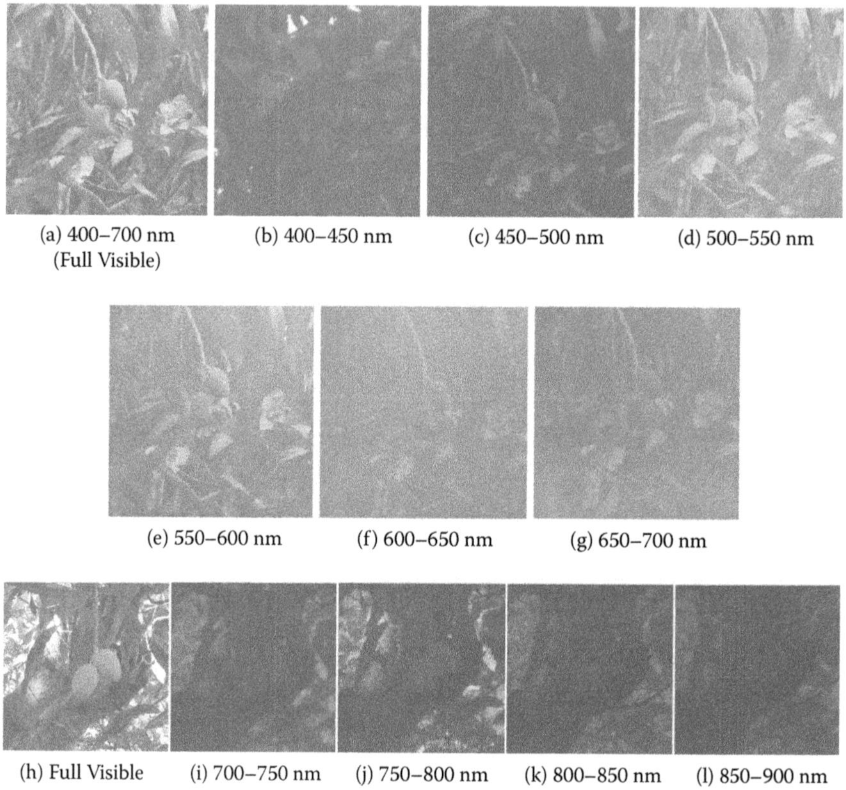

Figure 16.5 **(See color insert.)** Image of mango tree (fruit and foliage) acquired using a CCD (RGB) camera with filters: (a) 400–700 nm, (b) 400–450 nm, (c) 450–500 nm, (d) 500–550 nm, (e) 550–600 nm, (f) 600–650 nm, (g) 650–700 nm, (h) 400–700 nm, (i) 700–750 nm, (j) 750–800 nm, (k) 800–850 nm, (l) 850–900 nm.

modulation), in contrast to the LIDAR system of point-by-point imaging with a laser beam. At present, the sensing equipment for these techniques is expensive, relative to CCD imaging, and computational requirements are high. Thus, these technologies have not been implemented for estimation of fruit crop yield.

16.4 *In-field issues*

The success of an attempt to use machine vision to count fruit load on trees will be determined not only by the imaging technology used, but also by a number of additional in-field parameters, including the nature of the plant material, both fruit and canopy, the environment (lighting), and the detector system used.

16.4.1 *Plant material*

The color, shape, and surface texture of the fruit and its degree of difference from the background canopy (leaves, flowers, and stems) will define the difficulty of the detection system. Fruit and foliage may share characteristics with the background features in terms of color range, or edge shape. To state the issue in obvious terms, the detection of bright orange citrus fruit against a dark green background is more straightforward than detection of green beans against a background of green foliage.

Another significant issue is occlusion of fruit, i.e., partial obscuration in the image by foliage, other fruits, etc. Different canopy structures will produce different levels of occlusion. A tree pruned to create a "flatter" planar canopy will have advantages for machine vision, and robotic harvesting.

A third issue is that of multiple flowering events on a given tree, producing fruit of different maturities. Generally, the orchard manager seeks an estimate of the load of a particular flowering event, which will translate to a particular harvest event, rather than a total canopy fruit load estimate.

16.4.2 *Image field of view*

Most of the papers have used images that vary markedly in the distance from detector to tree. Some have utilized a close view where few fruit were visible (e.g., <10) and the background relatively uniform. Others have used images taken some distance from the tree, capturing fruit with varied lighting and background within the tree canopy. While the close view is appropriate to test and develop an image processing solution, eventual deployment of the image processing technique within orchard management requires imaging of the whole canopy.

16.4.3 *Lighting conditions and camera platform*

If detection is conducted during daylight hours, lighting and thermal conditions will vary widely within a tree, within an orchard, and over time, depending on the season, time of day, and local weather conditions. For example, ambient light could involve strongly directional bright sunlight through to muted lighting (dawn, dusk, or overcast conditions) (Figure 16.6). If detection is conducted during night hours, lighting will depend on the artificial source employed.

Another variable between reports of attempts to apply machine vision to fruit load estimation is the platform used to carry the imager through the orchard. Images have been collected from ground-based vehicles (farm utility vehicles or specially built vehicles) (Payne et al., 2013b) and from remotely piloted mini-helicopters (MacArthur et al., 2006).

(a) 7 am (b) 1 pm (c) 2 am

Figure 16.6 Images of mango trees (fruit and foliage) acquired using a CCD (RGB) camera under tropical conditions at different times: (a) 7 a.m., long shadows; (b) 1 p.m., strong directional lighting and short shadows; and (c) 2 a.m., artificially lit by LED spotlights, minimal background and evenly lit fruit.

16.4.4 Reality check

Published studies differ widely with respect to the number of samples used in development of a method, and the use of (or failure to use) truly independent data (image) sets in validation of the model. Ideally, a method should be tested against canopy images from different seasons and growing conditions. As a general rule, the greater the variety of conditions, the harder to develop an image processing approach that presents consistent results. Further, the work often involves assessment of the machine count of fruit within an image, presenting data such as percentage of fruit detected within an image, percentage errors, the coefficient of determination of the linear correlation (R^2), and the root mean square of errors between machine vision and manual counts. However, the success of an imaging algorithm in detection and counting of fruit within an image must be coupled to a consideration of how well the image represents field reality. Field reality might be defined as the actual end of season harvested yield, or the in-field count of fruit on the tree.

16.5 Image processing—RGB

While there are several technical challenges in any imaging application (e.g., lighting, lens choice), the major challenge of the automated crop yield estimation problem lies in processing of the images to detect (segment) fruit. While there is a wide array of approaches, there are commonalities: (1) use of a preprocessing stage, basically to compensate for undesirable characteristics of the image or to enhance the desirable ones; (2) a segmentation stage, in which the image is separated into pixels likely to be associated with fruit or background; and (3) a fruit detection stage, when the remaining pixels are assessed, grouped, and usually counted as individual fruit blobs. A classification of both segmentation and fruit detection stages, with identification of combinations employed in published reports (Figure 16.7), is discussed in the following sections.

16.5.1 *Preprocessing stage*

While most papers report segmentation of raw images, several approaches have been reported to adjust image intensity, to accommodate the effect of variable lighting at the time of image acquisition, prior to segmentation. Chinchuluun and Lee (2006) converted the original photo to grayscale, blurred the image using a disk filter mask with a radius of 25 pixels, and then quantized the image into 16 gray levels. Each of the red, green, and blue components of the color image was then enhanced using gamma correction, producing a much more even lighting result. Kurtulmus et al. (2011) preprocessed images to reduce the impact of variable illumination using histogram equalization and a logarithm transform to the intensity component of the HSI color space.

Another preprocessing technique was reported by Jin-jing et al. (2009), involving a vector median filter. An n*n window was passed across each image, using the RGB values of each pixel to calculate an average vector for the window. A vector within the window with the least distance to the average vector was then determined, and the center pixel in the area was replaced with this vector. This process had the effect of making the apples stand out, and deemphasizing the background.

16.5.2 *Segmentation stage*

Segmentation of the pixels in an image involves categorization of each pixel as a member of a class, in this case fruit, as opposed to foliage or other background structures. There are a number of approaches that have been taken by different researchers, including use of color, texture, and shape (contour). These approaches are summarized on the left-hand side of Figure 16.7 and discussed below.

16.5.2.1 *Segmentation using RGB color*

Color is a characteristic used for identification of fruit and can demarcate the fruit from the background (Table 16.1). The examples of application of segmentation based on RGB input include correlation with white apple flower number (Aggelopoulou et al., 2011), blueberry fruit number at harvest (Chang et al., 2012; Swain et al., 2010; Zaman et al., 2008), and orange fruit number 1 month prior to harvest (MacArthur et al., 2006). An excellent result was reported for the blueberry fruit count (R^2 = 0.98 for automated count against actual), usable results were reported for the apple flower count, and a poor result was obtained with orange fruit count (R^2 = 0.37).

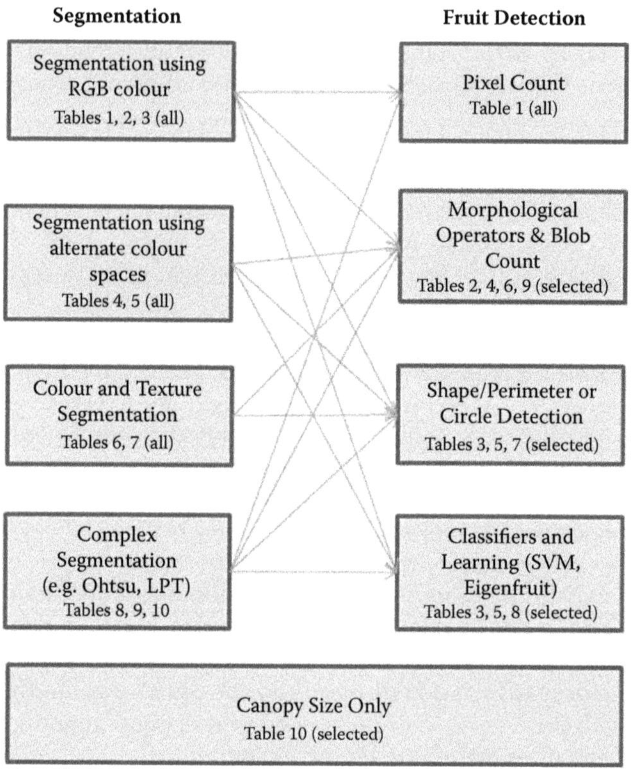

Figure 16.7 Classification of processing approaches based on two stage algorithms: (1) segmentation, the splitting of fruit and background pixels, and (2) fruit detection, the processing of remaining pixels into identifiable fruit fragments.

16.5.2.2 *Segmentation using processed RGB information*

A number of studies have documented the use of sophisticated processing of the RGB information to perform segmentation (Tables 16.2 and 16.3). Bulanon et al. (2002) reported the use of red color difference distribution for the segmentation of Fuji apples. This approach was continued for segmentation of orange fruit by Hannan et al. (2009). A bimodal distribution of red chromaticity values enabled separation of fruit and background pixels by the use of a threshold value, independent of lighting conditions. Bulanon et al. (2004) considered the use of all three chromaticity components for Fuji apples, an approach continued for segmentation of Valencia orange fruit (Bulanon et al., 2007, 2009a, 2009c). Images of Fuji apples were segmented within a tree canopy image using difference indexing of the chromaticity model, with nearest-neighbor clustering (Wenhua et al., 2009). Segmentation of images of yellow and red apples based on a global mixture of Gaussians in the RGB color space, using RGB video footage,

Table 16.1 Results of Segmentation Using RGB Color Thresholding Approaches with Crop Harvest Tonnage as the Reference

Fruit/cultivar/stage	Camera/platform/environment	Segmentation method	Blob and count method	Sample size	Reference value	Remarks	References
Apple Red Chief Flowering stage	Handheld commercial RGB camera—Canon PowerShot A510 (3.2 megapixels, 2048 × 1536 pixelx images, focal length 5.8–23.2 mm) Black cloth behind tree for contrast Images taken between 11:00 and 13:00	Segment using RGB threshold set based on calibration set	Pixel count of thresholded flower pixels	60 samples 53 validation	Actual yield	18% error in predicted yield Standard deviation of 3.47 t/ha	Aggelopoulou et al. (2011)
Blueberry At harvest	Digital Canon color camera (10 megapixels, 24-bit digital color camera) Camera mounted downward on tripod Partially cloudy (11 images), otherwise sunny	Segment on normalized blue chromaticity coordinate, using a fixed value (75)	Morphological operation, pixel count	60 sample locations in 2 fields (30 blueberry, 30 bare spot/weedy)	Actual yield	Field 1, $R^2 = 0.98$ Field 2, $R^2 = 0.99$	Zaman et al. (2008)

Continued

Table 16.1 (Continued) Results of Segmentation Using RGB Color Thresholding Approaches with Crop Harvest Tonnage as the Reference

Fruit/cultivar/ stage	Camera/platform/ environment	Segmentation method	Blob and count method	Sample size	Reference value	Remarks	References
Wild blueberry At harvest	Mounted Canon digital color camera (10 megapixels, 24-bit digital color camera) Farm motorized vehicle	Segment on normalized blue chromaticity coordinate, using a fixed value (75)	Morphological operation, pixel count	Two fields	Actual yield	Field 1, $R^2 = 0.90$ Field 2, $R^2 = 0.97$	Swain et al. (2010), Zaman et al. (2010)
Wild blueberry At harvest	CMOS sensor camera (lEye 1220SE/C color cameras, Imaging Development System, Inc., Woburn, MA) Specialized farm motorized vehicle developed	Segment on normalized blue chromaticity coordinate, using a fixed value (80)	Morphological operation, pixel count	Two fields	Actual yield	$R^2 = 0.95$ $p < 0.001$ RMSE = 0.23 mg/ha	Chang et al. (2012)
Orange Hamlin	Color digital camera (Videre model STH-MDCS-VAR-C stereo vision sensor) Mini-helicopter	Segment using a mixture of Gaussians to represent the orange class model	Orange pixels	15 trees	Actual yield	$R^2 = 0.373$	MacArthur et al. (2006)

Table 16.2 Summary of Processed RGB Segmentation Approaches Combined with Morphological Operators and with Fruit Number per Image as the Reference

Fruit/cultivar/ stage	Camera/platform/ environment	Segmentation method	Blob and count method	Sample size	Reference value	Remarks	References
Apple Fuji Harvesting season	Color CCD camera	Segment on the red color difference plane, using a minimum distance classifier	Blob count	15 trees, 4 images each (different lighting conditions), i.e., 60 images (20 calibration, 40 validation)	Manual image count	Best—cloud 95.4%/0.9% Worst—backlit 89.1%/18.2%	Bulanon et al. (2012)
Apple Fuji Harvesting season	Digital color CCD camera Natural daylight conditions—front lighting, back lighting, fruit in the shade, cloudy	Segment on trichromaticity components (*r*, *g*, and *b*), using a linear decision function	Remove noise using a low-pass filter Label blobs and calculate area, maximum and minimum axes Count blobs	As above	Manual image count	Front lit conditions—93.1% success, 1.3% false detections	Bulanon et al. (2004)

Continued

Table 16.2 (Continued) Summary of Processed RGB Segmentation Approaches Combined with Morphological Operators and with Fruit Number per Image as the Reference

Fruit/cultivar/ stage	Camera/platform/ environment	Segmentation method	Blob and count method	Sample size	Reference value	Remarks	References
Apple Gold Blush (yellow) Dixie Red (red) Ace Spur (red)	Digital color CCD camera, video stream (SonyDFW-x700 IEEE-1394 8-bit 1024 × 768 color CCD camera, Tokyo, Japan) Black curtain to restrict view of other trees	Segment using a global mixture of Gaussians (GMOG) method, based on mixture of Gaussians (RGB color space)	Count blobs	10 images	Manual image count	85–96% of fruit correctly detected (red and yellow) Between 1 and 6 false positives per image (113–213 fruit per image)	Tabb et al. (2006)
Apple Fuji	Stereo picking robot with binocular stereo vision system consisting of two digital cameras (Canon SX110 IS, 1/2.3 in. CCD, 3456 × 2592 pixels, focal length of 6–60 mm, f/2.8 to f/4.3)	Segment using difference indexing of chromaticity model, using nearest-neighbor clustering	Morphological filtering (remove areas smaller than the mean of all areas, open, close, fill holes) Additional obscured fruit processing (line)	10 images	Manual image count	Detection rate was greater than 90%; errors < 10%	Wenhua et al. (2009)

Table 16.3 Results of Processed RGB Segmentation Approaches Combined with Complex Fruit Detection and Fruit Number per Image as the Reference

Fruit/cultivar/stage	Camera/platform/environment	Segmentation method	Blob and count method	Sample size	Reference value	Remarks	References
Orange Harvesting season	Digital color camera (Sony Cybershot, DSC-P8) in automatic mode Sunny day conditions	Segment on normalized red chromacity coordinate, threshold determined by ROC graph calculated from 20 test images	Region labeling, size filtering, perimeter extraction and perimeter-based detection Separated fruit via shape analysis (circle detection)	20 inner canopy images (10 scenes, 2 lighting conditions) 25 near-view images 60 far-view images (5 trees)	Manual image count	Inner canopy, 99% detected, 3 false detections (of 82) Near-view, 93%, 9 false detections (of 131) Far-view, 90% detected, 8 false detections (of 219)	Hannan et al. (2009)
Apple Fuji Harvest	Color CCD video camera (images selected) Natural daylight conditions	Image enhancement performed using a vector median filter Segmentation using custom seeded region growing (seeds chosen on RGB color)	Morphological filter to reduce noise; support vector machine (based on color and shape features) to recognize apples	In-field test	Fruit detection by harvester in field	89% detected, 11% false detections	Jin-jing et al. (2009), Jiet al. (2011)

Continued

Table 16.3 (Continued) Results of Processed RGB Segmentation Approaches Combined with Complex Fruit Detection and Fruit Number per Image as the Reference

Fruit/cultivar/ stage	Camera/ platform/ environment	Segmentation method	Blob and count method	Sample size	Reference value	Remarks	References
Orange Valencia Robotic	Commercial digital camera (Sony Cybershot DSC-P8) Natural lighting conditions Camera in two positions relative to region of interest (0.5 cubic m), and six orthogonal views	Segment on r chromaticity component using an automatic threshold selection method (dependent on r intensity distribution)	Morphological operations to remove noise and fill holes Label blobs and extract features such as perimeter and area Perform circle detection to detect fruit centers	15 sets (6 views—all directions) from 5 trees 10 sets (9 views—from front) from 5 trees	Manual on-tree ROI count	Front view best of all views (50% fruit viewable) 6 front views best result of all (90% fruit viewable) Image processing algorithm achieved 87%	Bulanon et al. (2007, 2009a, 2009c)
Pomegranate	Commercial digital camera	Segment using a modified difference index	Circle detection	20 images	Manual image count	0–9.1% errors per image	Akin et al. (2012)

was performed by Tabb et al. (2006). Jin-jing et al. (2009) segmented images using a seeded region approach, where the seeds were chosen on RGB color information. Akin et al. (2012) segmented canopy images using a modified difference index (Table 16.3). All the approaches achieved high detection rates (above 90%), with low error rates.

16.5.2.3 Segmentation using alternate color spaces

To achieve segmentation, characteristics of color spaces (e.g., grayscale, HSI, HLS, Lab), rather than the raw RGB data, have been used by several workers (Tables 16.4 and 16.5). Chhabra et al. (2011) segmented images of mango on grayscale, using a predetermined threshold. Seeded k-means clustering on the hue (H) and saturation (S) components of the HSI color space for oranges was used by Chinchuluun and Lee (2006). Annamalai and Lee (2003) determined an optimal H threshold for segmentation of tangelo fruit by trial and error. They continued this approach with oranges (Annamalai et al., 2004). Kurtulmus et al. (2011) used Cb, Cr, and H components to threshold images for identification of immature green citrus fruit, although information on how the threshold values were determined was not provided. Regunathan and Lee (2005) differentiated Valencia orange fruit from the background canopy using H and S from the HLS color space, employing three different classifier approaches: Bayesian, neural network, and Fisher's linear discriminant analysis. Considerable preprocessing was conducted before using the CIE Lab color space to segment images of kiwifruit on the vine into fruit, leaf, branch, and dark background regions using a minimum distance classifier (Wijethunga et al., 2008) and a neural network self-organizing map (Wijethunga et al., 2009). Plebe and Grasso (2001) also segmented images using a neural network, using modified HSV space images. However, there was no clear advantage in the use of any given color space. Chhabra et al. (2011) and Wijethunga et al. (2008) achieved fruit detection rates of 90% with low errors for mango and kiwifruit variety Gold. Mid-range results were reported by Plebe and Grasso (2001), Annamalai and Lee (2003), Annamalai et al. (2004), Chinchuluun and Lee (2006), and Kurtulmus et al. (2011).

16.5.2.4 Segmentation using color and texture

Color space and texture characteristics have also been combined to achieve segmentation of fruit (Tables 16.6 and 16.7). Stajnko et al. (2009) segmented images of both green and red apple fruit on trees using HSI and texture characteristics of contrast, uniformity, and entropy. Later, this group reported the use of the hue component to perform an initial segmentation, as input for further sophisticated texture-based segmentation, involving spatial-frequency representation and a support vector machine classification (Rakun et al., 2011). Zhao et al. (2005) undertook segmentation based on the redness buffer (3R – (G + B)) of an image, using this

Table 16.4 Results of Alternative Color Space Segmentation, Combined with Morphological Operators, and with Fruit Numbers or Crop Harvest as the Reference

Fruit/cultivar/stage	Camera/platform/environment	Segmentation method	Blob and count method	Sample size	Reference value	Remarks	References
Tangelo Orlando End of citrus harvest season	Stationary mode using an RGB digital camera (model S100, Canon) with 640 × 480 pixels) Images taken in natural daylight	Segment using hue saturation (HSI plane) threshold based on calibration set	Morphological operations, blob count	90 images—of these, 59 were close view (20 calibration, 39 validation) and 31 were normal view (11 calibration, 20 validation)	Manual image count	Normal view, $R^2 = 0.76$, error rate per photo between 1.01% and 78.51%	Annamalai and Lee (2003)
Oranges Hamlin Harvesting season	Color analogue camera (model FCB-EX780S, Sony, New York, NY) Mounted on moving truck with images taken in stationary mode Natural daylight lighting conditions	Segment using hue saturation (HSI plane) threshold based on calibration set	Morphological operations, size-based blob count	98 trees, 354 images (329 validation images)	Manual image count Harvest yield	$R^2 = 0.79$ compared with image count $R^2 = 0.53$ compared with harvest yield	Annamalai et al. (2004)

Citrus 2 months prior to harvest	Two CCD cameras (model FCB-EX780S, Sony) Trailer mounted in stationary mode Natural outdoor illumination	Local gamma correction to correct illumination k-means clustering in the HSI color model (four clusters, set centroid values); segment using hue saturation	Morphological operation, marker-based watershed algorithm, blob count	512 images/32 trees 336 training images/21 trees; 176 validation images/11 trees	Manual image count Actual yield	$R^2 = 0.83$ compared with image count $R^2 = 0.64$ compared with actual yield	Chinchuluun and Lee (2006)
Orange Valencia	Color camera (model FCB-EX780S, Sony)/stationary mode/ultrasonic sensor/natural daylight conditions over 4 days	Segment pixels into four categories using hue and saturation (from HLS plane) and three classifiers (Bayesian, neural network, Fisher's linear discriminant)	Morphological filters to remove noise and fill holes Watershed transformation to separate touching blobs Blob count	37 training, 37 validation images	Manual image count	The root mean square errors for the three classifiers were: Bayesian 4.2, neural network 2.6, and Fisher 7.2	Regunathan and Lee (2005)

Table 16.5 Results of Alternative Color Space Segmentation, Combined with Complex Fruit Detection, with Fruit Numbers or In-Field Count as the Reference

Fruit/cultivar/stage	Camera/platform/environment	Segmentation method	Blob and count method	Sample size	Reference value	Remarks	References
Mango Ripe/unripe	Color digital camera (Sony TX color digital camera with 3008 × 2000 pixels (16 bits) Daylight conditions	Histogram equalization on grayscale image Segment using predetermined threshold	Morphological operations to filter noise and remove small unwanted objects; marker-controlled watershed transform to separate overlapping fruit Blobs input for neural network performing classification, feature analysis, and extraction	96 fruit (70% training, 15% validation, 15% testing)	Manual image count	90% accurate results 10% error	Chhabra et al. (2011)
Tangelo Orlando Immature	Color digital camera (Powershot SD880IS, Canon) with a resolution of 3648 × 2736 pixels) Sunny and shady in natural daylight conditions	Preprocessing to enhance image illumination (histogram equalization and logarithm transform) applied to HSI color space Segment image using Cb, Cr and H components, using preset threshold	Calculate "eigenfruit" vector for intensity, saturation, and circular Gabor texture (threshold determined adaptively), perform majority voting Merge and count detected blobs	96 images (32 training, 64 validation)	Manual image count	75.3% correctly detected 27.3% false positives 24.7% missed fruit	Kurtulmus et al. (2011)

Orange	Picking robot (including custom-built stereo vision)	Translation of image to a modified HSV space (combined HS only projected along a single plane perpendicular to the normal HS place) Segmentation using a multilayer perceptron, trained by back propagation or errors algorithm	Detect edges and detect closed curves and orange center Further 3D processing for robot arm	721 oranges in field	In-field count	Up to 85% of fruit harvested (cloudy day, no wind)	Plebe and Grasso (2001)
Kiwifruit Gold Green	Color digital camera (Lincoln Ventures Ltd., a Lumenera Le259 camera) Fixed to camera facing upwards Night with artificial lighting (fluorescent lights)	Preprocessing (center of window, contrast darker, lighter, emphasis blue and reduce noise) Four regions (fruit, leaves, branches, dark background) selected, mean and variance of each region in the L*a*b color space determined Segmented using a minimum distance classifier	Regional maximas used to obtain fruit counts	128 gold images (50 for model, 78 for validation) 72 green images (30 for model, 42 for validation)	Manual image count	90% gold, 60% green, average percentage error (under- or overestimation —7–37%)	Wijethunga et al. (2008)
Kiwifruit 4 months, 3 months, and 2 weeks prior to harvest	Color digital camera Fixed camera facing upwards Night with fluorescent or halogen lighting	Preprocessing as above Normalize a and b for L*a*b* color space Segment using SOM neural network	Pixel counts only	Three sets of five images	Preliminary work only		Wijethunga et al. (2009)

Table 16.6 Status of Color and Texture Approaches, Combined with Morphological Operators, with Fruit Numbers as the Reference

Fruit/cultivar/ stage	Camera/platform/ environment	Segmentation method	Blob and count method	Sample size	Reference value	Remarks	References
Mango Calypso Harvest	RGB camera (Canon 50D SLR camera using a 28–135 mm zoom lens) Tripod mounted Overcast daylight conditions	Color segmentation on normalized difference index (NDI), variance filter, filtering on Cr and Cb layers (YCbCr)	Blob count filtered on size	15 calibration images 555 validation images	Manual image count	$R^2 = 0.74$ RMSE = 7.7 51% detected Error rate 2.4%	Payne et al. (2013)
Mango Calypso 6 weeks prior to harvest, or "stone hardening"	RGB camera mounted on vehicle (Canon 50D SLR camera using a 28–135 mm zoom lens) Stationary photos Evening conditions 4 LED spotlights mounted around and 1 m from camera	Variance filter, border-limited mean (mango sized), Hessian filter, filtering on Cb layer, and Cr/Cb layer ratios	Blob count filtered on size	4 sets: Calibration (100 images) Validation 1 (10 images, same season, same trees) Validation 2 (74 images, same season, different trees) Validation 3 (42 images, next season, different orchard and trees)	Manual image count	Best set: $R^2 = 0.63$, 78.3% detected, error rate 10.6% Worst set: $R^2 = 0.63$, 56% detected, error rate 18.5%	Payne et al. (2014)

Table 16.7 Results of Color and Texture Approaches, Combined with Complex Fruit Detection, with Fruit Numbers or Pixel Count as the Reference

Fruit/cultivar/ stage	Camera/platform/ environment	Segmentation method	Blob and count method	Sample size	Reference value	Remarks	References
Apple Golden Delicious/ Jonagold Throughout vegetation period (June–September)	CCD camera (2288 × 1712 pixels taken with a digital camera type, Olympus 765UZ)	Segment based on hue, saturation, and intensity using empirically determined thresholds Segment based on the magnitude of red color intensities using thresholds determined for texture (contrast, uniformity, entropy)	Circle detection (using CHT algorithm) for clearly visible fruits Template matching (circular template)	20 trees	Manual image count	89% detected, 2.2% false detections	Stajnko et al. (2009)
Apple Gala/Golden Delicious	RGB camera Multiple viewpoints (20 ± 5° intervals) Daytime on a clear, sunny day	Segment based on hue (HSI plane) using generous threshold Segment based on texture (spatial-frequency texture classification/ SVM to binarize photos) Segment using 3D reconstruction based on projective reconstruction		Preliminary samples	Pixel count	25–46% pixels detected	Rakun et al. (2011)

Continued

Table 16.7 (Continued) Results of Color and Texture Approaches, Combined with Complex Fruit Detection, with Fruit Numbers or Pixel Count as the Reference

Fruit/cultivar/ stage	Camera/platform/ environment	Segmentation method	Blob and count method	Sample size	Reference value	Remarks	References
Apple Red/Green	RGB camera	Calculate texture of a region on the redness buffer of an image Detect edges and discard open edge contours Threshold regions based on average texture values Repeat following morphological functions on initial texture image	Split large contours using a circle fitting algorithm Count closed contours	Considered single photos only	Manual image count	Detected 18 of 20 apples	Zhao et al. (2005)
Apple Golden Delicious	Color CCD camera (Fujifilm FinePix S8000fd) Natural daylight conditions Middle of the day or close to sunset	Segmentation using a K-nearest-neighbor (KNN) classifier using color (RBG) and smoothness (variance of 6 × 6 area)	Detect seed areas and expand Perform shape analysis considering arcs, linear segments, and anamorphic segments Group contours into circles, and count each circle as an apple	Daytime set of 17 images (8 calibration, 9 validation) Dusk set of 17 images (9 calibration, 8 validation)	Manual image count	More than 85% fruit detected In dusk conditions, 95% fruit detected, with error rate of <5%	Linker et al. (2011, 2012), Cohen et al. (2011)

value as an index of the texture of a region. RGB and texture features were combined and used as an input into a classifier for segmentation of green apple fruit (Linker et al., 2012). Payne et al. (2013) performed color segmentation using the normalized difference index (NDI) of red and green layers. A 3 × 3 variance filter (texture filtering) and a filter based on the YCbCr color space were used in this segmentation approach (Payne et al., 2014). This algorithm was altered for fruit with less blush (reddish color), by dropping the NDI filtering, altering the YCbCr filtering, and adding both border-limited mean and Hessian filters.

Among these studies, the best results were obtained by the group working with green apples in diffuse light (dusk), with 95% of fruit detected with an error rate of less than 5% (Cohen et al., 2011; Linker et al., 2011, 2012). Stajnko et al. (2009) achieved 89% detection of apple fruit with 2.2% false detections. Only 51% of fruit was detected by Payne et al. (2013), with an error rate of 2.2%, while in later work the detection rate was increased to 78.3%, but errors were also increased to over 10% (Payne et al., 2014).

16.5.3 Blob identification and counting stage

Once the segmentation is over, the user is left with a subset of pixels of the image to process. A simple approach is to correlate pixel count to fruit number per image or orchard yield (Table 16.1). Due to incomplete and imperfect segmentation results, this subset of fruit pixels will represent portions of fruit and some falsely detected pixels (leaves or other background features). A disadvantage of this method is that it conveys no information on fruit size. Alternatively, a large range of approaches have been employed to group the segmented fruit pixels into blobs that represent countable fruit.

16.5.3.1 Morphological techniques

A majority of papers used morphological techniques, such as erosion and dilation, and the watershed algorithm (Tables 16.2, 16.4, 16.6, 16.9). These approaches assume that smaller groupings of pixels are misidentifications, and remove such groupings from further consideration. While the assumption is not always correct, the correction of false positive identifications generally far outweighs the removal of true positives. In some cases, filtering on the area or size of the remaining blobs was also carried out.

16.5.3.2 Regional maxima

Wijethunga et al. (2008, 2009) noted that kiwifruit had clear regional maxima (viewed from below, with fruit hanging from an overhead trellis, kiwifruit present a "bull's-eye" of floral tube material in the center of the fruit). These maxima were used to identify fruit with 90% detection achieved for the Gold variety, whereas only 60% detection was obtained with the Hayward variety.

16.5.3.3 Shape, perimeter, and circle detection

In human vision, our attention is drawn to objects with a signature outline as we identify fruit on a tree. Similarly, in machine vision the shape and perimeter of the segmented blobs can be used to identify fruit within an image. Hannan et al. (2009) removed smaller particles and then performed perimeter extraction. These perimeters were used to identify centers of fruit, including overlapping fruit (Table 16.3). Similarly, working with orange fruit, Plebe and Grasso (2001), Zhao et al. (2005), Bulanon et al. (2007), and Stajnko et al. (2009) extracted perimeters and performed circle detection. The group working with green apples (Linker et al., 2011, 2012; Cohen et al., 2011) used an approach in which the identified blobs were expanded into more likely fruit regions, and then an analysis of the perimeter of the resultant blobs was conducted. Circle detection approaches generally had good results, with high fruit detection rates of 85–95%, with low error rates (2.2–5%).

16.5.3.4 Classifiers and learning approaches

To quantify the number of fruit following image segmentation, a number of classifier and learning approaches have been introduced. Stajnko et al. (2004) used erosion and kernal filtering (3 × 3), and then input the selected fruit areas into a feature extraction, particle selection, and classifier process to identify fruit areas. Chhabra et al. (2011) performed a 3 × 3 filter over each image, applied morphological open and close operations, and then applied a watershed algorithm to separate overlapping fruit. In this study, 12 digital morphological features were employed to characterize and classify fruit within a support vector machine. Jin-jing et al. (2009) and Ji et al. (2011) also used a support vector machine to characterize and classify fruit within a canopy image. Majority voting was performed by Kurtulmus et al. (2011) after classifying fruit using an eigenfruit approach and a circular Gabor filter. However, these sophisticated approaches did not produce better results than those reported by using simpler procedures. Detection rates of fruit were reasonable to good, ranging between 75 and 90%.

16.6 Image processing—Spectral, thermal, and others

As stated above, thermal and spectral imaging has been used in the fruit load estimation study, either alone or in combination with RGB imaging.

These techniques are at an earlier stage of application compared to color image processing.

16.6.1 Thermal or thermal and RGB imaging

Bulanon and coworkers applied thermal imaging alone (Bulanon et al. 2008) and in combination with RGB images (Bulanon et al., 2009b) (Table 16.8). Segmentation was performed with Otsu and histogram tail methods. Further, they have trialed image fusion using a Laplacian pyramid transform and a fuzzy logic procedure. Overall, better results were achieved using the combination of thermal and RGB imaging (Stajnko et al., 2004; Wachs et al., 2010). However, the results reported by the various authors appear insufficient to justify the additional expense, maintenance, and processing involved in using a thermal camera.

16.6.2 Spectral imaging

The various reports on the use of spectral imaging to classify fruit pixels in a canopy image are summarized in Tables 16.9 and 16.10. Using spectral imaging and a combined ECHO algorithm and maximum likelihood classifier (MLC), Alchanatis et al. (2007) achieved high detection rates (87%) along with high false detection rates (15%). Similar results were achieved by Bulanon et al. (2010) following principal component analysis (PCA) and pattern recognition techniques. Discriminant analysis of hyperspectral images, using Otsu segmentation with morphological operations, was found to be effective in differentiation at a pixel level between fruit and leaf with 85% correct classification (Kane and Lee, 2006). However, significant in-field problems were reported, such as target shifting, light/shadow changes, raw image saturation, and multiple leaf reflections, with the effect of degrading the quality of the collected image significantly. Okamoto and Lee (2009) also reported an in-field application in which spectral imaging and linear discriminant analysis on pixels failed to distinguish more difficult background from that of obscured fruit. In summary, spectral imaging is a powerful technique; however, evidence remains lacking that the technique offers significant improvement in fruit classification compared to RGB imaging.

16.6.3 Canopy estimations

Several groups have attempted to relate fruit load to a measure of a canopy character. For example, Zaman et al. (2006) related tree fruit yield to a measure of canopy size achieved using ultrasonic sensors (Table 16.10). Ye et al. (2006) trialed the use of overhead hyperspectral imagery-associated

Table 16.8 Results of Thermal Approaches, with Fruit Numbers or Pixel Identification as the Reference

Fruit/cultivar/ stage	Camera/platform/ environment	Segmentation method	Blob and count method	Sample size	Reference value	Remarks	References
Orange Hamlin	Thermal infrared camera (ThermaCAM P65HS infrared camera, FLIR, Inc.) 24 h period Dew point meter readings Color RGB images + halogen light (nighttime) for reference	Complex approach Compensate images using measured environmental factors Segment using Otsu (not effective) or histogram tail method (effective 4 p.m.–midnight)	Pixel count	4 trees, images over day	Pixel identification	Successful segmentation; 70% of pixels identified 6% false detection	Bulanon et al. (2008)
Orange Hamlin	Color digital camera (D-490ZOOM, Olympus, Japan) and a thermal camera (ThermaCAM P65HS, FLIR Systems, Inc., North Billerica, MA)	Image fusion performed using Laplacian pyramid transform (LPT) and fuzzy logic (fuzzy logic better performance)		96 pairs of visible and thermal images for each ROI	Pixel identification	Fuzzy logic performed better than LPT	Bulanon et al. (2009b)

Apple Golden Delicious 5 stages—after fruit drop to collecting fruit	RGB camera and thermal camera AGEMA 570 (FLIR, Inc.) Late afternoon	Convert thermal images to bitmap format Threshold using normalized difference index (NDI)	Morphological erosion Extract features, select particles, and use classifier evaluation to select apple blobs	600 images (120 images per stage, 20 images per sunny side, 20 images per shady side)	Manual image count	R^2 from 0.83 to 0.88	Stajnko et al. (2004)
Green apples Golden Delicious Granny Smith	Thermal infrared and color image modalities, thermal infrared camera (SC2000, FLIR, Inc., Sweden)	Method 1: Register thermal and color image, segment in L*a*b space using k-means clustering (173 training set); segment in thermal space using set threshold (determined from manual selection) morphological operations (fill holes, open); apply round shape-structuring element and watershed transform; apply Hough transform to detect circles Method 2: Haar classifier to detect apples; train classifier with majority voting		180 images	Manual image count	Combined approach 74% detection	Wachs et al. (2010)

Table 16.9 Results of Spectral and Ultrasonic Approaches, Combined with Morphological Operators, with Fruit Numbers or Pixel Identification as the Reference

Fruit/cultivar/ stage	Camera/platform/ environment	Segmentation method	Blob and count method	Sample size	Reference value	Remarks	References
Apple Golden Delicious First stage of growth (July/ 4 cm) Just before harvest (September/ 8 cm)	Tripod mounted B&W camera with an acousto-optical tunable filter (AOTF, Brimrose) Images taken between 11:00 and 16:00, bright sunlight (July) and partly overcast (September)	Complex (ECHO algorithm/ maximum likelihood classifier (MLC))	Morphological operations, watershed algorithm, blob count	13 samples (533 spectral images)	Manual image count	Correct detection 87% False detection 14.9%	Alchanatis et al. (2007), Safren et al. (2007)

Orange Valencia Ready for harvest	Monochrome camera with filter wheel carrying six optical band pass filters (PixeLINK PL-B955HU, ThorLabs, six band pass filters) Natural daylight conditions	Principal component analysis (PCA) to reduce dimensionality of data Pattern recognition techniques to classify pixels into fruit or background class (Linear discriminant classifier, Artificial Neural Networks)	Modified watershed transform for fruit declustering Blob analysis	6 multispectral images	Manual image count	Artificial neural network offered 89.3% detection with 16.6% false positive rate LDC offered 76.8% with 19.5% false detection	Bulanon et al. (2010)
Green citrus Tangelo, Valencia, and Hamlin	Hyperspectral camera 369–1042 nm (ImSpector V10, Spectral Imaging Ltd.)	Pixel discrimination based on a linear discriminant analysis + spatial image processing steps (noise reduction filtering, labeling, and area thresholding)	Morphological functions, including median filter and erode Analyze particles including 300 pixel size threshold	37 training images, 37 validation images	Pixel identification Manual image count	70–85% correct pixel identification 80–89% foreground fruit identified	Okamoto and Lee (2009)

Table 16.10 Results of Spectral and Ultrasonic Approaches, Combined with Complex Fruit Detection, with Fruit Numbers or Crop Harvest Tonnage as the Reference

Fruit/cultivar/stage	Camera/platform/environment	Approach	Sample size	Reference value	Remarks	References
Green citrus Orlando, Tangelo, Hamlin, Valencia	Spectrophotometer measuring diffuse reflectance 200–2500 nm range Fruit moisture content Samples only	Discriminant analysis in lab to determine optimal wavelengths for leaf/fruit separation	Samples only	Pixel identification	R^2 as high as 1.0	Kane and Lee (2006)
Citrus	Monochromatic near-infrared camera with optical band pass filters (NIR InGaAs camera, FLIR Systems) Sunny conditions Difficult but usual field conditions (wind, sunlight changes)	Histogram stretching Smoothing filters Validated index relating 1064, 1150, 1572 nm Otsu segmentation method Morphological operations applied	36 trees, 500 images	Pixel identification	84.5% correct pixel identification $R^2 = 0.74$ when predicting the actual number of citrus pixels	Kane and Lee (2007)

Citrus Multiple times through season	Airborne visible and NIR (407–898 nm) Airborne Imaging Spectrometer for Applications (AISA) Eagle system (Pasco Co. Ltd., Tokyo, Japan)	Range of techniques, including: • Back propagation network to relate average canopy reflectance to citrus yield • TBVI calculated for red and NIR • NDVI, SR (simple ratio), PRI (photochemical reflectance index)	31 sample trees	Fruit yield	193 models achieved a R^2 value between 0.6 and 0.8 (May)	Ye et al. (2006)
Orange Valencia	Automatic ultrasonic system (tree canopy volume) Sensor-based automatic yield monitoring system	No processing required	40 plots	Fruit yield	20 plots, $R^2 = 0.80$ for tree size compared with fruit yield 20 plots, $R^2 = 0.42$ for actual tree yield compared with predicated fruit yield	Zaman et al. (2006)

spectral characteristics to tree fruit yield. In both cases, the results are mixed and worthy of further consideration.

16.7 Comparative performances of imaging techniques

There are a diverse array of imaging and image processing techniques reported for the application of estimation of fruit number on a tree canopy; however, it is difficult to recommend a best approach. We therefore provide a broad brush classification of the papers and results to allow for discussion of the current state of play in the field, and likely future developments. Published reports have been qualitatively assessed on two scales, fruit detection rate and robustness of estimation (Table 16.11). In our view, the reports categorized to the lower parts of the matrix (e.g., in quadrant 3) are not poor. Our assessment criteria are directly related to the commercial applicability of the approach, and a paper was graded with low score where robustness was not tested across independent image sets. These assessments are encoded into the following recommendation section (Section 16.8).

16.7.1 Fruit detection rate

The findings were grouped into one of three categories, based on the reported detection rate (percent of fruit in image detected), the false positive rate, and the R^2 of the machine to manual count:

- Category 1: Repeatable results: $R^2 > 0.9$, percentage of fruit detected > 90%, percentage errors < 10%.
- Category 2: Valuable progress: $R^2 > 0.6$ to < 0.9, percentage of fruit detected > 60% to < 90%, percentage errors > 10% to < 30%.
- Category 3: Poor: $R^2 < 0.6$, percentage of fruit detected < 50%, percentage errors > 10%.

16.7.2 Robustness

To achieve high detection and low false positive rates in a given set of images it is essential to adjust the classification procedures. However, overfitting of a classification routine is a common occurrence. The true test of a procedure lies in its use across a number of populations representing the range of conditions to be found in a commercial practice. A procedure that is stable in performance across such a range is said to be robust. To achieve this, the algorithm should be established using a set of calibration images. Then it has to be tested on sets of images acquired

Table 16.11 High-Level Comparison of Results with a View to Repeatability for Commercial Crop Vision Applications, including the Crop Considered

		Repeatability		
	Commercial	High	Medium	Low
Robustness	High		**Payne et al. (2014)/mango**	
	Medium		Aggelopoulou et al. (2011)/apple Annamalai et al. (2004)/oranges Chinchuluun and Lee (2006)/citrus Kurtulmus et al. (2011)/tangelo Stajnko et al. (2009)/apple Regunathan and Lee (2005)/orange Payne et al. (2013)/mango *Stajnko et al. (2004)/apple *Wachs et al. (2010)/apple *Kane and Lee (2007)/citrus *Okamoto and Lee (2009)/citrus *Zaman et al. (2006)/orange *Ye et al. (2006)/citrus	Bulanon et al. (2002)/apple Bulanon et al. (2004)/apple Cohen et al. (2011), Linker et al. (2011, 2012)/apple Wijethunga et al. (2009)/kiwifruit
	Low	MacArthur et al. (2006)/orange	Annamalai and Lee (2003)/tangelo Zhao et al. (2005)/apple Plebe and Grasso (2001)/orange Jin-jing et al. (2009)/apple *Bulanon et al. (2008)/orange *Alchanatis et al. (2007), Safren et al. (2007)/apple *Bulanon et al. (2010)/orange	Zaman et al. (2008)/blueberry Swain et al. (2010)/blueberry Chang et al. (2012)/blueberry Bulanon et al. (2007, 2009a, 2009c)/orange Hannan et al. (2009)/orange Chhabra et al. (2011)/mango Wenhua et al. (2009)/apple Tabb et al. (2006)/apple *Kane and Lee (2006)/citrus

Note: Repeatability levels are categorized as high (repeatable with like sets), medium (possibly repeatable with like sets), or low (not repeatable with another set). Commercial robustness levels are categorized as high (good representation of field conditions), medium (representative of field conditions, but not sufficient to be repeatable at commercial scale), or low (trial on a small scale). Items in the top right square are techniques that have been well tested and which produced viable, repeatable results (none currently existing). Papers marked with asterisks used multi- or hyperspectral imaging sensors. The method used to allocate papers to categories is described in Section 16.7.

from trees with different canopy structures, different orchards, and different seasons. If the prediction result is poor, the algorithm should be adjusted using the test populations, and then it should be retested using new test sets. The following categories were considered for robustness:

Category 1: Sufficient: Reports in this category have trial sets acquired under a range of orchard conditions and seasons, acquired at a crop stage when yield estimation is of commercial significance. Images of the whole canopy were used rather than a small area of the canopy.

Category 2: Progress: Images of a given orchard are not sufficient to test robustness.

Category 3: Poor: Images are unlikely to represent the diversity of the orchard and acquired at a time not relevant to crop yield estimation or with narrow views of the tree canopy.

16.8 *Summary and recommendations*

The potential for fruit load estimation by machine vision has been recognized for several decades. However, the technology has not seen commercial adoption, and the generally reported potential of 85% recognition of fruit within a canopy image with less than 5% errors seems overly optimistic in practice. Consistent results are likely to come with attention to the need to create images that maximize contrast between fruit and foliage, through selection of sensor type and image analysis algorithms, and attention to lighting and canopy conditions to maximize fruit contrast. The use of hyperspectral imaging should be revisited as equipment costs decrease.

16.8.1 *Imaging sensors*

Of the 46 studies assessed, 32 used digital cameras storing images in RGB format, 4 used stereo imaging equipment, 1 used RGB video, 5 used thermal imaging, 5 used spectral imaging, and 1 used ultrasound. As noted above, no technology gave a marked improvement over RGB imagery, so the RGB camera is recommended as the sensor of choice for the application of fruit load assessment, considering the relative cost of equipment.

16.8.2 *Time of day*

In most studies, images were acquired during daylight hours. However, natural lighting conditions are variable, ranging from strong directional lighting to diffuse light, creating variations in shadowing and color presentation (Payne et al., 2013). Indeed, in recognition of this issue, Aggelopoulou et al. (2011) employed black cloth behind apple trees. A number of studies acquired images under overcast conditions, and this generally assisted fruit detection by limiting shadowing and creating more even lighting. However, less than complete cloud cover will result in unpredictable lighting. Wijethunga et al. (2009) and Payne et al. (2013) acquired canopy images at night using artificial lighting for kiwifruit and mango and reduced the rates of false detections. We therefore recommend image acquisition at night using artificial lighting.

16.8.3 Image processing

It is not possible to recommend one particular RGB image processing approach as the best for further development, even within the context of a particular crop (e.g., citrus). It may be that the suite of image processing tools already available (in ImageJ and MATLAB®) is adequate for separation of fruit from the background for a given fruit type, with use of an expert system to guide selection of the tool and its parameters for the particular application. Alternatively, new image processing approaches based on neural network and classification techniques may offer significant improvements in the stages of segmentation and blob counting. In either case, the next step in moving the technology from laboratory to field use is the assembly of algorithms within software with a graphic user interface (GUI) to allow ease of use.

We recommend maintenance of a public repository of images of different fruit types to test new approaches of segmentation and blob counting for calibration. This repository needs to contain multiple sets of images of trees from different agronomic and lighting conditions, to be used in validation of a given approach. The authors maintain such a set of images of mango fruit trees that can be accessed on request.

16.8.4 Incorporation to a horticultural system

Success in segmentation of fruit within a canopy image represents the start of a process of implementation, not its conclusion. Almost all published work to date compares machine vision with manual counts of fruit visible within the image, and very few groups have reported the estimation of whole tree and orchard fruit yield from 2D images. Future studies should address this issue, with consideration of the influence of canopy structure, and manipulations of canopy form to minimize physical overlap of leaves over fruit, and of fruit color, shape, and texture with other tree features.

Further, the logistics of collection of canopy images across entire orchards (e.g., of 70,000 trees) requires consideration. Commercial in-field use of a machine vision-based crop yield system will also require implementation of a number of ancillary capabilities. For example, (1) geolocalization of tree-specific data will likely require GPS or radio frequency identification (RFID) capability, (2) variation in canopy size and interrow spacing will likely require use of multiple cameras and selection of appropriate lenses, or use of video imaging where resolution allows, and (3) the need for consistent lighting providing contrast between fruit and foliage will likely require LED illumination at specific wavelengths. Appropriate decision support software will need to be commissioned to allow comparison of images.

Acknowledgment

Support from Horticulture Australia Limited (HAL) under project HAL Agreement HG08075 is acknowledged. We thank Phil Purdy of Integrated Spectronics, Sydney, for supply of the hyperspectral image (Figures 16.2 and 16.3).

References

Aggelopoulou, A.D., Bochtis, D., Fountas, S., Swain, K.C., Gemtos, T.A., and Nanos, G.D., Yield prediction in apple orchards based on image processing, *Precis. Agric.*, 12, 448–456, 2011.

Akin, C., Kirci, M., Gunes, E.O., and Cakir, Y., Detection of pomegranate fruits on tree using image processing, in *1st International Conference on Agro-Geoinformatics, Agro-Geoinformatics 2012*, Shanghai, China, August 2–4, 2012, pp. 1–4.

Alchanatis, V., Safren, O., Levi, O., and Ostrovsky, V., Apple yield mapping using hyperspectral machine vision, in *6th European Conference on Precision Agriculture*, ed. J.V. Stafford, Skiathos, Greece, June 3–6, 2007, pp. 555–562.

Annamalai, P., and Lee, W.S., Citrus yield mapping system using machine vision, in *ASAE Annual International Meeting*, Las Vegas, NV, July 27–30, 2003, Paper 031002.

Annamalai, P., Lee W.S., and Burks, T.F., Color vision system for estimating citrus yield in real-time, in *ASAE Annual International Meeting*, Ottawa, Canada, August 1–4, 2004, Paper 043054.

Bulanon, D.M., Burks, T.F., and Alchanatis, V., Study on fruit visibility for robotic harvesting, in *ASAE Annual International Meeting*, Minneapolis, MN, June 17–20, 2007, Paper 073124.

Bulanon, D.M., Burks, T.F., and Alchanatis, V., Study on temporal variation in citrus canopy using thermal imaging for citrus fruit detection, *Biosyst. Eng.*, 101, 161–171, 2008.

Bulanon, D.M., Burks, T.F., and Alchanatis, V., Fruit visibility analysis for robotic citrus harvesting, *Trans. ASABE*, 52, 277–283, 2009a.

Bulanon, D.M., Burks, T.F., and Alchanatis, V., Image fusion of visible and thermal images for fruit detection, *Biosyst. Eng.*, 103, 12–22, 2009b.

Bulanon, D.M., Burks, T.F., and Alchanatis, V., Improving fruit detection for robotic fruit harvesting, *Acta Hortic.*, 824, 329–336, 2009c.

Bulanon, D.M., Burks, T.F., and Alchanatis, V.,A multispectral imaging system for citrus fruit detection, *Env. Control. Biol.*, 48, 81–91, 2010.

Bulanon, D.M., Kataoka, T., Okamoto H., and Hata, S. Development of a real-time machine vision system for the apple harvesting robot, in *IEEE Society for Instrument and Control Engineers 2004 Annual Conference*, Sapora, Japan, August 4–6, 2004, vol. 1, pp. 595–598.

Bulanon, D.M., Kataoka, T., Ota, Y., and Hiroma, T., A segmentation algorithm for the automatic recognition of Fuji apples at harvest, *Biosyst. Eng.*, 83, 405–412, 2002.

Chang, Y.K., Zaman, Q., Farooque, A.A., Schumann, A.W., and Percival, D.C., An automated yield monitoring system II for commercial wild blueberry double-head harvester, *Comput. Electron. Agric.*, 81, 97–103, 2012.

Chhabra, M., Gupta, A., Mehrotra, P., and Reel, S., 2011. Automated detection of fully and partially ripened mango by machine vision, in *Proceedings of the International Conference on Soft Computing for Problem Solving, Advances in Intelligent and Soft Computing*, December 20–22, 2011, vol. 131, pp. 153–164.

Chinchuluun, R., and Lee, W.S., Citrus yield mapping system in natural outdoor scenes using the watershed transform, in *ASAE Annual International Meeting*, Portland, OR, July 9–12, 2006, Paper 063010.

Cohen, O., Linker, R., and Naor, A., Estimation of the number of apples in color images recorded in orchards, in *IFIP advances in information and communication technology, computer and computing technologies in agriculture IV*, 2011, vol. 334, Springer, Berlin, pp. 630–642.

Hannan, M.W., Burks, T.F., and Bulanon, D.M., A machine vision algorithm combining adaptive segmentation and shape analysis for orange fruit detection, *Agric. Eng. Int. CIGR eJournal*, XI, manuscript 1281, 2009.

Ji, W, Zhao, D., Cheng, F., Xu, B., Zhang, Y., and Wang, J., Automatic recognition vision system guided for apple harvesting robot, *Comput. Electric. Eng.*, 38, 1186–1195, 2012.

Jiménez, A.R., Ceres, R., and Pons, J.L., A survey of computer vision methods for locating fruit on trees, *Trans. Am. Soc. Ag. Eng.*, 43, 1911–1920, 2000.

Jin-jing, W., De-an, Z., Wei, J., Jun-jun, T., and Ying, Z., Application of support vector machine to apple recognition using in apple harvesting robot, in *International Conference on Information and Automation, ICIA '09*, June 22–24, 2009, pp. 1110–1115.

Kane, K.E., and Lee, W.S., Spectral sensing of different citrus varieties for precision agriculture, in *ASAE Annual International Meeting*, Portland, OR, July 9–12, 2006, Paper 061065.

Kane, K.E., and Lee, W.S. Multispectral imaging for in-field green citrus identification, in *ASAE Annual International Meeting*, Minneapolis, MN, June 17–20, 2007, Paper 073025.

Kurtulmus, F., Lee, W.S., and Vardar, A., Green citrus detection using "eigenfruit," color and circular Gabor texture features under natural outdoor conditions, *Comput. Electron. Agric.*, 78, 140–149, 2011.

Li, P., Lee, S.H., and Hsu, H.Y., Review on fruit harvesting method for potential use of automatic fruit harvesting systems, *Procedia Eng.*, 23, 351–366, 2011.

Linker, R., Cohen, O., and Naor, A., Determination of the number of green apples on trees using color imaging, in *ASAE Annual International Meeting*, Louisville, KY, August 7–10, 2011, Paper 1110603.

Linker, R., Cohen, O., and Naor, A., Determination of the number of green apples in RGB images recorded in orchards, *Comput. Electron. Agric.*, 81, 45–57, 2012.

MacArthur, D.K., Schueller, J.K., Lee, W.S., Crane, C.D., MacArthur, E.Z., and Parsons, L.R., Remotely-piloted helicopter citrus yield map estimation, *in ASAE Annual International Meeting*, Portland, OR, July 9–12, 2006, Paper 063096.

Okamoto, H., and Lee, W.S. Green citrus detection using hyperspectral imaging, *Comput. Electron. Agric.*, 66, 201–208, 2009.

Payne, A., Walsh, K.B., Subedi, P.P., and Jarvis, D., Estimation of mango crop yield using image analysis—segmentation method, *Comput. Electron. Agric.*, 91, 57–64, 2013a.

Payne, A., Walsh, K., Subedi, P., and Jarvis, D., Estimating mango crop yield using image analysis using fruit at stone hardening stage and night time imaging, *Comput. Electron. Agric.*, 100, 160–167, 2014.

Plebe, A., and Grasso, G., Localization of spherical fruits for robotic harvesting, *Mach. Vision Appl.*, 13, 70–79, 2001.

Rakun, J., Stajnko, D., and Zazula, D., Detecting fruits in natural scenes by using spatial-frequency based texture analysis and multiview geometry, *Comput. Electron. Agric.*, 76, 80–88, 2011.

Regunathan, M., and Lee, W.S., Citrus fruit identification and size determination using machine vision and ultrasonic sensors, in *ASAE Annual International Meeting*, Tampa, FL, March 5–9, 2005, Paper 053017.

Safren, O., Alchanatis, V., Ostrovsky, V., and Levi, O., Detection of green apples in hyperspecctral images of apple-tree foliage using machine vision, *Trans. ASABE*, 50, 2303–2313, 2007.

Stajnko, D., Lakota, M., and Hočevar, M., Estimation of number and diameter of apple fruits in an orchard during the growing season by thermal imaging, *Comput. Electron. Agric.*, 42, 31–42, 2004.

Stajnko, D., Rakun, J., and Blanke, M.M., Modeling apple fruit yield using image analysis for fruit color, shape and texture, *Eur. J. Hortic. Sci.*, 74, 260–267, 2009.

Swain, K.C., Zaman, Q.U., Schumann, A.W., Percival, D.C., and Bochtis, D.D., Computer vision system for wild blueberry fruit yield mapping, *Biosyst. Eng.*, 106, 389–394, 2010.

Tabb, A., Peterson, D., and Park, J., Segmentation of apple fruit from video via background modeling, in *ASAE Annual International Meeting*, Portland, OR, July 9–12, 2006, Paper 063060.

Wachs, J.P., Stern, H.I., Burks, T.F., and Alchanatis, V., Low and high-level visual feature-based apple detection from multi-modal images, *Precis. Agric.*, 11, 1–19, 2010.

Wenhua, M., Baoping, J., Jicheng, Z., Xiaochao, Z., and Xiaoan, H., Apple location method for the apple harvesting robot, in *2nd International Congress on Image and Signal Processing*, CISP '09, October 17–19, 2009, pp. 1–5.

Wijethunga, P., Samarasinghe, S., Kulasiri, D., and Woodhead, I., Digital image analysis based automated kiwifruit counting technique, in *23rd International Conference on Image and Vision Computing*, New Zealand, IVCNZ 2008, November 26–28, 2008, pp. 1–6.

Wijethunga, P., Samarasinghe, S., Kulasiri D., and Woodhead, I., Towards a generalized color image segmentation for kiwifruit detection, in *24th International Conference on Image and Vision Computing*, New Zealand, IVCNZ 2009, November 23–25, 2009, pp. 62–66.

Ye, X., Sakai, K., Garciano, L.O., Asada, S., and Sasao, A., Estimation of citrus yield from airborne hyperspectral images using a neural network model, *Ecol. Model.*, 198, 426–432, 2006.

Zaman, Q.U., Percival, D.C., Gordon, R.J., and Schumann, A.W., Estimation of wild blueberry fruit yield using digital color photography, *Acta Hortic.*, 824, 57–65, 2008.

Zaman, Q.U., Schumann, A.W., and Hostler, H.K., Estimation of citrus fruit yield using ultrasonically-sensed tree size, *Appl. Eng. Agric.*, 22, 39–44, 2006.

Zaman, Q.U., Schumann, A.W., Percival, D.C., Read, S., and Esau, T.J., Automated, low-cost yield mapping of wild blueberry fruit, *Appl. Eng. Agric.*, 26, 225–232, 2010.

Zhao, J., Tow, J., and Katupitiya, J., On-tree fruit recognition using texture properties and color data, in *2005 IEEE/RSJ International Conference on Intelligent Robots and Systems*, August 2–6, 2005, pp. 263–268.

Index

G

J

K

L

M

N

O

P

For Product Safety Concerns and Information please contact our EU representative GPSR@taylorandfrancis.com
Taylor & Francis Verlag GmbH, Kaufingerstraße 24, 80331 München, Germany

www.ingramcontent.com/pod-product-compliance
Ingram Content Group UK Ltd.
Pitfield, Milton Keynes, MK11 3LW, UK
UKHW022213190926
475702UK00003B/341
9781466583016